Elements of Seismic Hazard in Mines

Aleksander J. Mendecki

Elements of Seismic Hazard in Mines

Springer

Aleksander J. Mendecki
Research & Development
Institute of Mine Seismology
Somerset West, Western Cape, South Africa

ISBN 978-3-031-93238-0   ISBN 978-3-031-93239-7   (eBook)
https://doi.org/10.1007/978-3-031-93239-7

This work was supported by Institute of Mine Seismology.

© The Editor(s) (if applicable) and The Author(s) 2025. This book is an open access publication.

Open Access This book is licensed under the terms of the Creative Commons Attribution 4.0 International License (http://creativecommons.org/licenses/by/4.0/), which permits use, sharing, adaptation, distribution and reproduction in any medium or format, as long as you give appropriate credit to the original author(s) and the source, provide a link to the Creative Commons license and indicate if changes were made.
The images or other third party material in this book are included in the book's Creative Commons license, unless indicated otherwise in a credit line to the material. If material is not included in the book's Creative Commons license and your intended use is not permitted by statutory regulation or exceeds the permitted use, you will need to obtain permission directly from the copyright holder.
The use of general descriptive names, registered names, trademarks, service marks, etc. in this publication does not imply, even in the absence of a specific statement, that such names are exempt from the relevant protective laws and regulations and therefore free for general use.
The publisher, the authors and the editors are safe to assume that the advice and information in this book are believed to be true and accurate at the date of publication. Neither the publisher nor the authors or the editors give a warranty, expressed or implied, with respect to the material contained herein or for any errors or omissions that may have been made. The publisher remains neutral with regard to jurisdictional claims in published maps and institutional affiliations.

Cover illustration: Presentation of one of at least two conjugate sets of Ortlepp shears with more than 90 mm of normal displacement. © Gerrie van Aswegen (2008)

This Springer imprint is published by the registered company Springer Nature Switzerland AG
The registered company address is: Gewerbestrasse 11, 6330 Cham, Switzerland

If disposing of this product, please recycle the paper.

Preface

About the Book The title of this book implies that it can only be selective or incomplete. It is not a textbook nor a review of the current state of research nor does it cover all issues relevant to seismic hazard in mines. The book summarises the author's main area of expertise and a few decades of experience in the selected subjects, without any claim that the described methods are the most suitable to the problem at hand. There are other published methods for seismic hazard assessment in mines that would deliver good results, therefore I refrained from critical analysis.

The book grew from the paper "Mine Seismology: Glossary of Selected Terms", which was published in 2013 in the Proceedings of the 8th International Symposium on Rockbursts and Seismicity in Mines. The idea behind the Glossary was to serve both as quick reference material and as a primer on selected terms used in mine seismology. As the scope of the Glossary broadened and the volume increased the character of the glossary changed to a *Mine Seismology Reference Book: Seismic Hazard* published in 2016. Both the glossary and the reference book are available for free download from the Institute of Mine Seismology website. This book also draws on *Seismic Monitoring in Mines* (Mendecki, 1997) and on *A Guide to Routine Seismic Monitoring in Mines* (Mendecki et al., 1999).

It is complementary to more comprehensive books on earthquake seismology, e.g. Aki and Richards (2002) and Shearer (2009), on critical phenomena in the natural sciences relevant to seismology, e.g. Keilis-Borok and Soloviev (2003) and Sornette (2004), on earthquake hazard and risk analysis, e.g. McGuire (2004) and Baker et al. (2021), and on mine seismology, e.g. Gibowicz and Kijko (1994). A very useful source of practical knowledge on mine seismology are the Proceedings of the International Symposia on Rockbursts and Seismicity in Mines and the relevant papers published by the South African Institute of Mining and Metallurgy. I would also like to recommend Ortlepp's book, *Rock Fracture and Rockbursts: An Illustrative Study*, that shows the complexity of the damage mechanism of rockbursts, and remind us of our limited understanding of the subject. As Ortlepp states "This book is essentially a phenomenological study intended to increase understanding rather than to discuss methods of control". He then pointedly writes that the structure of the book was determined by the "Recognition of the close association between rock fracture and rockbursting and adherence to the dichotomy between source mechanism and damage mechanism". This study should be on the bookshelf of every mine seismologist and rock mechanics expert.

Seismology is a very diverse science, ranging from mathematics being the language of science, geophysics, classical rock mechanics, fracture mechanics, statistical physics and geology. All that brings in a plethora of new terms into the seismological parlance. Consequently, the language of earthquake and mine seismology includes terms which may not be familiar to some geotechnical and mining engineers. The proliferation of different earthquake magnitude scales managed to confuse even this presumably simple concept. Therefore, in this book the common logarithm of seismic potency, $\log P$, is used as a measure of magnitude. Scalar seismic potency is directly observable from seismic spectra and it is a product of two parameters, $P = \Delta\epsilon \cdot V$, where the $\Delta\epsilon$ is the strain change at source during the seismic event and V is the source volume. Seismic potency and radiated seismic energy are two fundamental independent parameters describing the seismic source.

One of the objectives of this book is to introduce, define, and explain some of these terms and concepts. Another one is to clarify some of the misconceptions, in particular relating to the analysis and interpretation of mining-related seismicity and to seismic hazard assessment for mines. I hope the book will help to explain some of the intricacies and limitations of the power law size distribution that, when plotted as a straight line on a log-log scale, looks deceptively simple. Finally, it emphasises the need for testing the quality and integrity of the data before any analysis and assessment is attempted, the importance of quantifying observations, estimating the associated uncertainties and presenting results in meaningful plots.

In the first draft version of this book I used different data sets to demonstrate different practical examples. Then, recalling Keilis-Borok remark that "For any given, not too absurd, hypothesis one can find a self-consistent body of data", I decided to use as much as is viable the same data set in my practical examples. I think it works rather well because it introduces some degree of objectivity and it demonstrates the limitations imposed on us by the available data.

The first chapter of the book, "Preliminaries," starts with the authors views on different aspects of mine seismology, states the main differences between earthquakes and mine induced seismicity, lists the major factors influencing seismic hazard in mines, and proceeds to the general formulation of the Seismic Hazard Management Plan for Seismic Monitoring. Such plan should specify: (1) the objectives of seismic monitoring for a particular mine, (2) what the mine needs to have in terms of technology and skill, and (3) what the mine needs to do daily, weekly, monthly, and yearly to achieve the stated objectives. At the end of this chapter, almost as an appendix, is a short note on seismic magnitude scales.

Chapter 2, "Seismic Sources in Rock," starts with the glossary of terms used to describe seismic sources, and a short description of seismic potency, radiated seismic energy, and associated parameters like apparent stress, energy index, and apparent volume. There is a short description of the static solution for a circular crack subjected to a uniform shear strain given by Eshelby, that explains the basic source scaling relations, and a section on the circular ω^2 source model by Brune. While Brune's model fits the spectra of seismic events recorded in deep high stress mines, it does not work as well in many mines with a weaker inhomogeneous rock mass and/or at intermediate depth and in caving mines. Seismic events in such an environment radiate less energy per unit of deformation at source. Therefore Chap. 2 introduces more general point source models suggested by Beresnev and Atkinson that allow for "softer spectra", e.g., $\omega^{2.5}$ or ω^3. It also describes the final static deformation and strains induced by the double-couple source, gives the radiation pattern for the near-, intermediate-, and far-fields, and derives the expression for the surface of such a source for a given inelastic strain drop.

Chapter 3, "Monitoring Rock Mass Stability," is a continuation of the seismic stability analysis suggested by Mendecki (1993). The assumption here is that an inhomogeneous rock mass subjected to loading displays certain seismic symptoms when approaching instability: (1) an overall softening, measured by a decrease in the average value of the apparent stress; (2) increased rate or accelerating deformation, measured by an increase in the activity rate and/or apparent volume; (3) an associated increase in correlation length where one would expect to observe an increase in the spatial distribution of seismic activity, measured, for example, by seismic diffusivity; and (4) a decrease in the dimensionality, or localisation, of seismic activity as can be measured by the shape factor. The objective of seismic stability analysis is not to predict instability or to manage seismic exposure in the short term, but to guide control measures to mitigate seismic hazard.

Chapter 4, "Size Distribution and Seismic Hazard," starts with a general description of extreme events characterised by heavy tailed distributions with a few very large values compared to the other values of the data set. Typical examples of power law scaling are the Gutenberg and Richter magnitude frequency relation and the Omori law for aftershock decay. However, both are open-ended distributions, and therefore have no characteristic scale, i.e. there is no upper limit on magnitude in the former and there is no end to the duration of aftershock sequences in the latter. Therefore, a need arises for a more physical description, either by introducing a soft transition to finite energy release or a sharp cut-off by a double-truncated distribution. Another important issue described in this chapter is the so-called m_{max} or $\log P_{max}$, which in earthquake seismology is the maximum magnitude, or potency, earthquake that a given seismogenic region can deliver. However, in mines the maximum possible size event scales with the footprint of the mine and with the degradation of rock mass stiffness, both increasing as mining progresses creating conditions conducive for ever larger events to occur. Therefore, the maximum size event associated with mining is not an ultimate number, but needs to be estimated periodically. It is the next record-breaking seismic potency, energy, or magnitude which can be estimated and that needs to be managed.

Chapter 5, "Time Distribution and Seismic Hazard," starts with the time characteristics of seismic activity, the coefficient of variation and proportional variability, and progresses to stationary Poisson processes, probabilities of exceedance, and mean recurrence times. In this section there are two practical examples. The first one compares seismic hazard between three mines with different volumes of rock extracted over the same time period at different depths and it is done in the $\log E$ domain. It shows the utility of plotting the cumulative potency vs the cumulative volume of rock extracted as an indication of seismic hazard. The second example compares the seismic hazard characteristics of two mostly overlapping seismic and rock extraction data sets selected from the same mine, and it is done in the $\log P$ domain. Having partial data on the volume of rock extraction, it postulates that apart from increasing extraction ratio, the more concentrated production blasting at greater depth may have contributed to the observed increase in seismic hazard.

The rest of this chapter is dedicated to the seismic rock mass response to step loading, mainly by production blasting and by larger seismic events. Seismic rock mass response to blasting is driven strongly by the stress level in rock surrounding the blast and by the volume of rock blasted. In most cases rock extraction by blasting induces and triggers seismic events immediately and in close proximity to the blast. In some cases the rate of activity follows typical aftershock sequences that can be described by the simple Omori law or by the stretched exponential relaxation function. Here, the non-stationary Poisson process needs to be invoked to assess the probabilities of having larger events during the aftershock sequences. Two examples of seismic hazard assessment after large events are presented.

In practice mines frequently group blasts to optimise the required exclusion time after blasting. It is important then to schedule blasting sequences to limit the overlap of seismic relaxation. If larger blasts are too frequent and/or too close to each other, i.e. if the next blast is still during the excitation phase of the previous one, it may push a larger and larger part of the system into the sub-critical stage. There is also the possibility that such blasting may induce a larger seismic event that would not have happened otherwise, as opposed to advancing the clock for events that are almost ready to be triggered. This issue is addressed in the section Seismic Rock Mass Response and Blasting Sequence, where the defined Proximity Index may help to test whether the preferred time differences and distances between blasts sequence of a planned sequence are at an acceptable level.

Chapter 6, "Ground Motion Hazard," starts with the general description of transient strains and stresses and the dynamic stress concentration factors for a plane harmonic incident P- and S-wave wave hitting a tunnel. Then it gives a rudimentary explanation of the ground motion (GM) at seismic source, describes parameters that measure the intensity of GM, and presents a case study on amplification of GM at the skin of an excavation. Understanding, quantifying, and forecasting GM motion in the near and intermediate field of seismic radiation is the most important issue in mine-induced seismicity since most damage caused by seismic events is observed in excavations very close to their sources. Since the maximum ground velocity at source is controlled by the strength of the rock mass, small and large events produce similar ground velocity at source, but large events affect a substantial volume of rock, hence the probability of hitting a vulnerable structure is considerably higher. In smaller mines the strong ground velocities and displacements associated with larger events may affect the entire infrastructure.

The next section is dedicated to the Ground Motion Prediction Equation (GMPE) and its applications, e.g., plotting the expected GM at strategic locations in real time, seismic fragility curves that show the probability that tunnel support may be damaged under different seismic loads given its remaining deformation capacity, and the GM alert program, called GMAP. The next section describes the real-time version of GMAP, called GMAS, that does not require the GMPE and estimated source parameters to issue an alert.

The next section describes Mapping Seismic Ground Motion Hazard, which in earthquake seismology is called the probabilistic seismic hazard assessment and its limitations, the major being the uncertainty in the spatial distribution of the expected seismicity.

The last section, Modelling Seismic Hazard, describes what is called the deterministic hazard, i.e., kinematic modelling of GM produced by potential seismic sources defined by their expected maximum potency or energy, placed at the most likely locations that can produce the highest intensity of GM motion at a given site. Examples of modelling GM produced by extended and complex sources are given.

The Appendix, "Basic Concepts in Probability and Statistics," gives a simple overview of the subject with emphasis on statistical dependence. Having a number of parameters that may be relevant to the subject at hand, we would like to select a subset that is most informative, mainly those that are independent. However, some of the relevant parameters may not be totally independent and then we need to select a set of parameters that are least dependent.

The book is intended for mine seismologists, mine geotechnical engineers, and consultants who are, or planning to be, involved in advising mines on how to manage the seismic rock mass response to mining. It may also be useful to Mining and Geotechnical Engineering students.

Somerset West, South Africa   Aleksander J. Mendecki

Acknowledgments

While writing this book I benefited from discussions with my colleagues from the Institute of Mine Seismology. My late friend Assen Ilchev deserves a special mention. He was a great person, mathematical physicist, and polymath; having a conversation with Assen was like attending University. Our discussions always started with physics, followed by the history of science, and, since Assen was a great cook, we always ended up with food and wine. Very topical, since the offices of our Institute are at the Equitania wine farm that produces a very good blend of Cabernet Sauvignon and Cabernet Franc. Assen was born in Bulgaria, and told me that in Bulgarian folklore there are 70000 songs praising red wines and only one about white wines and it goes like "Eh ti, bjalo vino, zashto ne si cherveno?", meaning "Oh you white wine, why aren't you red?". Assen's great insight into physics and his mathematical rigour made my task of writing this book far more demanding, but it will certainly benefit readers.

I would also like to thank Gerrie van Aswegen, an expert in structural geology and mine seismology, with particular experience in forensic analysis of processes leading to and resulting from larger seismic events. During our numerous underground visits to deep level gold mines in South Africa Gerrie drew my attention to the dynamic brittle shear fractures associated with seismic sources described by Dave Ortlepp, which otherwise I would certainly miss. He also argued that these fractures are sufficiently unique to warrant a special name—Ortlepp shears (van Aswegen, 2008). Gerrie also tested the stability concept described in Chap. 3 and offered critical comments that helped to improve the method. He reviewed and provided comments on Chaps. 1, 3, and 5.

Gerrie also provided the photograph on the front cover that shows one of at least two conjugate sets of Ortlepp shears. The right hand shear in the photograph shows more than 90 mm of normal displacement. These structures formed dynamically as part of an extremely damaging mine seismic event that closed a 37 m wide panel in a 110 m wide stope. The stope was advancing along the strike of the Carbon Leader reef in gold mine some 70 km west of Johannesburg, South Africa. The reef dips ~26° southerly and the stope was advancing westerly. The seismic event had a complex source mechanism, being some 15% DC (from the Ortlepp shears) and 85 implosive (stope closure). The Ortlepp shear was exposed when a 10 m wide panel was mined up-dip in front of the damaged panel, yielding a perfect cross section of the fractures and shears ahead of the damaged stope.

Peter Mountfort perused the manuscript and his comments enabled me to see things more clearly. Peter is an expert in seismic sensors and systems and he reviewed all references to sensors and the quality of waveforms I used in this book. He is also an excellent software programmer and I frequently asked for help when struggling with my code. I would also like to acknowledge countless thought-provoking discussions I had over the last 15 years with Cornel du Toit on the subject of probability theory and statistical seismology.

Artur Cichowicz reviewed and provided valuable comments on Chap. 2, Seismic Sources, and on Chap. 6 Ground Motion Hazard. I'm also grateful to Savka Dineva who asked pertinent question regarding stability analysis. I hope for more critical comments by readers and for that I'm already grateful.

I was fortunate to interact personally with, learn from, and be influenced by prominent seismologists: Keiti Aki, Vladimir Keilis-Borok, Yehuda Ben-Zion, Roel Snieder, Didier Sornette, Slawomir Gibowicz, Roman Teisseyre, Art McGarr, Artur Cichowicz, Steve Spottiswoode, Andzej Kijko, Stanislaw Lasocki, Vaclav Vavrycuk, Jan Sileny, and Cezar Trifu. I learned by working and interacting with rock mechanics researchers: David Ortlepp, Peter Kaiser, Rob Bewick, Dick Stacey, Richard Brummer, Paul Young, Aleksander Linkov, Vladimir Lyakhovsky, Matthew Handley, William Joughin, and Ernesto Villaescusa. I also benefited greatly by interacting with prominent geotechnical and mining experts: Eduardo Rojas, Matthew Sullivan, Alan Moss, German Flores, Lourens Scheepers, Koos Bosman, David Cuello, Javier Vallejos, Sergio Gaete, David Beck, David Finn, Cristian Orrego, and Chris Chester.

During 1990s I made over 100 underground visits to South African deep gold mines, initially with David Ortlepp and then with Gerrie van Aswegen. Most of these visits were after severe rockbursts, unfortunately many of them with fatal consequences. The intensity of the damage and a realisation of the consequences focus my effort on developing useful seismological tools to mitigate rockburst risk in seismically active mines.

Bibliography

Aki, K., & Richards, P. G. (2002). *Quantitative seismology* (2nd ed.). University Science Books.
Baker, J. W., Bradley, B. A., & Stafford, P. J. (2021). *Seismic hazard and risk analysis*. Cambridge University Press.
Beresnev, I., & Atkinson, G. M. (1997). Modeling finite-fault radiation from the omega-n spectrum. *Bulletin of the Seismological Society of America, 87*(1), 67–84.

Brune, J. N. (1970). Tectonic stress and the spectra of seismic shear waves from earthquakes. *Journal of Geophysical Research, 75*(26), 4997–5009.

Eshelby, J. D. (1957). The determination of the elastic field of an ellipsoidal inclusion and related problems. *Proceedings of the Royal Society of London, Series A, Mathematical and Physical Sciences, 241*(1226), 376–396.

Gibowicz, S. J., & Kijko, A. (1994). *An introduction to mining seismology.* Academic Perss.

Keilis-Borok, V. I. (1964). Seismology and logics. In H. Odishaw (Ed.), *Research in geophysics* (Vol. 2, pp. 61–79). MIT Press.

Keilis-Borok, V. I., & Soloviev, A. A. (2003). *Nonlinear dynamics of the Lithosphere and earthquake prediction*, Springer.

McGuire, R. K. (2004). *Seismic Hazard and risk analysis*, Earthquake Engineering Research Institute.

Mendecki, A. J. (1993). Real time quantitative seismology in mines: Keynote Address. In R. P. Young (Ed.), *Proceedings 3rd International Symposium on Rockbursts and Seismicity in Mines, Kingston, ON, Canada* (pp. 287–295), Balkema, Rotterdam.

Mendecki, A. J. (1997). *Seismic monitoring in mines* (1 ed., 262 pp.). Chapman and Hall.

Mendecki, A. J., van Aswegen, G., & Mountfort, P. (1999). A guide to routine seismic monitoring in mines. In A. J. Jager & J. A. Ryder (Eds.), *A Handbook on Rock Engineering Practice for Tabular Hard Rock Mines* (chap. 9, pp. 287–309). The Safety in Mines Research Advisory Committee.

Ortlepp, W. D. (1997). *Rock fracture and rockbursts - An illustrative study.* Monograph Series M9 (126 pp.). South Afican Institute of Mining and Metallurgy.

Shearer, P. M. (2009). *Introduction to seismology* (2 ed., 396 pp.). Cambridge University Press.

Sornette, D. (2004). *Critical phenomena in natural sciences, chaos, fractals, selforganization and disorder: Concepts and tools*, synergetics. Springer.

van Aswegen, G. (2008). Ortlepp Shears - dynamic brittle shears of South African gold mines. In Y. Potvin (Ed.), *Proceedings 1st Southern Hemisphere International Rock Mechanics Symposium, Perth* (pp. 111–120).

Contents

1 Preliminaries... 1
   1.1   Progress and Limitations in Seismic Monitoring in Mines............ 1
   1.2   Earthquakes and Mine Induced Seismicity................................ 9
   1.3   Seismic Hazard Factors in Mines... 14
   1.4   Probabilistic Hazard: Long, Intermediate, and Short Term........... 15
   1.5   Seismic Hazard in the Time and Volume Mined Domain.............. 17
        1.5.1   Seismogenic Volume, Time Span of Data, and
                 De-clustering.. 19
        1.5.2   Testing the Quality and Integrity of Data...................... 20
   1.6   Seismic Hazard Management Plan: Seismic Monitoring................ 22
        1.6.1   Objectives of Seismic Monitoring in Mines.................... 23
        1.6.2   System Requirements.. 25
        1.6.3   Recommended Analysis.. 29
   1.7   A Short Note on Magnitude Scales... 31
References... 36

2 Seismic Sources in Rock.. 41
   2.1   Introduction... 41
   2.2   Seismic Sources: Glossary of Terms.. 43
   2.3   Seismic Potency and Potency Tensor....................................... 46
   2.4   Radiated Seismic Energy.. 49
   2.5   Apparent Stress, Energy Index, and Apparent Volume................. 51
   2.6   Circular Crack: Static Solution by Eshelby................................ 54
   2.7   Circular Crack Model by Brune.. 56
   2.8   More General Point Source Models 60
        2.8.1   Source Time Functions and Spectra............................ 62
        2.8.2   Far Field in Time Domain and Spectra........................ 64
        2.8.3   Q Corrections and Site Effects................................... 68
   2.9   Frequency Range $\log P$ and $\log E$...................................... 72
   2.10 Source Parameters from Spectra: Examples.............................. 78

	2.11	Final Static Deformation for Double-Couple Source....................	83
		2.11.1 Radiation Patterns...	83
		2.11.2 Final Static Displacement and Induced Strain................	85
	References...		90

3 Monitoring Rock Mass Stability.. 95
   3.1 Note on Time, Order, Disorder, and Stability............................. 95
   3.2 Stability of Deformation and Stability of a System 99
   3.3 Seismic Softening and Accelerating Deformation........................ 102
   3.4 Statistical Parameters of Co-seismic Deformation 104
      3.4.1 Seismic, Strain, Strain Rate, Stress, and Stiffness 104
      3.4.2 Seismic Viscosity, Relaxation Time, and Deborah
            Number... 105
      3.4.3 Seismic Diffusivity 106
      3.4.4 Seismic Schmidt Number.. 108
      3.4.5 Shape Factor—Sphericity... 108
   3.5 Seismic Stability Analysis.. 109
      3.5.1 Assumptions and Data Selection................................. 109
      3.5.2 Stability Example... 110
      3.5.3 Stability—General Comments..................................... 115
   References... 116

4 Size Distribution and Seismic Hazard... 119
   4.1 Fat Tails, Power Laws, Extreme, and Unexpected Events.............. 119
   4.2 Open-Ended Power Law (OE).. 121
      4.2.1 Selection of P_{min} 123
      4.2.2 Estimation of α and β of the OE Relation...................... 124
   4.3 Open-Ended Tapered Power Law (OET)..................................... 125
   4.4 Upper Truncated Power Law (UT) 126
      4.4.1 Estimation of α and β of the UT Relation 127
      4.4.2 Comments on Data Selection and Parameter
            Estimation 128
   4.5 Confidence Limits... 129
      4.5.1 Confidence Limits for β 130
      4.5.2 Confidence Limits for the Power Law Fit 130
   4.6 Utility of Power Law Distributions... 131
      4.6.1 Limitations and Benefits... 131
      4.6.2 Missing Potency and β... 132
      4.6.3 Power Law and log(Energy) vs. log(Potency)
            Relation 133
      4.6.4 Power Law and Stress Transfer 136
      4.6.5 Power Law and Information Entropy........................... 137
      4.6.6 Power Law Exponent, Rock Properties, Stress,
            and Stiffness 138
   4.7 Expected Maximum Event Size.. 140
      4.7.1 What Is P_{max} or m_{max} 140

	4.7.2	Balance of the Effective Volume Mined and P_{max} 142

- 4.7.2 Balance of the Effective Volume Mined and P_{max} 142
- 4.7.3 Order Statistics: P_{max}—The Upper Limit to the Next Largest Event 144
- 4.7.4 Record Statistics—The Next Record Breaking Event 146
- 4.7.5 The Expected Next Record Breaking Potency 147
- 4.7.6 Rank Plot 149
- References 151

5 Time Distribution and Seismic Hazard 155
- 5.1 Time Characteristics of Seismic Activity 156
  - 5.1.1 Coefficient of Variation and Proportional Variability 157
- 5.2 Intermediate Term Hazard 160
  - 5.2.1 Homogeneous Poisson Distribution 161
  - 5.2.2 Probabilities of Exceedance and Mean Recurrence Times 162
- 5.3 Empirical Probabilities from the Observed Recurrence Times 166
- 5.4 Example: Seismic Hazard Difference Between Three Mines 167
- 5.5 Example: Seismic Hazard Difference in Time 172
- 5.6 Seismic Response to Step Loading: Short Term Hazard 179
  - 5.6.1 Introduction 179
  - 5.6.2 Seismicity Rate Change 180
  - 5.6.3 Omori Relaxation Function 183
  - 5.6.4 Omori Distribution, Parameters, and Simple Omori 184
  - 5.6.5 Stretched Exponential Relaxation Function 186
  - 5.6.6 Stretched Exponential Distribution and Its Parameters 187
  - 5.6.7 Hazard Function, Conditional Probability, and Stretched Exponential 189
  - 5.6.8 Information Entropy and Stretched Exponential 190
  - 5.6.9 Non-stationary Poisson Process 191
  - 5.6.10 Aftershock Sequence After $\log P = 2.61$ Event 193
  - 5.6.11 Aftershock Sequence After $\log P = 5.47$ Event 197
- 5.7 Seismic Rock Mass Response to Blasting Sequence 200
  - 5.7.1 Introduction 200
  - 5.7.2 Scaled Volume and Proximity Index of Blasts 202
  - 5.7.3 Example: Proximity of Blasting and Seismic Response 203
- References 210

6 Ground Motion Hazard 213
- 6.1 Seismic Waves, Transient Strains, and Stresses 214
- 6.2 Static and Dynamic Stresses Around a Circular Tunnel 218
- 6.3 Ground Motion at Source 223
- 6.4 Ground Motion Characteristics 226
  - 6.4.1 Engineering Characteristics of Ground Motion 227
- 6.5 Ground Motion Amplification at the Skin of Excavation 231
  - 6.5.1 Ejection Velocity 236

	6.6	Simple GMPE and Its Utility	238
		6.6.1 Introduction	238
		6.6.2 Simple GMPE for PGV and CAD (Based on Mendecki, 2019)	241
		6.6.3 SGMPE Example	244
		6.6.4 Perceptibility of Ground Motion	247
		6.6.5 Damage Inspection and Cumulative CAD Plots	247
		6.6.6 Probability $\Pr(\geq PGV, R, \Delta T)$	250
		6.6.7 Seismic Fragility Curves and Damage Potential	251
		6.6.8 Seismic Ground Motion Alert Program—GMAP	253
	6.7	Seismic Ground Motion Alert System—GMAS	261
	6.8	Mapping Seismic Ground Motion Hazard	265
		6.8.1 Methodology	265
		6.8.2 Example: Data and Size Distribution	266
		6.8.3 GMPE and Survival Function, $\Pr(\geq GMP_x; P, R)$	268
		6.8.4 Distribution of Distances	269
		6.8.5 Probabilities and Hazard Maps	270
		6.8.6 Limitations	271
	6.9	Modelling Ground Motion—Deterministic Hazard	273
		6.9.1 Introduction	273
		6.9.2 Implementation and Examples	278
	References		287
A	**Basic Concepts in Probability and Statistics**		**293**
	A.1	Basic Concepts in Probability	293
		A.1.1 Random Variable	293
		A.1.2 Union and Conditional Probability	294
		A.1.3 Distribution Function and Bimodality	295
		A.1.4 Conditional Probability and Hazard Function	296
	A.2	Basic Concepts in Statistics	297
		A.2.1 Variance, Coefficient of Variation, and Covariance	298
	A.3	Statistical Dependence	299
		A.3.1 Correlation Coefficient	299
		A.3.2 Hoeffding Test	300
		A.3.3 Distance Correlation	300
		A.3.4 Information Entropy	301
		A.3.5 Mutual Information	303
		A.3.6 Example: Anscombe Test	304
		A.3.7 Example: Noisy Data	306
		A.3.8 Example: Source Parameters	307
	A.4	Correlation Integral and Correlation Dimension	308
	A.5	Random Walk and Diffusion	312
		A.5.1 Space-Time Distribution of Aftershocks	313
	References		314

About the Author

Aleksander J. Mendecki graduated with a PhD in Seismology from the Silesian University of Technology in Poland in 1981, where he also worked as an Assistant Professor. In 1985, he became the Head of the Seismology Department at the Anglo American Corporation in South Africa. In 1991, he was appointed as the Managing Director and the Head of Research at ISS International Limited (ISSI), and in 1994, he worked as the Managing Director of ISS Pacific (ISSP). In 2010, he founded the Institute of Mine Seismology in South Africa and in Australia, and with the acquisition of both ISSI and ISSP, he became their Chairman and Head of Research. His main areas of expertise are monitoring seismic rock mass response to mining, quantification of seismic sources and near-source ground motion, deterministic and probabilistic seismic hazard assessment in mines, and the application of quantitative seismology to rock mass stability. He is the editor and the main author of the book Seismic Monitoring in Mines published by Chapman & Hall, London, 1997, and the author of the book Mine Seismology Reference Book: Seismic Hazard published in 2016. He has published a number of papers in international journals and presented and published keynote lectures resulting from the International Symposia on Rockburst and Seismicity in Mines, Kingston, Canada, in 1993; Krakow, Poland, in 1997; Johannesburg, South Africa, in 2000; Dalian, China, 2009; in St Petersburg, Russia, in 2012; and Santiago de Chile, Chile, in 2017.

Chapter 1
Preliminaries

Abstract This chapter starts with the authors' views on different aspects of mine seismology, states the main differences between earthquakes and mine induced seismicity, lists the major factors influencing seismic hazard in mines, and proceeds to the general formulation of the seismic hazard management plan for seismic monitoring. Such plan should specify (1) the objectives of seismic monitoring for a particular mine, (2) what the mine needs to have in terms of technology and skill, and (3) what the mine needs to do daily, weekly, monthly, and yearly to achieve the stated objectives. At the end of this chapter, almost as an appendix, is a short note on seismic magnitude scales.

1.1 Progress and Limitations in Seismic Monitoring in Mines

Over the last 40 years, considerable progress has been made in seismic monitoring technology, the quantification of seismic sources, and the quantification and analysis of mining induced seismicity. However, not enough is said about the limitations of seismic sensors, shortcomings in data acquisition, our limited understanding of statistical mechanics, non-linear mechanics and rock mass dynamics, and the tendency to over-interpretation.

Seismic Sensors and Systems The world first digital and intelligent seismic monitoring system, i.e. with A/D conversion at the sensor site, and each site working independently and as a part of the network, was developed in South Africa and began routine operation in January 1988. There were 48 three-component sites mostly located underground with a few on surface to monitor seismic activity associated with gold mining in the Welkom area that spanned 20 km by 15 km (Mendecki et al., 1988). At that time more than a hundred thousand people worked daily at depths of 500 to 2500 m below surface applying scattered mining method, imposed by geological conditions to the tabular reefs, extracting 28 billion tonnes of ore and waste rock per year and producing almost 200 tonnes of gold.

Since then, a few hundred mines worldwide adopted digital seismic monitoring to manage seismic rock mass response to mining. Consequently, a number of geotechnical practitioners and geophysicists were exposed to, and trained in, the practical aspects of running seismic networks and in routine data interpretation. In the process, a new competence has emerged, namely that of mine seismologist.

Digital processing has benefited greatly from improvements in semiconductor technology over the past few decades, and, to a lesser extent, so has analogue signal conditioning and conversion. But transduction from ground motion to an electrical analogue, especially in the form of sensors which are rugged and small enough for installation in underground boreholes, has not. The result is that sensors are often the limiting factor in the performance of the seismic system as a whole. The principal reason is mass: Even in the noisy environment of a mine, with events occurring relatively close to the sensors, useful signals are small, and the simplest way to decrease sensor noise is to increase the mass, but the improvement in semiconductor price/performance has been attained largely by shrinking the size of the individual features. Applications with lower signal to noise ratio requirements, such as vehicle airbag triggers, use low-cost semiconductor MEMS, i.e. micro-electro-mechanical systems, accelerometers quite successfully. For mines, however, these lack dynamic range and can measure only the strongest ground motions, for example on the skin of an excavation during a damaging event. The passive geophone with coil and magnet introduced over 60 years ago is still holding its own in mine seismic monitoring. While they have limitations, we are not always able to utilise fully their capabilities mainly by incorrect installation and by applying an approximate transfer function that translates the voltage generated by the geophone into the instrumental ground velocities. In addition, there are transverse resonances (spurious frequencies) which make themselves felt at some sites but are not consistent enough from event to event to allow removal by site response correction.

Optical cavity accelerometers have been constructed which use optical cooling to reduce the thermal noise from a small mass with the high dynamic range of the electro-mechanical force balance accelerometers and wider bandwidth but are still relatively expensive.

To detect smaller events in mines, accelerometers or velocity sensors are generally used which attenuate low frequencies compared to displacement sensors. This is problematical since the measurement of seismic potency is taken from the low frequency asymptote of the displacement spectrum, which then requires integration of the velocity or acceleration ground motion recordings. The magnification of low frequencies inherent in the integration process, especially when compounded by the deconvolution of a velocity sensor response, must be carefully controlled to prevent the data from being swamped by noise. This can even lead to difficulties in picking the dominant frequency.

The new emerging technology in seismic monitoring is distributed acoustic sensing (DAS) that uses an optoelectronic instrument connected to an optical fibre to measure strain or strain rate at many positions along the fibre, effectively working as a seismic array (Lindsey & Martin, 2021). However, applications of DAS in underground mines are limited by a relatively high cost of interrogators, higher noise floor, the single-component nature of the recorded signals, and by the need for tight, uniform coupling of the fibre with the surrounding rock mass. Loosely coupled

1.1 Progress and Limitations in Seismic Monitoring in Mines

sections of fibre produce strong site effects in some cases larger than the actual seismic wave (duToit et al., 2022). But, if installed and grouted in longer boreholes, one can use it to constrain source location, to measure strain changes close to hydrofracturing sites (Luo et al., 2021), for seismic rock mass characterisation and exploration. It can also be used for tailings dams monitoring.

Seismic data on ground motion are grossly under-sampled in space and to a degree also in time. As I stated in Mendecki (1997b) ".. by analysing only good quality waveforms or well behaved seismic events one is utilising only a fraction of the time rock mass responds to mining seismically. In the case of 1000 seismic events recorded and analysed per day of an average duration, say 0.1 second each, one is listening to the rock mass for only for 0.1% of the day. Surely there is useful information lost during the remaining 99.9% of the time, since there are numerous coherent structures associated with convolved fracturing, unprocessable tremor type events and nonstochastic noise that constitutes a legitimate seismic rock mass response to mining".

Hence, there was a need to acquire and utilise continuous data. Over the last 10 years or so, progress in data communication facilitated a collection of continuous, as opposed to triggered, waveforms, and today, an average seismic network in a mine produces and acquires 10 billion samples of data per day, and largest systems produce 100 billion samples. Such a massive amount of data puts a great demand on our capability to process, characterise, and quantify it. The use of DAS will magnify the problem. Recent progress in machine learning and convolutional neural networks (CNN) for phase picking facilitates the development of seismic monitoring workflow platforms to seed databases with useful information derived from continuous waveforms, e.g. see Trugman et al. (2022) or Zhang (2022), and Arowsmith et al. (2022) for a review of big data seismology.

Quantitative Seismology Digital systems provide quality data that facilitate the quantification of ground motion parameters, seismic sources, and seismicity. Apart from the timing and location, a seismic source is quantified by seismic potency, or seismic moment, radiated seismic energy, and predominant frequency and is characterised by its source mechanism that decomposes the inelastic strain tensor into isotropic and deviatoric components.

The importance of the quality of seismological processing and the integrity of data can hardly be overestimated. The process starts with the location of seismic events, which is important for the following reasons. (1) It indicates the location of potential rockbursts. (2) All subsequent seismological processing, e.g. seismic source parameter and attenuation and velocity inversion, depends on location. (3) All subsequent interpretation of individual events depends on location, e.g. events far from active mining, close to excavations or, in general, in places not predicted by numerical modelling, may raise concern. (4) All subsequent interpretation of seismicity, e.g. spatial interaction, clustering—specifically localisation around planes, migration, spatial and temporal gradients of seismic parameters, and other patterns are judged by their location and timing. But accurate location is difficult because it relies on reconstructing the complex wave path from the source to the receiver and on picking the arrivals of the appropriate phases. Also, there are thousands of these events each day to be located and seismologically processed and the response time matters—information delayed may be information denied—with consequences that may exceed the cost of monitoring.

Seismicity is defined by a number of seismic events occurring within a given volume of rock over a certain period of time and is quantified using mainly the four, largely independent, quantities. (1) Average time between events. (2) Average distance between consecutive events. (3) Cumulative seismic potency or seismic moment. (4) Cumulative radiated energies. A quantitative description of seismic events and of seismicity allows one to derive information about their size and time distributions, spatial and temporal pattern formation, migration or diffusion, and about changes in the strain and stress regime and the rheological properties of the rock mass associated with seismic radiation. Although seismic waveforms do not provide direct information about absolute stresses, they do provide useful information about stress orientation and about the co-seismic spatial and temporal strain and stress changes.

In today's practice, we quantify seismic sources by fitting spectra of the observed waveforms to the source model which predicts a far-field displacement spectrum that is constant at low frequencies and proportional to the seismic potency or seismic moment (Keilis-Borok et al., 1960) and decays as the inverse squared power at higher frequencies, the so-called ω^2 model (Aki, 1967 and Brune, 1970). In mines, seismic events close to excavations and associated fractures zones have larger isotropic components, therefore are slower, radiate less energy at higher frequencies, and deliver displacement spectra that decay according to an $\omega^{2.5}$ or in some caving mines even an ω^3 model.

In general, a simple point source representation of seismic sources, whether crack-like or the pulse-like, does not include the co-seismic generation of rock damage in the source volume which alters the seismic radiation (Johnson & Sammis, 2001; Sammis et al., 2009). There is ongoing work on more realistic sources with volume component (Lyakhovsky et al., 2016; Kurzon et al., 2019; Johnson et al., 2021).

Production Data The seismic rock mass response to mining is mainly the result of stress changes due to rock extraction. The catalogue of seismic data is used to gain insight into people's exposure to seismicity and to quantify the evolving hazard as mining progresses. If seismic data is used in conjunction with data on rock extraction, i.e. the timing, location, and volume of the in situ rock extracted, then we may gain insight into possible causes of and guide the effort into prevention and control of potential rock mass instabilities that could result in rockbursts.

Complexity of the Problem On a macroscopic scale, all processes in nature are dissipative. Natural systems consist of a large number of elements which, at any given time, are not in the same state. Therefore, in order to accommodate the differences, a macroscopic system spontaneously generates local flow of energy and momentum in addition to those imposed by the external conditions. Close to equilibrium, the distribution of fluctuations is more or less random, the correlation time and the correlation length are short, and non-linearities are mostly hidden. Away from equilibrium, the system is more susceptible to the action of intermittent intrinsic instabilities that, due to their non-linear nature, are agents of spatial and temporal correlations. The finite values of spatial and temporal correlations measure the distance from equilibrium, and as this distance grows, the influence of non-linearities increases. When the range of spatial fluctuations increases, elements of the different parts of the system interact and the system can generate and maintain a reproducible relationship among its distant parts. In this process of self-organisation, the system creates spatial and temporal patterns that are not directly imposed by external forces. When the range of correlations becomes comparable with the system size, the resulting coherence, or order, may influence its behaviour qualitatively, and the system may become critical and undergo bifurcation, i.e. transition to another state (Nicolis & Prigogine, 1977). The approach to the critical point and the nature of the instability depends on the degree of disorder or heterogeneity in the system. Increase in disorder leads to a slower transition and diffused instability, while highly homogeneous systems may crack in one go with little precursory behaviour.

An important agent in the development of spatial and temporal correlations in overstressed rock is seismic activity itself. By breaking numerous asperities, seismic events smooth the system, allowing transfer of stresses over larger distances and thus paving the way for even larger events. Disorder, on the other hand, plays a stabilising role and, to a degree, can be engineered into the system by a scattered layout and/or by a scattered sequence of mining or blasting. Disordered directions of local stresses and a slower loading rate may also play a role, since it promotes healing and thus stress roughening.

In mines, the dynamics is excited mainly by the transient deformation associated with sudden rock extraction. Each such excitation causes a certain spatial and temporal pattern of events to develop that prevails only over a limited period of time after which the next excitation causes a new, and in many cases, compounded pattern of events. These responses change not only the rock mass properties but also the energy balance and redistribute that energy across a wide range of scales. As a consequence, it generates power laws of the size distribution of seismic events.

Modelling with Seismic Data Science does not deliver a universal truth underlying a given natural phenomenon, but it does provide models of reality with various degrees of approximation. When building a model, we are trying to understand the mechanism that generates the observed data and to express it mathematically in the simplest possible way to capture the essence of the problem. But one can always find more than one model that can reproduce the observed data, the problem known as non-uniqueness. Therefore, models cannot be validated since we do not know if we use the correct equations to get a solution. What we can do is to verify the model to make sure that the equations used are solved correctly. To add credibility to the numerical modelling, we should try to adjust the model parameters to match the available relevant observations. This process is known as calibration. However, we can never claim that the model is calibrated because of the sparsity of observations and because the dynamic systems are evolving and the same action may produce different outcomes in future. Frequently, after calibration, a model still delivers a wrong prediction, so there is a need to re-calibrate again and again. This is also the case when modelling the seismic rock mass response to mining, the more data we incorporate trying to reconcile the model with data the better, however, we can never claim that the model is calibrated. See an excellent paper on the subject of verification, validation, and calibration of numerical model by Oreskes et al. (1994).

Models are useful to analyse the influence of various aspects of the system on its behaviour that helps to understand the model and indirectly the problem at hand. Here, the quality of a model may be quantified by the following ratio: the number of features the model can explain divided by the number of parameters (Ben-Zion, 2017). Models are also used to gain insight into the future behaviour of a system. While the predictive power of models is limited, their utility can be enhanced by scenario based modelling, when we compare various plausible future scenarios with the same model parameters in order to formulate a better outcome.

Past seismic data offers a reasonable forecast of the intermediate future of the size distribution of seismicity, but it is less successful in forecasting the time distribution and even less regarding the space distribution. Numerical modelling, however, is better in forecasting the potential locations of larger events but not so their timing.

Training Traditionally, the training and practice of mining and geotechnical engineers was based on classical mechanics of conservative, and therefore reversible, systems at equilibrium, with focus on static linear elastic concepts. Statics is concerned with the equilibrium state of bodies under the action of forces. In mechanics, equilibrium is a particular state in which both the velocities and the accelerations of all the material points of a system are equal to zero, that means no "fluctuations", i.e. no movements relative to the rest frame of the centre of mass. Such mechanics prohibits the interaction of system components, and it excludes the possibility of emergent behaviour.

The assumption of linearity implies superposition, i.e. the end effect of the combined action of two different causes is merely the superposition of the effects of each cause taken individually. But in a non-linear system adding a small cause to one that is already present can induce effects surpassing by far the amplitude of the individual effects. Elasticity is well suited to explain the propagation of seismic waves in the far field, but it fails to explain the near and intermediate field deformation and the transfer of stresses and strains over larger distances.

A seismic source is a volume of rock where linear elasticity breaks down, and the processes leading to and resulting from such instability are driven by non-linear dynamics embedded in dissipative, and therefore irreversible, inhomogeneous, or disordered, systems which are far from equilibrium. While the static elastic approach offered, and will continue to offer, useful solutions, its validity is limited to systems at, or close to, equilibrium. The real practical benefit of the applied static elastic solutions is underpinned by the ability and the experience of geotechnical engineers to interpret results taking into account the limitations of the method in relation to the problem at hand.

Fat Tails, Power Laws, and Seismic Hazard Extreme events can be considered as large deviations from the average behaviour in an evolving system (Frisch & Sornette, 1997). Their recurrences are characterised by the thickness of the tail that defines the probabilities of having large events. The thicker the tail the higher the probability, so we should expect the unexpected. A thicker tail implies not only more larger events, i.e. higher hazard, but also a less predictable size of the next largest event. Unfortunately, it also means a higher probability of being surprised by the Black Swan—a big event with major effect, unexpected at first but rationalised by hindsight. Such made up explanations after the fact create misconceptions that causes of these events are understood and, hence, that they can be avoided or predicted.

It is therefore of great interest to find out how the smaller events interact in creating the conditions for larger events to occur. Failure can then be thought of as a scaling-up process in which failure at one scale is part of the damage accumulation of the larger instability.

Understanding the mechanisms generating power laws and their implications is one of the subjects of statistical seismology, which is a field of research at the interface of probability theory, stochastic modelling, and statistical physics. While physical models try to understand the process completely, the stochastic models estimate probability distributions of potential outcomes by allowing for random variation in one or more parameters. The main objective of statistical seismology though is to derive the large-scale laws of a physical system from the specification of the relevant microscopic elements and their interactions. One of the important practical applications of statistical seismology is the assessment of seismic hazard.

Prediction Versus Forecasting A prediction can be understood as a deterministic binary statement, true or false, about a future event that can be validated or falsified with a single observation. A forecast can be defined as a statement of probability about a future event. In his excellent book, Prigogine (1980) writes "Even in physics, as in sociology, only various possible scenarios can be predicted", which should be a guiding principle for practitioners involved in numerical modelling. One should however remember the calculated probability reflects what we know about the system, and not what the system is actually doing. An individual forecast can never be validated by a single observation and requires multiple observations to establish a degree of confidence.

The United Nations Disaster Relief Organization (UNDRO, 1979), defined hazard in general as the probability of occurrence, within a specific period of time in a given area, of a potentially damaging phenomenon.

Uncertainty Uncertainty is the existence of more than one possibility, and it is measured by a set of probabilities assigned to a set of possibilities. Risk is a state of uncertainty where some of the possibilities involve a loss, and is measured by assigning losses to some possible outcomes. Therefore, one may have uncertainty without risk but not risk without uncertainty. The notion that events are uncertain is both complicated and uncomfortable, and therefore we tend to underestimate uncertainty and consequently underestimate risk.

In general, uncertainties can be divided into two categories: (1) Aleatory uncertainty, (from the Latin word alea for dice), or purely stochastic variability, where the ambiguity in outcome is inherent in the nature of the process, and no amount of additional measurements can reduce the inherent randomness. (2) Epistemic uncertainty, or ignorance or a lack of complete knowledge of the process, which results in certain influential variables not being considered, and affects our ability to model it. It also includes insufficient and inaccurate measurement. The epistemic uncertainty can be diminished by taking more data, by using more accurate instrumentation, by better experimental design and acquiring better insight into specific behaviour with which to develop more accurate models. The guiding concept in dealing with epistemic uncertainty should be that less data means larger uncertainty. While the aleatory variability reflects the natural randomness of the monitored process, the epistemic variability is the result of the uncertainty in the expected outcome due to a lack of knowledge or inaccurate measurements.

One way to deal with epistemic uncertainty is to use logic trees, first introduced by Kulkarni et al. (1984). A logic tree consists of a series of branches that describe the alternative models and/or parameter values. At each branch, there is a set of branch tips that represent the alternative credible models or parameter values. The weights on the branch tips represent the judgment about the credibility of the alternative models. The branch tip weights must sum to unity at each branch point. Only epistemic uncertainty should be on the logic tree. A common error is to put aleatory variability on some of the branches. Logic trees reflect the degree of belief of the group of experts in the alternative models. However, evaluating the alternative models involves considering alternative representations, new observations, and, in some cases, data from analogous regions. This process is also subjective. An alternative approach is to develop the single best model, but this requires a consensus by experts, which in some cases may be more difficult than constructing a logic tree.

Interpretation Managing seismic hazard involves judgment under uncertainty. Judgment, frequently defined as "an intelligent use of experience" or "a sense of what is important", is a vague attribute. In making predictions and judgments under uncertainty, people do not always follow probability theory but frequently relying on a limited number of heuristic principles: rules of thumb, educated guesses, intuitive judgments, experience-based reasoning, or simply on common sense. These rules reduce the complex tasks of assessing probabilities to simpler judgmental operations, that may be useful, but also may lead to severe and systematic errors (Tversky & Kahneman, 1974; Kahneman, 2003). Our understanding of how judgment works is far from satisfactory, and the definition of common sense is different to different people. Seismic hazard assessment delivers probabilities assuming a given probability distribution. However, we do not observe probability distributions, only a series of events from the system that generates data. So we assume a given probability distribution, say Poissonian, on the basis of limited data.

Motivational Bias Human judgment under uncertainty is also affected by motivational bias, which may be even more critical than statistical misconceptions or errors. In some cases, an expert may be motivated by incentives to see things in a certain way. But most frequently an expert is defensively conservative or may want to suppress uncertainty in order to appear authoritative or knowledgeable, or has taken a strong stand in the past and wants to be consistent. In addition, there may be a conflict of interest between the loyalty demanded by the organisation the expert represents or even by the client he consults for, and the expert's objectivity (Hogarth, 1975; Vick, 2002). To alleviate some of these problems, it is advisable to subject the applied methodologies, procedures, and results of analysis to peer review.

Communication To be effective, mine seismologists and geotechnical engineers need to communicate their observations to mine management and convince them that their advice is credible. There is a difference between being vigilant, that helps to establish credibility, and being alarmist. The best advise here is, state the expected, monitor, and notify if the observed exceeds the expected. And when planning ahead, it can help to expect the unexpected.

1.2 Earthquakes and Mine Induced Seismicity

Use of Magnitude Different mines use different magnitude scales that, in many cases, differ significantly and, in some cases, are not consistent over time. Therefore, in this book, the common logarithm of seismic potency, $\log P$, is used as a measure of magnitude. Scalar seismic potency is a product of two parameters, $P = \Delta\epsilon \cdot V$, where the $\Delta\epsilon$ is the strain change at source during the seismic event and V is the source volume. In hard rock strains $\Delta\epsilon \leq 10^{-6}$ are considered elastic and strains $\Delta\epsilon \geq 10^{-3}$ crack intact rock. Therefore, a seismic event with $\log P = 1.0$, i.e. Hanks-Kanamori $m = 1.59$, is not just 10 m^3 of something, but such event creates approximately $V = 10^1/10^{-3} = 10^4$ m^3 of rock subjected to cracks, which is substantial and may, and frequently does, lead to damage if its source is close to an excavation.

$\log P$ is simple, appropriate for the range of sizes of seismic events recorded in mines and independent of rigidity, and thus seismic hazard may be objectively compared between different mines and between different periods of time for the same mine. Table 1.1 translates selected values of $\log P$ into Hanks and Kanamori magnitude, expressed here in terms of potency, $m = (2/3) \log P + 0.92$ or $\log P = (3/2) m - 1.38$ and to Nuttli magnitude used in some Canadian mines, $m_N = 0.97m + 0.59$, which gives $m_N = 0.65 \log P + 1.48$ (Sonley & Atkinson, 2005).

A one parameter description is inadequate to gain an insight into the stress and strain changes at seismic sources. For that, one must quantify the two independent source parameters: the seismic potency, P, and radiated energy, E. It gives an easy estimate of the apparent stress, $\sigma_A = 10^{\log E - \log P}$, in Pascals, which is a model independent measure of stress change at the source.

Tectonic Earthquakes Earthquake driving forces cannot be controlled. They are fairly constant over time and relatively slow compared to changes in stresses induced

Table 1.1 log P, Hanks and Kanamori, m, and Nuttli (Sonley & Atkinson), m_N, magnitudes

log P	−5.0	−4.0	−3.0	−2.0	−1.0	0.0	1.0	2.0	3.0	4.0	5.0
m	−2.41	−1.75	−1.08	−0.41	0.25	0.92	1.59	2.25	2.92	3.59	4.25
m_N	−1.77	−1.12	−0.47	0.18	0.83	1.48	2.13	2.78	3.43	4.08	4.73
m	−5.0	−4.0	−3.0	−2.0	−1.0	0.0	1.0	2.0	3.0	4.0	5.0
log P	−8.88	−7.38	−5.88	−4.38	−2.88	−1.38	0.12	1.62	3.12	4.62	6.12
m_N	−4.26	−3.29	−2.32	−1.35	−0.38	0.59	1.56	2.53	3.50	4.47	5.44

by mining. For example, different segments of San Andreas fault in California slip at the rate of a few to 40 mm per year. This slow loading facilitates the process of self-organisation that leads to a state at which the system develops reproducible relationships among its distant parts. Over time the system creates well defined spatial and temporal patterns that are not directly imposed by external forces.

There is a specific model, called self-organised criticality, SOC, developed by Bak et al. (1987) that combines the concepts of self-organisation and criticality to explain complexity. It assumes that under very general conditions, dynamical systems, when slowly driven, organise themselves into a state with a structure whose statistical properties can be described by simple power laws. The system becomes critical in the sense that all parts of the system influence each other and such systems fluctuate around a state of marginal stability. The SOC is a phenomenological definition, and the more constructive one would be the SDIDT—Slowly Driven Interaction-Dominated Threshold Systems (Jensen, 1998). The slow drive is needed to allow the system to relax from one metastable configuration to the other. The notion of interaction dominated means that the dynamics of the system is dominated by the mutual interaction between many degrees of freedom rather than by the intrinsic dynamics of the individual degrees of freedom. The effect of a threshold is to allow a large number of static metastable configurations. The threshold instability means that the incoming energy slowly builds up the profiles until the threshold is locally overcome and the system then self-organises itself via fast relaxation events that dissipate the energy excess across the system.

A state of SOC would imply a slowly driven system far from equilibrium with small fluctuations about the critical state over large timescales and sensitivity to minor perturbations that could trigger large events that can span the length scale of the system. The model self-organises to produce a power law in the frequency size distribution, despite having very simple rules governing interactions and no tuning parameters. In the case of tectonics, the SOC would therefore require the crust to be perpetually near a state of global failure, rendering individual events unpredictable both temporally and spatially.

Alternatively, seismicity may be described as a process undergoing self-organised sub-criticality, SOSC, as suggested by Al-Kindy and Main (2003). SOSC is characterised by the system being below the critical point but still maintaining power law statistics on a local scale. An SOSC state would also suggest a finite degree of statistical predictability in the dynamics of the system, contrary to a pure SOC.

While the long term seismic and aseismic loading are the main preparatory processes for larger earthquakes, their exact timing can be influenced by transient loading caused by surface waves from other earthquakes (dynamic triggering), and/or earth tides, atmospheric pressure or heavy precipitation amongst others. These loading sources all produce a similar amplitude stress change on a fault that is three to four orders of magnitude less than the ambient stress on the fault, and they occur on timescales from short periods on the order of 10 to 100 seconds to long periods, of the order of a year (Johnston, 2017).

Mining Induced Seismicity Induced seismicity results from stress changes that are at least on the same order as the ambient pre-mining stresses. Triggered seismicity results from stress changes that are considerably smaller than the ambient pre-mining stress, i.e. when the system is close to critical. Mining cannot induce large earthquakes but could potentially trigger such events (McGarr et al., 2002).

The seismic rock mass response to mining is a result of stress changes due to rock extraction, blasting, hydrofracturing, and, to a lesser extent, due to the relocation of the extracted rock to rock waste sites, and therefore it is not a spontaneous process. Mining induces stresses at a particular place, at a particular time, and at a particular rate which are all highly variable compared to the tectonic regime. The average rate of deformation induced by mining is at least two orders of magnitude greater than the average deformation rate of tectonic plates. The bulk of seismic activity in mines starts with rock extraction, increases with the extraction ratio of the ore body, and, with the exception of large-scale mining taking place over a number of years, stops rather quickly with the cessation of mining.

Figure 1.1 left shows the cumulative number of events with $\log P \geq -2.0$ during 1435 hours before and 2200 after the main shock (MS) of $\log P \geq 2.61$ ($m \geq 2.66$) where size of the event scales with the radius of source volume and the colour indicates the distance of the event from the main shock.

The first aftershock with $\log P \geq -2.0$ was recorded 4, and the second 6.4 seconds after the MS. Within the first 16 hours after the main shock, there were 165 events that give the activity rate just over 10 events/hour and within the first

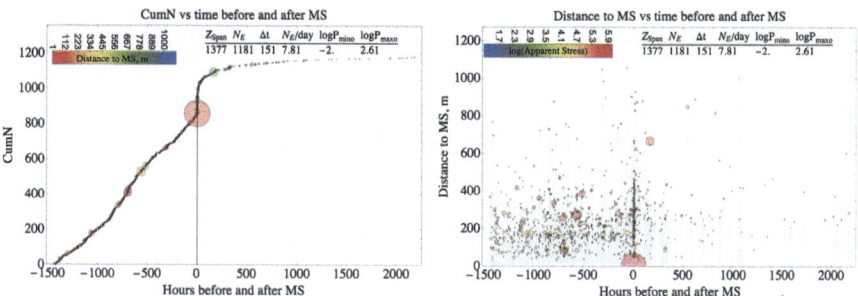

Fig. 1.1 Cumulative number *(left)* and distances *(right)* of seismic events to the MS with $\log P = 2.61$

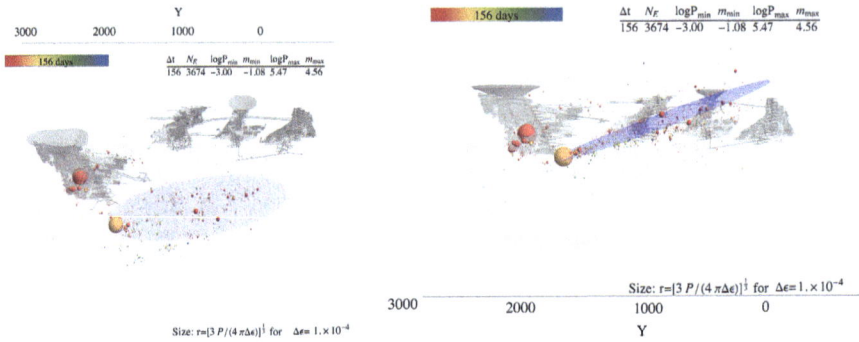

Fig. 1.2 Top *(left)* and section view *(right)* of the mine, including the nearest three mining operations. The main shock is shown in light grey and aftershocks coloured by the time of their occurrence

171 hours the rate of activity dropped to 1.4 events/hour. Figure 1.1 right shows the distances from the MS during the same time where colour scales with the logarithm of apparent stress, $\log \sigma_A = \log(E/P)$. The bulk of the immediate aftershocks spread almost 500 m from the MS.

With a very few exceptions, aftershock sequences after larger seismic events in mines or after major production blasts are not as well developed as they are after tectonic earthquakes, mainly due to faster relaxation facilitated by the presence of openings and extensive fracturing. However, there are a few exceptions where large events which are triggered by mining are actually driven by local tectonic stresses. In such cases, there are more aftershocks, and the aftershock sequences last longer and extend beyond the mining operations.

One such example is a $\log P = 5.47$ ($m = 4.57$) reverse slip event triggered by mining and driven mainly by tectonic sub-horizontal stress. Figure 1.2 shows the left corner and section views of the mine, including the three nearest mining operations. The main shock is marked by the large light grey sphere centred at the very bottom of the mine, and its aftershocks are coloured according to time. The size of each event represents the radius of the source volume taken as a sphere, $V = P/\Delta\epsilon$, where $\Delta\epsilon$ is the assumed strain change at the source, in this case $\Delta\epsilon = 1 \cdot 10^{-4}$ to make small aftershocks visible. The main shock was initiated off the bottom edge of the mine and propagated over one kilometre away from the mine along a sub-horizontal geological structure. The immediate aftershocks were at the bottom of the mine and away from the mine on the structure. The first aftershock with $\log P = 2.66$ one minute after the main shock located within the mine and the second largest aftershock, marked as dark yellow, with $\log P = 2.54$ occurred 37 days later.

Figure 1.3 left shows the cumulative number of events with $\log P \geq -3.0$ during 1338.7 hours before and 3730.5 hours after the MS shown here as the large red circle.

1.2 Earthquakes and Mine Induced Seismicity

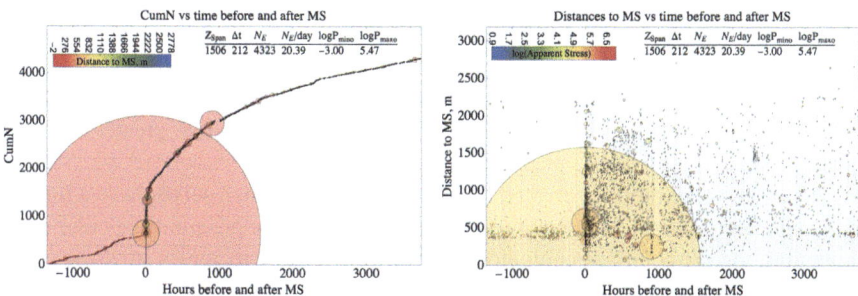

Fig. 1.3 Cumulative number *(left)* and distances to the main shock of seismic events with $\log P \geq -3.0$ before and after the main shock

Colour here scales with distance to the MS, and the size of the event here represents the radius of the source volume taken as a sphere, $V = P/\Delta\epsilon$, where $\Delta\epsilon$ is the assumed strain change at the source, in this case 10^{-2}. There was a steady rate of seismic activity, $\lambda_B = N(\log P \geq -3.0)/\Delta t_B = 0.484$, per hour before the main shock.

Figure 1.3 right shows distances of seismic events to the main shock over the same period of time where colour scales with the logarithm of apparent stress, $\log \sigma_A = \log(E/P)$. The spatial distribution of events before the MS is mostly concentrated at the mine, at distances of 300 to 500 m from the future MS, with only few events at distances over 500 m. From the very beginning of the aftershock sequence, the spatial distribution of aftershocks extended to over 2000 m from the MS. Approximately, 2900 hours after the MS one can see seismic activity at 200 m from the MS associated with the rehab of mining operations.

Smaller events in mines are highly correlated in space and in time with mining, while the largest events tend to occur after the extraction ratio and/or the depth of mining reach a critical level. The spatial distribution of larger events, however, is influenced by past mining layout and by the geological structures influenced by mining, and their temporal distribution is more random than that of small events.

Less frequently rockbursts result from a sudden shear fracture through pristine rock where no faults or discernible planar weaknesses existed before (Gay & Ortlepp, 1978). Certainly, to drive rupture through intact rock would require higher effective stress than to drive the same size event along pre-existing weakness, and therefore, it would produce higher ground velocities and therefore radiate more seismic energy.

Most frequently, the temporal distribution of larger events in mines is more random than that of small events. Small events are more clustered in time than larger events, while the time distribution of the larger events tends to a random or Poissonian distribution.

1.3 Seismic Hazard Factors in Mines

In mines, the rock mass dynamics is driven by the transient deformation associated with rock extraction, which is at least two orders of magnitude greater than the average slip rate of tectonic plates.

Where the excavation is created by blasting, the blasts themselves have a limited or local effect. The effects of mining are exacerbated by the proximity of older mine openings, the proximity of geological structures, and local tectonics stresses.

If the balance of the influx of energy and dissipation is not maintained, the system is susceptible to discharging its surplus energy in the form of larger seismic events, after which mining operations are frequently suspended and resumed only days or even months later.

Notwithstanding the complex nature of the inducement, seismic rock mass response to mining still develops reasonable size distributions and time relaxation patterns, which seismologists and geotechnical engineers use to manage seismic related risk.

Seismic hazard in mines is positively correlated with the following natural factors:

1. Virgin rock stress, which is a combination of depth and the local tectonic stress
2. The bulk strength of the rock mass
3. The degree of homogeneity, or smoothness, of the rock mass including geological features, specifically those with shear strength comparable to the shear stresses induced by mining excavations

In addition, there is a number of factors related to mining that may exacerbate the intensity of the seismic rock mass response to mining:

1. The extraction ratio
2. The spatial extent of the mined-out area or volume
3. The rate of mining and the spatial and temporal sequence of extraction
4. The additional stress induced by adjacent mining
5. The smoothness of the mine layout itself and in relation to the geological structures (Mendecki, 2012; Mendecki, 2013a)

Smoothness can be defined by the dimensionality of the object, as measured by its fractal dimension—the lower the fractal dimension the smoother the object. Fractals appear similar at any scale of observation. In mathematical terms, fractal objects exhibit fractional dimensionality, that is they are neither lines, nor surfaces or volumes. Their dimension falls in between the classical dimensions of Euclidean geometry. An object is fractal when its length L is a function of the length λ of the measuring device, $L \sim \lambda^{1-D}$, where D is the fractal dimension (Mandelbrot, 1967, 1975). If $N(\lambda)$ is the number of cubes of size λ needed to cover the object, then the box counting fractal dimension of an object can be estimated by $D = \ln N(\lambda) / \ln(1/\lambda)$ (Barnsley, 1988). Fractal dimension increases with the degree of irregularity, or raggedness of the object.

Seismic events originate and stop at inhomogeneities in the rock mass. By breaking numerous asperities, they smooth the system, allowing transfer of stresses over larger distances and thus paving the way for even larger events. Assuming that large seismic events are those small ones that were not stopped soon enough, then one way to manage seismic hazard is to engineer a mine layout that, together with the natural structures, is "rough" enough to limit the extent of ruptures. It has been argued that mining scenarios that introduce spatial heterogeneity, or roughness, may de-correlate the system and be less likely to develop larger dynamic instabilities (e.g. van Aswegen & Mendecki, 1999; Handley et al., 2000; Vieira et al., 2001; Mendecki, 2001 Figure 8; Mendecki, 2005; Durrheim et al., 2005).

1.4 Probabilistic Hazard: Long, Intermediate, and Short Term

Probabilistic seismic hazard analysis (PSHA) is a methodology that estimates how frequently a given level of a given ground motion parameter can be reached or exceeded at a point of interest, X, in future time ΔT, $\Pr[\geq GMP(X), \Delta T]$, where the GMP can be the peak ground velocity, PGV, or the cumulative absolute displacement, CAD.

The PSHA incorporates all potential sizes of seismic events, the frequency of their occurrence and source distances via the ground motion prediction equation (GMPE) to estimate the combined probability of exceedance at X for different time periods ΔT. It also disaggregates seismic hazard into its contribution from event size, distance, and the normalised residual, expressed in terms of the number of standard deviations from the median ground motion estimated by the assumed GMPE. The main objective of disaggregation is to show which component dominates seismic hazard at a given site.

For mines, the PSHA is frequently limited to the size distribution hazard, i.e. assessment of the probability that a potentially damaging event $\geq \log P$ will occur inside a given seismogenic volume of rock, ΔV, in future time ΔT, or while extracting a given volume of rock ΔV_m, i.e. $\Pr[\geq \log P(\Delta V), \Delta T \text{ or } \Delta V_m]$. To estimate this probability, we need to assume the size and the time distributions of seismic events for a given seismogenic volume of rock. The most frequently used size distribution is the upper truncated power law, and it is recommended that the upper limit for a given data set be estimated independently. As for time distribution it is frequently assumed that seismic activity can be described by a stationary Poisson process, whether in the time or in the volume mined domain, and this simplification is in part neutralised by making regular hazard estimates and comparing differences. In cases where there is a clear trend in seismic activity, one can try to extrapolate the parameters of the size distribution, and alternatively, one can apply the non-stationary Poisson process with a suitable intensity function.

This limiting PSHA in mines to $\Pr\left[\geq \log P\left(\Delta V\right), \Delta T \text{ or } \Delta V_m\right]$ is justified for smaller operations where the linear size of the target event, $\log P$, is comparable with the size of a mine. Moreover, most seismic systems for mines are designed to locate events and to estimate seismic source parameters, mainly potency P and radiated energy E, rather than to estimate the resulting ground motion at the skin of excavations. For this reason, seismic sensors are placed in boreholes, away from excavations, to avoid the very site effects that amplify ground motion at certain frequencies and contribute to damage. The ground motion prediction equations derived from such measurements will certainly underestimate seismic load and need to be corrected.

Seismic risk is a product of the probability of occurrence and the potential liability or vulnerability, $\mathcal{L}(X)$, at a given site, $\text{Risk}(\geq v, \Delta T) = \int_X \{\Pr[\geq v(X), \Delta T] \cdot \mathcal{L}(X)\} dX$. Since probability is dimensionless, seismic risk is expressed in the units of liability $\mathcal{L}$. Information on liability or vulnerability is usually more difficult to quantify than the information on seismic hazard, because it involves not only the material damage, injury and in some cases loss of life, but also the cost of business disruption, which is difficult to predict.

The intermediate term hazard would then cover a period of up to one Year, and results are presented as the expected probabilities versus different size events for selected time periods, e.g. 1, 3, 6, and 12 months. The long term hazard in a mining context should cover a period between one year and the expected life span of the mine but, because the uncertainties increase with time, 5 years may be sufficient. For the long term hazard, one can estimate the expected probabilities associated with different size events, $\log P$ or, for specific size events that are associated with selected recurrence times, $\log P\left(\bar{t}_{1y}\right)$, $\log P\left(\bar{t}_{2y}\right)$, $\log P\left(\bar{t}_{3y}\right)$, and $\log P\left(\bar{t}_{5y}\right)$, and for the estimated next record breaking $\log P_{nrb}$, versus time. Note that the data set used for the long term hazard assessment may frequently be the same as for the intermediate one. In such a case, the only difference is the presentation of results in the time domain, as opposed to the magnitude or $\log P$ domain, and the confidence limits.

The short term hazard would cover a period of less than a month and include the estimates of interval (or re-entry) probabilities immediately after sudden loading by larger seismic events or after production blasts.

In many cases, seismic hazard can only be estimated by means of numerical modelling, since there may not be enough seismic data available to perform such analysis. However, the results of numerical studies should be confirmed regularly as soon as data becomes available.

Limitations There is an exchange of views in the literature on the utility of PSHA. The spectrum of views is wide, from suggestions to drop it altogether (e.g. Mulargia et al., 2017) to the more pragmatic, stating that the shortcomings of the method do not invalidate the existence of the hazard curve, which comprises the basic assumption for PSHA (Anderson & Biasi, 2016).

There is a difference in the nature of ground motion hazard to underground structures due to seismic events in mines and to surface structures due to earthquakes. The main differences are the distances involved. The near-source ground motion due to small and larger seismic events is similar, but the hypocentral distances to underground excavations in mines are very small, say from a few metres to a few hundred metres. Earthquakes are usually many kilometres away. Therefore, earthquake engineers are mainly concerned with ground motions from larger earthquakes, but these earthquakes are less frequent and their activity rates are less certain. Recurrence times for large events in mines are also uncertain. One can then expect that the hazard maps for mines can better estimate the potential for less severe damage caused by smaller and medium size events than the infrequent large events.

This is why mines prone to larger events may wish to supplement probabilistic analysis with deterministic analysis, i.e. ground motion simulation. Such a simulation involves kinematic modelling of ground motion produced by sources defined by their expected maximum potency, or magnitude, and placed at the most likely locations that can produce the strongest level of peak ground velocity at a given site or sites. The likely locations of these sources may be determined by numerical modelling of the induced shear stresses on geological structures. In a kinematic model, the source process is defined by the spatial and temporal distribution of the slip vector, the local slip velocity function, and the rupture velocity, without taking into consideration the forces and stresses acting at the source, see Mendecki and Lötter (2011) and Mendecki (2016).

Probabilistic and deterministic methods for hazard assessment have advantages and disadvantages. Probabilistic methods can be viewed as inclusive of all deterministic events with a non-zero probability of occurrence. In this context, a proper deterministic method that models a particular larger event should ensure that that event is realistic, i.e. with a finite probability of occurrence. This points to the complementary nature of deterministic and probabilistic analyses: Deterministic events can be checked with a probabilistic analysis to ensure that the event is probable, and probabilistic analyses can be checked with deterministic events to see that rational, realistic hypotheses of concern have been included in the analyses (McGuire, 2001). However, the deterministic analysis can better account for specifics, i.e. the path and the site effects associated with a given strategic infrastructure.

1.5 Seismic Hazard in the Time and Volume Mined Domain

The mean recurrence interval between events above a certain potency estimated over the period of time Δt is $\bar{t} (\geq P) = \Delta t / N (\geq P)$, where Δt is the time span of data used and $N (\geq P)$ is the number of events not smaller than P over that time. Seismic activity rate for these events is then $1/\bar{t} (\geq P)$. The mean volume mined between events above a certain size during extraction of V_m volume of rock is $\bar{V} (\geq P) = V_m / N (\geq P)$. The number of events, $N (\geq P)$, per unit of volume of rock extraction is $1/\bar{V} (\geq P)$. Note that the terms "mean inter-event time", "mean recurrence interval" may be deceptive since in many cases the dispersion from the mean, as measured by the standard deviation, is comparable with the mean value. If the standard deviation of the observed recurrence intervals is less than 50% of the mean recurrence interval, then the seismic behaviour may be assumed to be periodic rather than episodic, and calculated probability estimates can be considered reasonable. For a very short time interval into the future, $\Delta T < \bar{t}(\geq P)$, the probability of having an event with potency not smaller than P can be estimated as $\Pr(\geq P) \simeq \Delta T/\bar{t}(\geq P)$. The calculated recurrence interval or the mean volume mined that stretch well beyond the span of the data set Δt or V_m should be treated with caution.

Example Data sets from three mines, A, B, and C, were collected over the same two year period, $\Delta t = 678$ days, all related to tabular mining with vertical principal stresses but with different geological structures, mining layouts, extraction ratios, depths, and rates of mining. Mine A practiced long-wall mining of a highly extracted tabular reef and mine B the sequential grid method. In both mines the rock extraction took place at practically the same depth of 3300 m. The extraction ratio of the tabular reef in mine A is over 80%, in mine B is 70%, and in mine C 60%. A simple numerical elastic model shows that due to the higher extraction ratio the mean vertical stress calculated over the un-mined areas in mine A is 1.7 times higher than in mine B. Mine C practiced scattered mining, imposed by the presence of larger faults, at an average depth of 1755 m.

Figure 1.4 compares cumulative volume mined and cumulative seismic potency in the three mines. Mine B delivered the highest production, production rate, and the highest seismic potency release. However, mine A delivered the highest observed potency release per volume mined, $\sum P/V_m$. Compared to A and B, the seismic potency release in mine C is low. Note that the production in B is 4.3 times higher than in A, however, the potency release per day is only 1.4 times higher, and the potency release per unit of volume mined is 3.1 times lower. Clearly, the most informative parameter of the inherent hazard is $\sum P/V_m$, and if it is unacceptably high, then slower pace of mining only delays the inevitable.

Seismic hazard induced by mining should therefore be expressed as a function of time, e.g. by the mean inter-event times, and as a function of rock extraction, e.g. by the mean inter-event volumes mined. During times when mining is suspended,

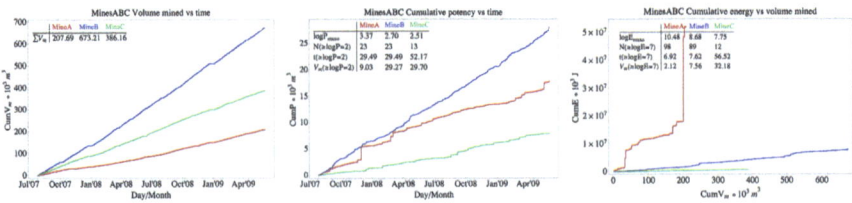

Fig. 1.4 All cumulative: volume mined versus time *(left)*, potency release versus time *(centre)*, and potency release versus volume mined *(right)* for mines A, B, and C over 678 days

1.5 Seismic Hazard in the Time and Volume Mined Domain

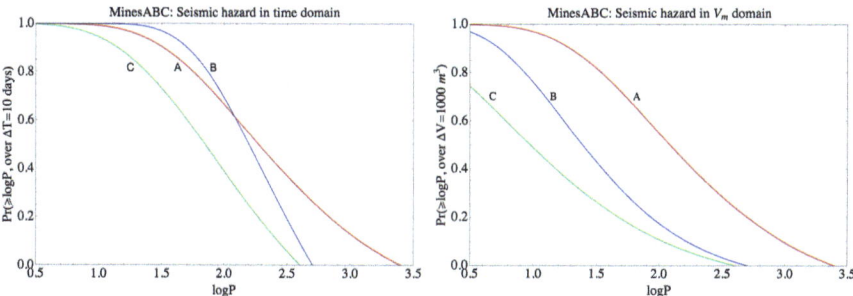

Fig. 1.5 Seismic hazard for the three mines in the time domain *(left)*, and in the volume mined domain *(right)*

seismic activity decays very quickly and the mean inter-event times increase, indicating lower hazard. The mean inter-event volumes mined, however, stay the same since the hazard potential did not change, and when mining is resumed, it will be very much the same as it was before.

Figure 1.5 left shows seismic hazard estimated for the three mines A, B, and C in the time domain for $\Delta T = 10$ days, assuming their current rate of mining. Seismic hazard at mine C is clearly the lowest, although converging with mine B for large potency events. Note the crossover point between mines A and B below which seismic hazard for data set B is higher. Figure 1.5 right shows seismic hazard potential, estimated in the volume mine domain, assuming all three mines extract the same volume of rock $\Delta V_m = 1000$ m^3, otherwise for the same set of parameters. Clearly, hazard potential at mine A is the highest of all three mines, then at mine B, but again slowly converging with mine C for large potency events.

1.5.1 Seismogenic Volume, Time Span of Data, and De-clustering

Probabilistic hazard analysis is based on data selected from a given volume of rock, ΔV, over a certain period of time, Δt. The volume ΔV should include all seismic events that are interdependent. Many properties of induced seismic activity indicate a hierarchical organisation, suggesting a connection among events in space and in time. The range of spatial interaction is at least equal to the size of the largest event. Kijko et al. (1993) detected similarities in probability patterns between clusters of seismic activity separated by more than 1000 m, although they did not consider production. If the linear size of a stand-alone mine is comparable with the size of the largest expected event, then ΔV should be defined by the volume of seismic activity within and around the mine. For the same reason, two or more adjacent mines may be treated as one seismogenic volume. It is useful to test the spatial extent of the seismic response to larger production blasts or to larger seismic events which, in some cases, may cover the entire mine.

Similar considerations need to be applied to Δt, which should span as far back as possible. Since mining scenarios may change, it is advisable to select the past data that is most relevant to future mining. The calculated probabilities over future time ΔT larger than Δt should be treated with caution. One should avoid sliding time windows and rather resort to cumulative windows and then normalise to compare. The future seismic rock mass response to mining is reasonably informed by the data on the size distribution, but it is less informed by the data on the time distribution, and even less by its space distribution. There is very little information about the possible location of future larger events in the past data. To gain insight into the spatial distribution of seismic hazard, it is better to resort to numerical stress modelling.

De-clustering Conceptually, one can try to separate seismic activity into two components: events that are independent (background events or main shocks) and events that are triggered either by static or dynamic stress changes. This process of discrimination is called de-clustering, e.g. Gardner and Knopoff (1974), Reasenberg (1985), Bottiglieri et al. (2009). In mines, the problem is aggravated by blasting and/or hydrofracturing that induce the bulk of seismic activity. Smaller blasts frequently produce waveforms difficult to be discriminated.

The objective of de-clustering is to make the working data set more stationary in time and more random or disordered in space, i.e. more Poissonian, so it can be treated independently in time and/or in space. Because there is no physical difference between foreshocks, mainshocks, and aftershocks, clusters of events are usually defined by their proximity in space and/or by the fact that they occur at rates greater than the seismicity rate averaged over longer duration. Since most of the "triggered" events are small, de-clustering would lower the exponent of the power law overstating seismic hazard for larger events (Mizrahi et al., 2021). However, de-clustering is an ill-posed problem and does not have a unique solution, and therefore, it is subjective. In most mines, seismic activity is strongly dependent, and de-clustering would either decimate the data set leaving not much to work with, or it would only partially de-cluster leaving a distorted data set. There is a case to be made for reintroducing aftershocks and foreshocks into the probabilistic hazard assessment (Taroni, 2024). A careful selection of the seismogenic volume in space and in time, selecting slightly higher $\log P_{min}$, and a more frequent assessment of time varying hazard, may offer a more objective solution. One can also account for a trend by extrapolating the parameters of the size distribution beyond the observed volume mined or the observed time span.

1.5.2 Testing the Quality and Integrity of Data

This is an important and frequently underestimated, or even overlooked, process that can influence the results of seismic hazard analysis considerably. It is also a process that takes time and that should be performed by an experienced mine seismologist.

1.5 Seismic Hazard in the Time and Volume Mined Domain

Firstly, one needs to select all available data, regardless of their quality. This original data set should be stored for future reference.

Secondly, one should recognise, document, and then remove all artifacts, e.g. blasts, ore-pass noise, test pulses, spikes, misassociated, and split events, and identify gaps in data recordings. Some split events, i.e. larger complex events separated by seismologist during processing into two or more simple ones, may significantly affect the time, and to a degree, the size distribution analysis. It is useful to plot the spatial distribution of seismic events, the activity rate versus time and the activity versus time of day or day of week. It is useful to plot the peak ground velocity (PGV), its associated frequency, and distance for each seismic station. Some blasts generate higher frequency PGV at close stations. One should test for amplitude saturation caused either by exceeding the output range of an amplifier—which produces squared off waveforms—or by exceeding the travel limit of the inertial mass within the geophone, known as displacement clipping. Seismic sensors used in mine networks can only sustain a few mm of displacement, and, for a given peak ground velocity, the peak ground displacement, PGD, increases as the associated frequency f decreases, approximately $PGD = PGV/(2\pi f)$. The displacement clipping produces a sharp reversal in velocity and is difficult to spot on velocity waveforms, but it can be seen more easily on displacement traces as a sharp triangular peak reaching the maximum displacement. If not recognised, amplitude saturations will distort seismic source parameter calculations and ground motion analysis.

Thirdly, one should test the quality of seismological processing, specifically seismic potency, P, radiated energy, E, and corner frequency, f_0. Here it is useful to test the relation between seismic potency and radiated energy on a log-log plot, $\log E$ versus $\log P$, the relation between the S- and P-wave energy, $\log E_S$ versus $\log E_P$, the relation between the S- and P-wave corner frequencies, $\log f_{0S}$ versus $\log f_{0P}$, the relation between the S-wave corner frequency and potency, $\log f_{0S}$ versus $\log P$, and check for outliers and patterns. Source parameters are also influenced by mislocation, and if the hypocentral distances are overestimated so are the source parameters, in some cases enormously. Plotting the recorded PGVs at each site versus distance for different $\log P$ in colour helps to identify events with suspect source parameters and/or poor location.

One should also examine in detail the quality of seismological processing of at least the 100 largest events, since they have a great influence on the results. Here it is important to test the influence of near-field recordings on results of source parameter calculations, mainly on seismic potency, energy, and corner frequency.

The potency and energy range of seismic events that a system can recover is limited by its frequency range (f_1, f_2), which is mainly determined by the capabilities of the seismic sensors. For example, in hard rock with $v_S = 3.6$ km/s and $\mu = 30$ GPa, the largest event for which the system can recover 85% of seismic potency is $P \simeq 7.41 \Delta\sigma / (10 f_1)^3$, which for a reasonably high stress drop of $\Delta\sigma = 3$ MPa and $f_1 = 3$ Hz (i.e. for 4.5 Hz sensor) gives $\log P = 2.9$ (Mendecki, 2013b). It is advisable to secure at least 80% recovery of seismic potency. Note that the estimates of seismic potency and radiated energy can be strongly influenced by their dependence on the applied correction for attenuation, Q. Apart from location and source parameters, one should try to resolve the source mechanisms of larger events and their spatial association with geological structures. One should also check if any changes made to the seismological processing software over time introduced step changes in the results.

Since seismic hazard in mines is driven by rock extraction, it is also useful to examine plots of cumulative potency versus time, cumulative potency versus volume mined in relation to cumulative volume mined versus time, see Fig. 1.4.

It needs to be recognised that the definition of a seismogenic volume and the selection of the appropriate data set introduce a degree of subjectivity.

1.6 Seismic Hazard Management Plan: Seismic Monitoring

The objective of this section is to facilitate preparation of the seismic monitoring part of seismic hazard management plans (SHMP) for seismically active mines. As mining goes deeper and the footprint and the extraction ratio of the ore bodies increases, more underground mines will identify seismicity as being the principal hazard, i.e. hazard that has a reasonable potential to result in multiple deaths in a single incident or a series of recurring incidents.

The process starts, or should start, before mining commences with deriving the expected seismic rock mass response to mining, which constitutes the first reference seismic hazard. In such a case, there is no seismic system, therefore no data, so the expected hazard depends on an expert opinion taking into account the nature of ore body, its geological setting, the planned mine layout, the results of numerical modelling of the expected stresses and strains, and by taking into account the experience of other mines in similar conditions. However, in many cases at the inception, mines did not expect to experience seismic problems, and therefore, frequently the first hazard assessment is done at a later stage when either underground workers reported "rock noises" or after the first incident of damage caused by the perceived seismic event. Most seismically active mines monitor seismic rock mass response to mining, and in these cases seismic hazard should be assessed quantitatively in terms of probabilities of exceedance of certain magnitudes and the resulting ground motion.

The seismic monitoring part of a good SHMP should be logical, consistent, and quantitative. All criteria should be defined to facilitate quantification and each routine task and action, timed and costed. The plan should also provide a framework to ensure compliance with the local health and safety regulations. At the same time, it should be realistic, i.e. should not be overly ambitious to make sure that the mine will be able to deliver on its own commitment. It should be reviewed every 6 months, or every time the seismic rock mass response to mining delivers the "unexpected".

In preparation, it is useful to follow a four-step process accepted by the industry. (1) Identify the type and the nature of seismic related hazards, e.g. slip on geological structures, bursting of pillars, or coal, gas, and rock outburst from a face in coal mines. (2) Assess the risk, i.e. the expected consequences, associated with each type of hazard as measured by the magnitude and the probability or frequency of occurrence, and then score it in the company accepted risk matrix. (3) Specify measures to be taken to manage seismic hazard and minimise the risk, e.g. changes to mine layout and/or the sequence of mining, rate of mining, introduction or changes to regional or local support, introduction of preconditioning, etc. (4) Reassess seismic hazard and review control measures. The seismic part of the SHMP should include the following:

1. The specific objectives of seismic monitoring for the mine.
2. What the mine needs to have to achieve the stated objectives, in terms of seismic monitoring technology and people and skill involved?
3. What the mine needs to do to achieve the stated objectives, i.e. a list of tasks the mine needs to do daily, weekly, monthly, and yearly to achieve the stated objectives?

Having the seismic part of the SHMP, one can prepare a seismic trigger action response plan that defines conditions, called triggers, that need to be continuously checked for and the respective actions to follow when those conditions occur. Actions are triggered when seismic behaviour deviates from the expected and could become hazardous. The SHMP and TARPs should be reviewed twice a year or every time seismic rock mass response to mining exceeds expectations in a significant way.

1.6.1 Objectives of Seismic Monitoring in Mines

Routine seismic monitoring in mines enables the quantification of exposure to seismicity and provides a logistical tool to guide the effort into prevention, control, and alert of potential rock mass instabilities that could result in damage, injury, or loss of life. The scope of seismic monitoring depends on the expected seismic rock mass response to mining. Mines with very low seismic hazard may opt for a limited scope of monitoring in order to register any potential changes in seismic response to mining in time and in space. As seismic activity increases, the scope of monitoring needs to be revised.

The objectives of seismic monitoring are an integral part of the seismic hazard risk management strategy adopted by mine management. The overall objective of seismic monitoring is to measure if the current seismic response to mining is as expected.

One can define the following general objectives of monitoring the seismic response of the rock mass to mining (modified after Mendecki, 1997a and Mendecki et al., 1999):

1. **Rescue.** To detect and locate potentially damaging seismic events, to alert management, and to assist in rescue operations.

2. **Prevention.** To confirm the rock mass stability related assumptions made during the mine design process and enable an audit of, and corrections to, the particulars of a given design while mining, for example:

    (a) Monitor seismic behaviour and the mechanism of seismic events associated with known geological structures and test for unknown geological structures.
    (b) Resolving spatial, temporal, and geometrical characteristics of co-seismic inelastic strain in selected areas of the mine to reconcile with corresponding numerical model(s) and to constrain numerical modelling of future mining scenarios.
    (c) Recording strong ground motion in solid rock and at the skin of excavations to quantify site amplification and assist in support design.
    (d) Monitoring the propagation of a caving and the development of the yield front in caving mines.
    (e) Quantify seismic rock mass response to production by plotting the the cumulative potency versus the cumulative volume of rock extraction, see Fig. 1.4 right.
    (f) Confirm the expected level of ground motion produced by larger events in solid rock and at the skin of excavations to confirm the support design, and monitor the consumption of deformation capacity of support due to seismicity.
    (g) Model the expected seismicity associated with different scenarios of future mining.
    (h) To monitor and to quantify an increase in seismic activity caused by larger seismic events or by production blasts in order to define the area-specific temporal exclusion zones and to guide the re-entry into the working places.

3. **Hazard Assessment.** To quantify the expected exposure to seismicity in terms of the probability of occurrence, within a specific period of time in a given area, of potentially damaging events, and potentially damaging ground motion in the intermediate and long term. The latest hazard results should be compared with previous ones, differences discussed, explained, and documented.

4. **Alerts.** To detect strong and unexpected changes in the spatial and/or temporal behaviour of seismic parameters that could lead to instability and affect working places immediately or in the short term.

5. **Back analysis.** To improve both the mine design and the seismic monitoring processes. All seismic events regardless of size that caused fatality, injury, or damage need to be back analysed thoroughly. It is also advisable to back analyse all near misses, i.e. seismic events that did not result in consequences—but had the potential to do so. Results of back analysis should form the basis for a regular critical review of the applied seismic risk management strategy, guidelines, and procedures.

A quantitative description of seismic events and seismicity is considered as a minimum requirement to accomplish these objectives.

1.6.2 System Requirements

The transducers, data acquisition hardware, and processing software which comprise seismic monitoring systems can best be characterised in terms of the amplitudes and frequencies of the ground motion which they can faithfully represent, and the average rate at which events may be recorded and processed.

One should also specify the minimum location accuracy and the minimum and maximum potency and energy that the system can recover from waveforms.

Dynamic Range in Amplitude The dynamic range in amplitude is commonly defined as the ratio of the highest amplitude that can be measured as the RMS noise level, which is considered to equal the amplitude of the smallest detectable signal. The ratio $R = A_\text{peak}/N_\text{RMS}$ is often expressed in decibels $\text{dB} = 20 \cdot \log_{10} R$. If the environmental noise within the mine is within 10^{-7} to 10^{-6} m/s and the expected maximum amplitude of ground velocity is 1 m/s, then to measure it we need the dynamic range of at least 120 dB or more. Miniature geophones can easily measure the ground noise 10^{-7} and less, but peak amplitudes are limited by internal displacement. A 4.5 Hz geophone with 0.7 damping and 4 mm peak-to-peak travel can accommodate 56 mm/s at the natural frequency, giving a minimum dynamic range of 115 dB. A 15 Hz geophone can tolerate higher tilt angles, which in turn reduces the available travel to as little as 0.5 mm, and lower damping is often used to enhance sensitivity at lower frequencies, all of which could reduce the peak velocity to 16 mm/s or less, with a corresponding dynamic range of 104 dB.

One can increase the signal range and the travel limits of geophones by overdamping. With normal damping of 0.7, the frequency response is flat to ground velocity above the natural frequency, f_n, and is proportional to f_n^2 below, which is caused by a double pole at that frequency. As the damping increases beyond 1, the poles separate, in such a way that the product of the pole frequencies remains constant. Between these poles, the velocity response is proportional to frequency, effectively making it flat to acceleration over this frequency range. For 4.5 Hz geophones, the maximum damping which can reasonably be achieved is 3.4, which means the acceleration response covers the frequency band from 0.7 Hz to 30 Hz. In this configuration, the ADC voltage clip limit is raised by a factor of 5 to 0.5 m/s, which is then slightly greater than the minimum internal displacement clip limit. The spectra need to be corrected for this response when calculating source parameters.

Sensors and Frequency Range The potency and energy range of seismic events that a system can recover is limited by its frequency range (f_1, f_2), which is mainly determined by the capabilities of seismic sensors. For the conventional ω^2 model at $f_1 = 0.2 f_0$, we can recover 96%, at $0.42 f_0$ 85%, and at the corner frequency only 50% of seismic potency, respectively. The conventional 4.5 Hz sensors are capable of recording frequencies down to $f_1 = 3$ Hz, and therefore, for seismic events with stress drop $\Delta \sigma = 5 \cdot 10^5$ Pa, they can reliably recover 85% of potency up to $\log P = 2.05$ or $m = 2.32$. The omnidirectional 14 Hz geophones are capable of recording reliably down to $f_1 = 10$ Hz and can recover 85% of potency up to $\log P = 0.5$ or

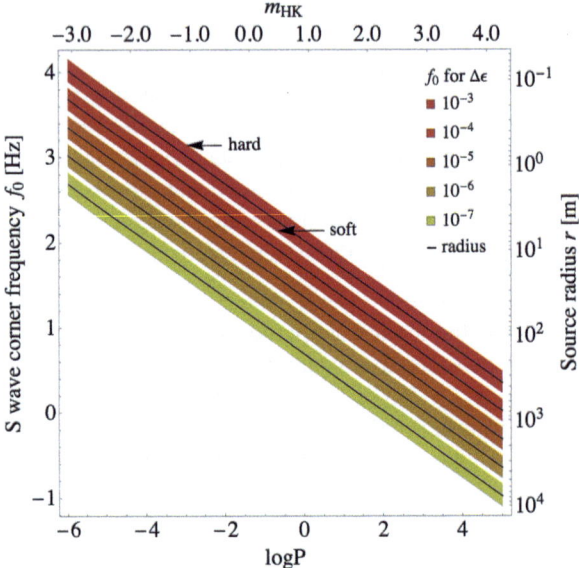

Fig. 1.6 Expected S-wave corner frequency and source radius as a function of log P for different strain drops and rock type (after Mountfort & Mendecki, 1997)

$m = 1.28$. If we assume that both sensors are capable of recording frequencies up to $f_2 = 2000$ Hz, then they can recover 85% of seismic energy down to $\log P = -2.3$ or $m = -0.58$. While the underestimation of seismic potency does not have a significant effect on the potency-based magnitude, it has a notable effect on the estimation of the energy index, apparent stress, and apparent volume.

Assuming the relation between the S-wave corner frequency, f_0, S-wave velocity, v_S, and source radius, $r = 0.3 v_S / f_0$, (Brune et al., 1979), and that $P = 16 \Delta \epsilon r^3 / 7$ (Eshelby, 1957), we can derive the following simple relation: $f_0 = 0.395 \sqrt[3]{\Delta \epsilon / P}$. Since $v_S = (\mu/\rho)^{1/2}$, one can construct a nomogram, see Fig. 1.6, representing the relations between these variables for hard rocks, defined here by $\mu = 37$ GPa, $\rho = 2700$ kg/m^3, and $v_S = 3700$ m/s (top of the band) and for soft rocks by $\mu = 7.2$ GPa, $\rho = 1800$ kg/m^3, and $v_S = 2000$ m/s (bottom of the band).

Figure 1.7 left shows the recovery of seismic potency as a function of the ratio of available frequencies at the lower end of the spectrum f_1, to corner frequency f_0 for $\omega^{1.5}$, ω^2, and ω^3 models. Figure 1.7 right shows that the ω^2 model produces 18% of energy below the corner frequency (left of $f_2/f_0 = 1$), the $\omega^{1.5}$ model less than 10%, while the ω^3 almost 50%. The energy recovery at Fig. 1.7 right is calculated for $f_1 = 0.2 f_0$ and f_2 varying from $0.2 f_0$ (0% recovery) to $10 f_0$ (87% recovery). Therefore for ω^2 model, the smallest event for which we can recover at least 85% of energy is $P \simeq 0.1 \Delta\sigma / (f_2/10)^3$, which for $\Delta\sigma = 3$ MPa and $f_2 = 1000$ Hz is $\log P = -0.5$.

1.6 Seismic Hazard Management Plan: Seismic Monitoring

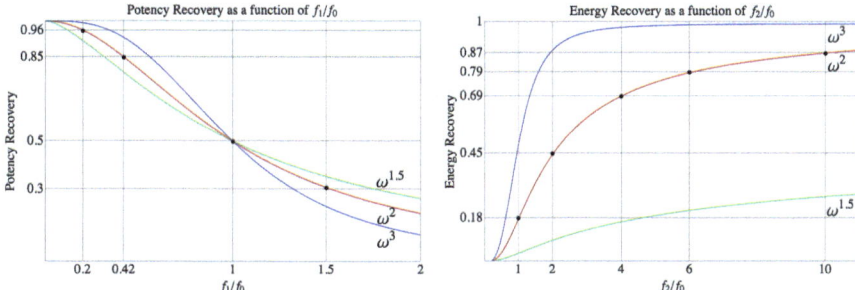

Fig. 1.7 Recovery of seismic potency as a function of f_1/f_0 *(left)* and radiated energy $E\,(0.2f_0, f_2)/E\,(f_1 = 0, \infty)$ as a function of f_2/f_0 *(right)* for the ω^2 model in red, the $\omega^{1.5}$ in green, and the ω^3 in blue. To secure a finite energy, the recovery for the $\omega^{1.5}$ model is defined as $E\,(0.2f_0, f_2)/E\,(f_1 = 0.01f_0, f_2 = 100f_0)$. The large black dots on the ω^2 model indicate particular recoveries of potency and energy. After Mendecki (2013b)

A mine wide system equipped with 4.5 Hz sensors should have at least one site capable of recording three-component strong ground motions of at least 0.5 m/s at lower frequencies.

Sensor Orientation There are two aspects to sensor orientation: Firstly, if the lower natural frequency 4.5 Hz geophones are not installed precisely vertically or horizontally, they do not function correctly; secondly, the true directions of ground motion must be found for each event to assist in location and to estimate its mechanism. The 4.5 Hz geophones are installed in vertical or in horizontal boreholes up to 10 m from the skin of excavations; therefore, it is relatively easy to secure proper orientation. The omnidirectional higher frequency geophones are frequently installed in longer boreholes, and it is recommended to install orientation sensing electronics in the borehole sensors.

Seismic Data and Data Access Data recorded by the seismic system should be available in open formats and be easily accessible:

1. Event Data. The user should be able to define a template format that can be exported on a regular basis. It should contain at least the following data per event: event time (UTC and local), location (X, Y, Z), seismic potency for P-wave and S-wave, radiated energy for P-wave and S-wave, corner frequency, for P-wave and S-wave, the number of triggered stations, and the number of accepted stations.
2. Trigger Data. For each triggered station, the following data should be logged: Trigger ID, Trigger Time, PGV or PGA, and Duration. This should be exportable in ASCII or CSV.
3. Seismogram Data. Seismograms of triggered or continuous data should be exportable to an open format such as ASCII, miniSEED, or SEG-Y. Such seismogram data should have clear indications of how gain factors are defined and which filtering or processing, if any, has been applied to the exported data. Additional metadata such as site coordinates, sensor type, including sensor

characteristics and configuration of the associated sites, should be exportable to simple textual formats. Timing information such as sampling rate and reference times per seismogram should be clear and well documented.

Rock Extraction Data Rock extraction data is an important parameter to constrain the interpretation of the seismic rock mass response to mining. During times when mining is suspended, seismic activity decays very quickly, and the mean inter-event times increase, indicating lower hazard. The mean inter-event volumes mined, however, stay the same since the hazard potential did not change, and when mining is resumed, it will be very much the same as it was before. For this reason, seismic hazard induced by mining should be expressed as a function of time and as a function of rock extraction. For caving mines, rock extraction data should be separated into undercutting and drawing or mucking.

Sensitivity and Location Accuracy The location of a seismic event is assumed to be a point within the seismic source that triggered the set of seismic sites used to locate it. The interpretation of location, if accurate, depends on the nature of the rupture process at the source—if a slow or weak rupture starts at a certain point, the closest site(s) may record waves radiated from that very point while others may only record waves generated later in the rupture process by a higher stress drop patch of the same source. One needs to be specific in determining the arrival times if the location of rupture initiation is sought, otherwise the location will be a statistical average of different parts of the same source.

Since the source of a seismic event has a finite size, the attainable location accuracy of all seismic events in a given area should be within the typical size of an m_{min} or $\log P_{min}$ event that defines the sensitivity of the seismic network, i.e. the magnitude above which the system records all events by the minimum number of stations to secure the required accuracy of location.

Seismological Processing and Source Parameters A seismic event is considered to be described quantitatively when apart from its timing, t, and location, $X = (x, y, z)$, reliable values are obtained for at least two independent parameters pertaining to the seismic source, namely seismic potency, P, and radiated seismic energy, E. Mining induced seismic events are complex, and underground excavations are frequently a part of the source volume; it is therefore useful to invert for source mechanism and to decompose it into the isotropic and deviatoric components.

Seismicity is defined for a given volume, ΔV, over a certain time, Δt, and it can be quantified by the basic four, largely independent, quantities: (1) average time between events, $\bar{t}$, (2) average distance, including source sizes, between consecutive events, $\bar{X}$, (3) cumulative seismic potency, $\sum P$, and (4) cumulative radiated energy, $\sum E$. Other parameters, e.g. apparent stress, $\sigma_A = E/P$, and apparent volume, $V_A = \mu P^2/E$, can be derived from seismic potency and energy. From these four independent quantities, we can derive a number of statistical parameters related to co-seismic deformation and associated changes in the strain rate, stress,

1.6 Seismic Hazard Management Plan: Seismic Monitoring

and rheology of the process. These parameters are described in the chapter on Monitoring Rock Mass Stability.

To quantify the ground motion hazard, one needs to develop the ground motion prediction equation (GMPE), in case of smaller mines for the entire mine or, in case of larger mines, for different mining areas.

1.6.3 Recommended Analysis

A seismic monitoring system provides a large amount of potentially useful data. To accomplish the stated objectives, it is recommended to define routine tasks that need to be performed regularly. These tasks may vary depending on the degree of seismic exposure. However, any unexpected or unusual seismic behaviour needs to be recognised and communicated. Mines with elevated seismic hazard after production blasts or after larger seismic events should institute a higher resolution seismic monitoring as a part of the re-entry decision-making process. This may involve lowering trigger levels and/or monitoring individual triggers or monitoring continuous ground motions at selected sites. This can be done in addition to monitoring triggered events. The intermediate and long term seismic hazard assessment should be done at least once a year and also every time the maximum observed event size has been exceeded.

Examples of Daily Tasks

- Check the system performance.
- Identify possible outliers of seismological processing and return them for reprocessing.
- Check for strong and unexpected changes in seismic activity. Unexpected activity close to excavations and events far from active mining, close to a shaft or, in general, in places not predicted by numerical modelling, should be noted.
- Compare the time of day plot of seismic activity with the average over the last week. Be mindful of the linear size and orientation of these events, since their spatial influence in certain directions may be considerable greater than that indicated by the routine plots of events as dots or spheres on a mine plan.
- Note any unusual activity after production blasts and after larger seismic events.

Examples of Weekly Tasks

- Note any recurrent system performance issues during the last week.
- Test the integrity of data, e.g. temporal gaps, the presence of blasts and ore-pass noise.
- Review locations and magnitudes of events recorded over the last week in terms of their distances from excavations. Persistent activity close to excavations and events far from active mining, close to a shaft or, in general, in places not predicted by numerical modelling, may raise concern.

- Examine if these unusual locations of seismic events are correct.
- Update plots of cumulative potency versus time and versus the cumulative volume of rock extraction and compare them with a short term history, e.g. with the mean value for the last week or month.

Examples of Monthly Tasks

- Note any persistent system performance issues during the past month.
- Review locations and magnitudes of events recorded over the last month in terms of their distances from excavations. Persistent and unexpected activity close to excavations and events far from active mining, close to a shaft or, in general, in places not predicted by numerical modelling, should be reconciled with the expected behaviour and explained.
- Update plots of the cumulative potency versus time and versus the cumulative volume of rock extraction. An increase in the rate of the cumulative potency versus cumulative volume mined, or versus time, may signify an unstable seismic deformation, and an accelerated potency release may indicate a temporary loss of control over the seismic rock mass response to mining.
- Check if time of day and day of week plots of seismic activity are as expected.
- Update the list with details of the 10 largest events and the list of record breaking events at the mine.
- Update the list with details of the 10 largest recorded ground motions at each monitoring site.
- Watch for amplitude saturation caused either by exceeding the output range of an amplifier or by exceeding the travel limit of the inertial mass within the geophone.
- Undertake advanced analysis of larger or damaging seismic events, i.e. location uncertainty, source mechanism, the character of aftershock activity, and the spatial distribution of strong ground motion.
- Test for any possible precursory behaviour to these events, e.g. activity rate and the cumulative apparent volume versus energy index plots.

Examples of Yearly Tasks

- Quantify the performance of the seismic system and compare with previous years.
- Reassess the configuration and the sensitivity of the seismic network.
- Examine the location accuracy of seismic events.
- The recommended location accuracy in a given area should be within the typical size of an event of that magnitude which defines the sensitivity of the seismic network for that area.
- Reassess the suitability of currently employed sensors to accurately record strong ground motion.
- Review lists of largest events, and record breaking events and largest recorded ground motions.
- Compare and reconcile seismic activity with the latest numerical stress model.

- Evaluate intermediate and long term seismic hazard and compare it with previous years.
- Review seismic responses to production blasts, threshold levels for ground motion alerts and for re-entry, and the re-entry protocol.
- Back analyse larger seismic events and comment on their effect on excavations.
- Estimate the site effect on the skin of critical excavations and reconcile with the current support design.
- Review seismic hazard management plan for seismic monitoring.

Forensic Analysis of Larger Events This is a part of back analysis, and the main objective here is to understand the cause(s) of these events, their impact on mine safety, and the mine infrastructure. Such analysis includes the inversion of the mechanism of the main shock and its aftershocks, if any, to see if these events are associated with unknown geological structure, and the ground motion study to reconcile support performance with the observed and simulated seismic loading. Frequently larger events are complex, comprising a few sub-events, and it is important to establish their spatial distribution in respect to pillars and geological structures.

1.7 A Short Note on Magnitude Scales

In 1931, a Japanese seismologist Kiyoo Wadati constructed a chart of the logarithm of the maximum ground motion versus distance for 31 shallow, 5 intermediate, and 5 deep earthquakes and noted that the plots for different earthquakes formed parallel concave lines (Wadati, 1931). In 1934, Charles Richter constructed a similar diagram of peak ground motion versus distance for southern California, see Fig. 1.8, and used the fact that earthquakes of different size gave almost parallel curves to create the first earthquake magnitude scale. Richter published his work in January 1935 (Richter, 1935). This is how Richter described his initial observations.

I suggested that we might compare earthquakes in terms of the measured amplitudes recorded at the Wood-Anderson torsion seismographs in California with an appropriate correction for distance. Wood and I worked together on the latest events, but we found that we could not make satisfactory assumptions for the attenuation with distance. I found a paper by Professor K. Wadati of Japan in which he compared large earthquakes by plotting the maximum ground motion against distance to the epicentre. I tried a similar procedure for our stations, but the range between the largest and smallest magnitudes seemed unmanageably large. Dr. Beno Gutenberg then made the natural suggestion to plot the amplitudes logarithmically. I was lucky because logarithmic plots are a device of the devil. I saw that I could now rank the earthquakes one above the other. Also, quite unexpectedly, the attenuation curves were roughly parallel on the plot. By moving them vertically, a representative mean curve could be formed, and individual events were then characterised by individual logarithmic differences from the standard curve. This set of logarithmic

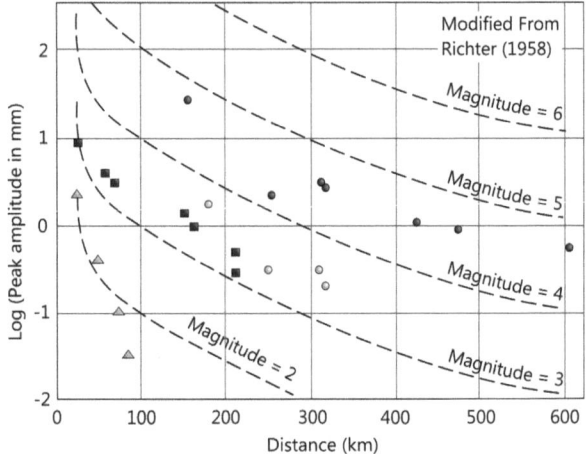

Fig. 1.8 Modified Richter attenuation graphs

differences thus became the numbers on a new instrumental scale. Very perceptively, Mr. Wood insisted that this new quantity should be given a distinctive name to contrast it with the intensity scale. My amateur interest in astronomy brought out the term "magnitude", which is used for the brightness of a star.

Richter defined the local magnitude of an event m_L at a given recording station as

$$m_L = \log\left[A(R)/A_0(R)\right] \quad \Rightarrow \quad A(R) = A_0(R) \cdot 10^{m_L}, \tag{1.1}$$

where A is the maximum zero-to-peak horizontal amplitude measured in mm on a Wood-Anderson torsion seismograph with magnification of 2080 at 0.8 second period (Uhrhammer & Collins, 1990), at epicentral distance R, and $A_0(R)$ is the reference maximum amplitude for the same distance. The local magnitude then is a relative measure of the strength of a seismic event, and a unit increase in magnitude corresponds to a ten-fold increase in the amplitude of ground displacement. The reference amplitude A_0[mm] as a function of distance R [km] was given by Richter in a table that can be approximated by the following formulae: $\log A_0 = 0.15 - 1.6 \log R$ for distances less than 200 km and $\log A_0 = 3.38 - 3.0 \log R$ for distances between 200 and 600 km. Richter arbitrarily chose the reference amplitudes so that the earthquakes he dealt with did not have negative magnitudes. At $R = 100$ km, $A_0 \simeq 0.001$ mm and on Wood-Anderson seismograph. The zero-to-peak amplitude for a magnitude $m_L = 3$ event would measure $A = 0.001 \cdot 10^3$, which is 1 mm or $\log A = 0$. To compute the displacement of ground motion $u(R)$, one must divide the amplitude measured on the instrument by its magnification, thus for $m_L = 3$ at 100 km $u = 1/2080 = 0.00048$ mm. The final magnitude of an event is taken as an average from a number of stations surrounding the event.

1.7 A Short Note on Magnitude Scales

The response of the Wood-Anderson seismometer, with nearly constant displacement amplification over the frequency range for local earthquakes in California, is implicitly included in Richter's definition of local magnitude. These instruments were of low sensitivity, capable of recording only horizontal ground motion, hence poorly suited to study local micro-earthquakes with well developed vertical components of ground motion. They have been mostly out of use since the late 1960s, and the local magnitudes are now computed from Wood-Anderson equivalent of modern high-gain seismographs (Brune & Allen, 1967; Bakun et al., 1978; Uhrhammer & Collins, 1990).

To cater for earthquakes at larger distances, the magnitude scale was later extended by the introduction of surface magnitudes. The surface wave magnitude was developed by Gutenberg (1945a) and is given by $m_S = \log(u/T) + c_1 \log \Delta + c_2$, where u is the maximum amplitude of Rayleigh waves in micrometres, T is the period, approximately 20 seconds, Δ is the distance in degrees, and c_1 and c_2 are calibration constants. Gutenberg (1945a) originally worked out $c_1 = 1.656$ and $c_2 = 1.818$, which applied for Pasadena, and in 1964 the International Association of Seismology and Physics of the Earth's Interior (IASPEI) adopted $c_1 = 1.66$ and $c_2 = 3.3$ proposed by Vanek et al. (1962). The m_S is suitable for shallow earthquakes that generate well developed surface waves and with source duration not much greater than 20 seconds.

To cater for earthquakes at all depths, Gutenberg (1945b) developed the body wave magnitude scale. The body wave magnitude is given by $m_b = \log(u/T) + q(\Delta, h)$, where T is the period associated with the maximum body wave amplitude, measured generally at $T = 1$ second, and $q(\Delta, h)$ is the calibrating function to correct for the epicentral distance Δ, depth h, and site effects. The body wave magnitude is more suitable for smaller earthquakes of shorter duration.

The body wave and the surface wave magnitudes coincide only for earthquakes with magnitude 6.5, for smaller events m_b is larger, and for greater events m_S is larger. Gutenberg and Richter (1956) related the three magnitude scales as follows: $m_b = 0.63 m_S + 2.5$, and $m_S = 1.27 (m_L - 1) - 0.016 m_L^2$.

There is also the m_{Lg} magnitude defined by Nuttli (1973) for regional distances based on the maximum amplitude of Lg waves, $m_{Lg} = \log u + 0.83 \log \Delta + c(\Delta - 0.09) \log e + 3.81$, where u in micrometres, Δ in degrees, and c is an attenuation coefficient different for different regions: $c = 0.07$ for the central USA and $c = 0.53$ for California. Lg waves are relatively short period, 1 to 6 seconds, have large arrival amplitude with predominantly transverse motion, and propagate along the surface with velocities close to the average shear velocity in the upper part of the continental crust (Aki & Richards, 2002). The Nuttli magnitude was used at one stage to quote the larger mine related events in Canada.

Before the advent of a high dynamic range, digital network waveforms of larger earthquakes recorded at distances less than 200 km were frequently saturated, and the maximum amplitude could not be measured. To alleviate the problem, Bisztricsany (1958) proposed to determine magnitude from the duration of long period surface waves. A basis for the duration magnitude was later given by Aki (1969) who pointed out that duration is virtually independent on distance within

100 km. One can therefore determine the magnitude of events which are only approximately located. Aki (1969) also suggested that the energy in the coda of signals from local events comes from back-scattered waves.

The duration magnitude m_D, determined from short period sensors, can be defined as $m_D = c_1 \log t_D + c_2 R + c_3$, where t_D is the duration of the waveform in seconds measured from the P-wave arrival and R is the distance in km. The total duration of an event is defined as the time interval between the onset of the first arrival and the point at which the signal falls and remains below the background noise level. The duration of ground motion depends, in part, on amplitude and thus may be influenced by site amplification, rock mass attenuation, radiation pattern of the source, and the background noise level. It is therefore important to estimate the duration magnitude at as many stations surrounding the event as possible, and the final magnitude is taken as an average over the stations. For small events, with signal duration dominated by S-P length, the duration magnitude may not be a stable estimate of event size because it does not measure properly the energy back-scattered from distant points. In such case, it is recommended to measure duration from the arrival of S-wave. In most cases quoted in the literature, the dependence of the duration magnitude on distance is rather weak. Today, the direct estimate of magnitude is performed routinely by national or regional seismological networks equipped with the low frequency sensors able to quantify larger earthquakes.

In most areas in the world, the low range of magnitudes of seismic events recorded in mines is not well covered by the regional or national seismological networks, and it is difficult to calibrate their records against local magnitudes. The modern seismic systems deployed in mines derive radiated energy and seismic potency, or seismic moment, from recorded waveforms. Since the underlying intention of different magnitude definitions is to offer an equivalent measure to earthquakes radiating the same amount of energy, one would like to scale the computed radiated energy with magnitude. Gutenberg and Richter (1956) derived an empirical relationship between the radiated seismic energy E and the surface wave magnitude of larger earthquakes,

$$\log E = 1.5m + 4.8 \quad \text{or} \quad m = (2/3 \log E - 3.2). \tag{1.2}$$

Choy and Boatwright (1995) calculated the radiated energy of 397 earthquakes with $m \geq 5.8$ by integrating the path and the radiation pattern corrected velocity-squared spectra and, assuming the slope 1.5, the least square regression fit yielded $\log E = 1.5m+4.4$. That indicates that, on average, the Gutenberg-Richter formula (1.2) may overestimate the radiated energy by a factor of 2.5. According to Eq. (1.2), a unit increase in magnitude corresponds to an approximately 30 times increase in radiated energy. Although the relation (1.2) is empirical, the implied scaling $m \sim \frac{2}{3} \log E$ can be explained for most moderate to large earthquakes in terms of a simple dislocation model with a constant stress drop, while very small earthquakes are likely to satisfy $m \sim \log E$ (Haskell, 1964; Kanamori & Anderson, 1975).

The radiated seismic energy of the P- or S-wave is proportional to the integral of the radiation pattern corrected far-field velocity pulse squared $\dot{u}^2(t)$ of duration

1.7 A Short Note on Magnitude Scales

t_s and in frequency domain to the square of the velocity power spectrum of the P- or S-wave $V_{P,S}^2(f)$,

$$E_{P,S} = \frac{8}{5}\pi\rho v_{S,P} R^2 \int_0^{t_s} \dot{u}^2(t)dt \quad \text{and} \quad E_{P,S} = 8\pi\rho v_{P,S} R^2 \int_0^\infty V_{P,S}^2(f)df, \quad (1.3)$$

where ρ is rock density, $v_{S,P}$ is S- or P-wave velocity, and R is the distance from the source. The total radiated energy $E = E_P + E_S$. The computation of the radiated seismic energy from waveforms requires a wide frequency bandwidth of the monitoring system, preferably from $0.2 \cdot f_0$ to $10 \cdot f_0$, where f_0 is the predominant frequency of a given event. Such bandwidth is frequently not available for smaller events either due to the limited capabilities of the sensors used or due to the insufficient sampling frequency of the data acquisition units. Moreover, the rate at which seismic events are produced in mines does not always allow for careful, time-consuming processing with proper corrections for attenuation and site effects. As a result, the estimates of radiated energy for smaller events are rather underestimated, in some cases by up to a factor of 10 that causes the error in magnitude by 0.7 units. In general, the energy-based magnitude scale struggles to represent small and large events adequately. The energy measure of the strength of an earthquake, $K = \log E$, was adopted in Russia (Rautian, 1960) and then in Polish coal mines (Gibowicz, 1963; Wierzchowska & Dubinski, 1973).

The energy available for seismic radiation during shear dislocation $\bar{u}$ over the area A driven by the average stress $\bar{\sigma}$ can be approximated by $E = \xi\bar{\sigma}P = \xi\bar{\sigma}M/\mu$, where ξ is the seismic efficiency. The product $\xi\bar{\sigma}$ is called apparent stress, $\sigma_A = E/P$, (Aki, 1966), which is a model independent measure of stress release at the source, and it does not does not depend on rigidity. Seismic sources similar in terms of their potency may differ by up to two orders of magnitude in radiated energies, reflecting stress differences at the place and at the time of their occurrence (Mendecki, 1993). Comparing $E = \sigma_A P = \sigma_A M/\mu$ with $m = (2/3) \log E - 3.2$ gives $m = (2/3)(\log P + \log \sigma_A) - 3.2$, which, assuming a constant apparent stress for larger earthquakes of 1.5 MPa and rigidity $\mu = 30$ GPa, gives

$$m = (2/3) \log M - 6.067 = (2/3) \log P + 0.92, \quad (1.4)$$

which is the magnitude-moment or magnitude-potency relation (Hanks & Kanamori, 1979).

The assumption of constant apparent stress implies a slope of 1.0 on the $\log E$ versus $\log P$ plot, which is not always supported by data. In a more general case, $\log E = d \log P + c$, where d is the slope and c the intercept that measures the $\log E$ released by a seismic event with $\log P = 0$. In such a case, apparent stress scales with potency as $\log \sigma_A = (d-1) \log P + c$, and for $d = 1$, apparent stress is independent of potency, $\sigma_A = 10^c$. For $d > 1$, apparent stress would increase with an increase in seismic potency. By combining $\log E = 1.5m + 4.8$ with $\log E = d \log P + c$, one obtains a more general magnitude-potency relation

$$m = (2/3)\, d \log P + (2/3)\, c - 3.2. \tag{1.5}$$

The slope d of the $\log E$ vs. $\log P$ plot of events recorded in mines is frequently reported to be higher than 1.0. Spottiswoode and McGarr (1975) calculated seismic moments, radiated energies and magnitudes of 24 events associated with deep level mining on the East Rand in SA from analog waveforms recorded at a surface site above the mine. They confirmed the relation (1.2) for events in the magnitude range $0 < m < 3$ and reported $\log M = 1.2m + 10.7$ which, for $\mu = 30\,\text{GPa}$, gives $m = 0.833 \log P - 0.186$ and, together with Eq. (1.2), gives $\log E = 1.25 \log P + 4.52$. Mendecki (1993) and van Aswegen and Butler (1993) analysed high dynamic range digital waveforms of thousands of events in the range, $-0.5 < m < 3.5$, recorded underground on West Rand, Klerksdorp, and Welkom gold mines. They reported $\log E = 1.5(\pm 0.1) \log(\mu P) - 10.5(\pm 0.5)$, which, for $\mu = 30\,\text{GPa}$, gives $\log E = 1.5 \log P + 5.22$ and $m = \log P + 0.28$.

A similar relation, $m = \log P + 0.72$, was obtained by Ben-Zion and Zhu (2002) for small earthquakes, $m < 3.5$, recorded by Abercrombie (1996) with 4.5 Hz sensors located in a deep borehole in California. For data sets with events in the magnitude range $1.0 < m < 6.0$, recorded by the TERRAscope/TriNet network, the linear fit is $m = 0.74 \log P + 0.98$, but the best overall fit is $m = 4.04\sqrt{\log P + 4.86} - 8.07$ (Ben-Zion & Zhu, 2002). This indicates that the non-linear term is required to describe the scaling between potency and magnitude values over a broad range of sizes with a single relation. A linear potency magnitude scaling relation can approximate only a limited range of data. The local magnitudes used by Ben-Zion and Zhu (2002) were determined independently by the Southern California Short Period Network.

References

Abercrombie, R. E. (1996). The magnitude-frequency distribution of earthquakes recorded with deep seismometers at Cajon Pass. *Tectonophysics, 261*, 1–7.
Aki, K. (1966). Generation and propagation of G waves from the Niigata earthquake of June 16, 1964. Part 2: Estimation of earthquake moment, released energy, and stress strain drop from the G-wave spectrum. *Bulletin Earthquake Research Institute Tokyo University, 44*, 73–88.
Aki, K. (1967). Scaling law of seismic spectrum. *Journal of Geophysical Research, 72*, 1217–1231.
Aki, K. (1969), Analysis of the seismic coda of the local earthquakes as scattered waves. *Journal of Geophysical Research, 74*(2), 615–631.
Aki, K., & Richards, P. G. (2002). *Quantitative seismology* (2nd ed.). University Science Books.
Al-Kindy, F. H., & Main, I. G. (2003). Testing self-organized criticality in the crust using entropy: A regionalized study of the CMT global earthquake catalogue. *Journal of Geophysical Research, 108*(B11), 2521. https://doi.org/10.1029/2002JB002230.
Anderson, J. G., & Biasi, G. P. (2016). What is the basic assumption for probabilistic seismic hazard assessment? *Seismological Research Letters, 87*(2A), 323–326. https://doi.org/10.1785/0220150232.
Arowsmith, S. J., Trugman, D. T., MacCarthy, J., Bergen, K. J., Lumley, D., & Magnani, B. (2022). Big data seismology. *Reviews of Geophysics, 60*(2), e2021RG000,769. https://doi.org/10.1029/2021RG000769.

References

Bak, P., Tang, C., & Wiesenfeld, K. (1987). Self-organized criticality: An explanation of 1/f noise. *Physical Review Letters, 59*, 381–384.

Bakun, W. H., Houck, S. T., & Lee, W. H. K. (1978). A direct comparison of "synthetic" and actual Wood-Anderson seismograms. *Bulletin of the Seismological Society of America, 68*(4), 1199–1202.

Barnsley, M. F. (1988). *Fractals everywhere*. Academic Press.

Ben-Zion, Y. (2017). On different approaches to modeling. *Journal of Geophysical Research: Solid Earth, 122*, 558–559. https://doi.org/10.1002/2016JB013922.

Ben-Zion, Y., & Zhu, L. (2002). Potency-magnitude scaling relation for southern California earthquakes with $1.0 < M < 7.0$. *Geophysical Journal International, 148*, F1–F5.

Bisztricsany, E. (1958). A new method for the determination of the magnitude of earthquakes. *Geophys. Kozlemen, 7*, 69–96.

Bottiglieri, M., Lippiello, E., Godano, C., & de Arcangelis, L. (2009). Identification and spatiotemporal organization of aftershocks. *Journal of Geophysical Research, 114*(B03303), 1–12. https://doi.org/10.1029/2008JB005941.

Brune, J. N. (1970). Tectonic stress and the spectra of seismic shear waves from earthquakes. *Journal of Geophysical Research, 75*(26), 4997–5009.

Brune, J. N., & Allen, C. R. (1967). A microearthquake survey of the San Andreas fault system in southern California. *Bulletin of the Seismological Society of America, 57*, 277–296.

Brune, J. N., Archuleta, R. J., & Hartzell, S. (1979). Far-field S-wave spectra, corner frequencies, and pulse shapes, *Journal of Geophysical Research, 84*(B5), 2262–2272. https://doi.org/10.1029/JB084iB05p02262.

Choy, G. L., & Boatwright, J. L. (1995). Global patterns of radiated seismic energy and apparent stress. *Journal of Geophysical Research, 100*(B9), 18,205–18,228.

Durrheim, R. J., Spottiswoode, S. M., Roberts, M. K. C., & van Z. Brink, A. (2005). Comparative seismology of the witwatersrand basin and bushveld complex and emerging technologies to manage the risk of rockbursting. *The Journal of The South African Institute of Mining and Metallurgy, 105*, 409–416.

duToit, H. J., Goldswain, G., & Olivier, G. (2022). Can das be used to monitor mining induced seismicity? *International Journal of Rock Mechanics and Mining Sciences, 155*(105127). https://doi.org/10.1016/j.ijrmms.2022.105127.

Eshelby, J. D. (1957). The determination of the elastic field of an ellipsoidal inclusion and related problems. *Proceedings of the Royal Society of London, Series A, Mathematical and Physical Sciences, 241*(1226), 376–396.

Frisch, U., & Sornette. D. (1997). Extreme deviations and applications. *Journal de Physique I, EDP Sciences, 7*(9), 1155–1171.

Gardner, J. K., & Knopoff, L. (1974). Is the sequence of earthquakes in Southern California, with aftershocks removed, Poissonian? *Bulletin of the Seismological Society of America, 64*(5), 1363–1367.

Gay, N. C., & Ortlepp, W. D. (1978). Anatomy of a mining-induced fault zone. *Geological Society of America Bulletin, 90*, 47–58.

Gibowicz, S. (1963). Magnitude and energy of subterrane shocks in Upper Silesia. *Studia Geophysica et Geodaetica, 7*(1), 1–37.

Gutenberg, B. (1945a). Amplitudes of surface waves and magnitudes of shallow earthquakes. *Bulletin of the Seismological Society of America, 35*, 3–12.

Gutenberg, B. (1945b). Amplitudes of P, PP and S and magnitudes of shallow earthquakes. *Bulletin of the Seismological Society of America, 35*, 57–69.

Gutenberg, B., & Richter, C. F. (1956). Earthquake magnitude, intensity, energy, and acceleration: (Second paper). *Bulletin of the Seismological Society of America, 46*(2), 105–145.

Handley, M. F., de Lange, J. A. J., Essrich, F., & Banning, J. A. (2000). A review of the sequential grid mining method employed at Elandsrand Gold Mine. *The Journal of The Southern African Institute of Mining and Metallurgy, 100*(3), 157–168.

Hanks, T. C., & Kanamori, H. (1979) A moment magnitude scale. *Journal of Geophysical Research, 84*, 2348–2350.

Haskell, N. (1964). Total energy and energy spectral density of elastic wave radiation from propagating faults. *Bulletin of the Seismological Society of America*, *56*, 1811–1842.

Hogarth, R. (1975). Cognitive processes and the assessment of subjective probability distribution. *Journal of the American Statistical Association*, *70*(350), 271–289. https://doi.org/10.2307/2285808.

Jensen, H. J. (1998). *Self-organized criticality: Emergent complex behavior in physical and biological systems*. Cambridge Lecture Notes in Physics 10 (1st ed.). Cambridge University Press.

Johnson, L. R., & Sammis, C. G. (2001). Effects of rock damage on seismic waves generated by explosions. *Pure and Applied Geophysics*, *158*, 1869–1908. https://doi.org/10.1007/PL00001136.

Johnson, S. E., Song, W. J., Vel, S. S., Song, B. R., & Gerbi, C. C. (2021). Energy partitioning, dynamic fragmentation, and off-fault damage in the earthquake source volume. *JGR Solid Earth*, *126*(e2021JB022616). https://doi.org/10.1029/2021JB022616.

Johnston, C. W. (2017). Stress modulation of earthquakes: A study of long and short period stress perturbations and the crustal response, Ph.D. thesis. University of California, Berkeley.

Kahneman, D. (2003). A perspective on judgment and choice - Mapping bounded rationality (Nobel Lecture). *American Psychologist*, *58*(9), 697–720. https://doi.org/10.1037/0003-066X.58.9.697.

Kanamori, H., & Anderson, D. L. (1975), Theoretical basis of some empirical relations in seismology. *Bulletin of the Seismological Society of America*, *65*(5), 1073–1095.

Keilis-Borok, V. I., Bessanova, E. N., Gotsadze, O. D., Kirilova, I. V., Kogan, S. D., Kikhtikova, T. I., Malinovskaya, C. N., Pavola, G. I., & Sarskii, A. A. (1960). Investigation of the mechanism of earthquakes (English translation). American Geophysical Union, Consultants Bureau, New York.

Kijko, A., Funk, C. F., & Brink, A. v Z. (1993). Identification of anomalous patterns in time dependent mine seismicity. In R. P. Young (Ed.), *Proceedings 3rd International Symposium on Rockburst and Seismicity in Mines, Kingston, ON, Canada* (pp. 205–210).

Kulkarni, R. B., Youngs, R. R., & Coppersmith, K. J. (1984). Assessment of confidence intervals for results of seismic hazard analysis. In *8th World Conference on Earthquake Engineering, San Francisco* (pp. 263–270).

Kurzon, I., Lyakhovsky, V., & Ben-Zion, Y. (2019). Dynamic rupture and seismic radiation in a damage–breakage rheology model. *Pure and Applied Geophysics*, *176*, 1003–1020. https://doi.org/10.1007/s00024-018-2060-1.

Lindsey, N. J., & Martin, E. R. (2021). Fibre-optic seismology. *Annual Review of Earth and Planetary Sciences*, *49*, 309–336.

Luo, B., Jin, G., & Stanek, F. (2021). Near-field strain in distributed acoustic sensing-based microseismic observation. *Geophysics*, *86*(5), 1SO–Z1. https://doi.org/10.1190/geo2021-0031.1.

Lyakhovsky, V., Ben-Zion, Y., Ilchev, A., & Mendecki, A. (2016). Dynamic rupture in a damage-breakage rheology model. *Geophysical Journal International*, *206*, 1126–1143. https://doi.org/10.1093/gji/ggw183.

Mandelbrot, B. B. (1967). How long is the coast of Britain? Statistical self-similarity and fractional dimension. *Science*, *156*, 636–638.

McGarr, A., Simpson, D., & Seeber, L. (2002). Case histories of induced and triggered seismicity. In W. H. K. Lee, H. Kanamori, P. C. Jennings, & C. Kisslinger (Eds.), *International Handbook of Earthquake and Engineering Seismology* (pp. 647–661). Academic Press.

McGuire, R. K. (2001), Deterministic vs probabilistic earthquake hazards risks. *Soil Dynamics and Earthquake Engineering*, *21*, 377–384.

Mendecki, A. J. (1993), Real time quantitative seismology in mines: Keynote Address. In R. P. Young (Ed.), *Proceedings 3rd International Symposium on Rockbursts and Seismicity in Mines, Kingston, ON, Canada* (pp. 287–295). Balkema.

Mendecki, A. J. (1997a). *Seismic monitoring in mines* (1 ed., 262 pp.) Chapman and Hall.

References

Mendecki, A. J. (1997b). Principles of monitoring seismic rock mass response to mining: Keynote Address. In S. J. Gibowicz & S. Lasocki (Eds.), *Proceedings 4th International Symposium on Rockbursts and Seismicity in Mines, Krakow, Poland* (pp. 69–80). Balkema.

Mendecki, A. J. (2001). Data-driven understanding of seismic rock mass response to mining: Keynote Address. In G. van Aswegen, R. J. Durrheim, & W. D. Ortlepp (Eds.), *Proceedings 5th International Symposium on Rockbursts and Seismicity in Mines, Johannesburg, South Africa* (pp. 1–9). South African Institute of Mining and Metallurgy.

Mendecki, A. J. (2005). Persistence of seismic rock mass response to mining. In Y. Potvin & M. R. Hudyma (Eds.), *Proceedings 6th International Symposium on Rockburst and Seismicity in Mines, Perth, Australia* (pp. 97–105). Australian Centre for Geomechanics.

Mendecki, A. J. (2012). Size distribution of seismic events in mines. In *Proceedings of the Australian Earthquake Engineering Society 2012 Conference, Queensland* (pp. 1–20).

Mendecki, A. J. (2013a). Characteristics of seismic hazard in mines: Keynote Lecture. In A. Malovichko & D. A. Malovichko (Eds.), *Proceedings 8th International Symposium on Rockbursts and Seismicity in Mines, St Petersburg-Moscow, Russia* (pp. 275–292). ISBN 978-5-903258-28-4.

Mendecki, A. J. (2013b). Frequency range, logE, logP and magnitude. In A. Malovichko & D. A. Malovichko (Eds.), *Proceedings 8th International Symposium on Rockbursts and Seismicity in Mines, St Petersburg-Moscow, Russia* (pp. 167–173). ISBN 978-5-903258-28-4.

Mendecki, A. J. (2016). *Mine seismology reference book: Seismic Hazard* (1 ed.). Institute of Mine Seismology. ISBN 978-0-9942943-0-2. www.imseismology.org/msrb/.

Mendecki, A. J., & Lötter, E. C. (2011). Modelling seismic hazard for mines. In *Australian Earthquake Engineering Society 2011 Conference, Barossa Valley*.

Mendecki, A. J., van Aswegen, G., Brown, J. N. R., & Hewlett, P. (1988). The Welkom seismological network. In C. Fairhurst (Ed.), *3rd International Symposium on Rockbursts and Seismicity in Mines, 08–10 June 1988, MN, USA* (pp. 237–244), Balkema, 1990.

Mendecki, A. J., van Aswegen, G., & Mountfort, P. (1999). A guide to routine seismic monitoring in mines. In A. J. Jager & J. A. Ryder (Eds.), *A Handbook on Rock Engineering Practice for Tabular Hard Rock Mines* (chap. 9, pp. 287–309). The Safety in Mines Research Advisory Committee.

Mizrahi, L., Nandan, S., & Wiemer, S. (2021). The effect of declustering on the size distribution of mainshocks. *Seismological Research Letters, 92*(4), 2333–2342.

Mountfort, P., & Mendecki, A. J. (1997). Seismic transducers. In A. J. Mendecki (Ed.), *Seismic Monitoring in Mines* (1 ed., chap. 1, pp. 1–20). Chapman and Hall.

Mulargia, F., Stark, P. B., & Geller, R. J. (2017). Why is probabilistic seismic hazard analysis (PSHA) still used? *Physics of the Earth and Planetary Interiors, 264*, 63–75.

Nicolis, G., & Prigogine, I. (1977). *Self-organization in nonequilibrium systems. From dissipative structures to order through fluctuations* (491 pp.). John Wiley and Sons.

Nuttli, O. W. (1973). Seismic wave attenuation and magnitude relations for east and north America. *Journal of Geophysical Research, 78*, 876–885.

Oreskes, N., Shrader-Frechette, K., & Belitz, K. (1994). Verification, validation, and confirmation of numerical models in the earth sciences. *Science, 263*, 641–646.

Prigogine, I. (1980). *From being to becoming. Time and complexity in the physical sciences*. W. H. Freeman and Company.

Rautian, T. G. (1960). Earthquake energy. *Transaction of the Joint Institute of Physics of the Earth, 9*, 35–114.

Reasenberg, P. (1985). Second-order moment of central California seismicity, 1969–1982. *Journal of Geophysical Research, 90*(B7), 5479–5495. https://doi.org/10.1029/JB090iB07p05479.

Richter, C. F. (1935). An instrumental earthquake magnitude scale. *Bulletin of the Seismological Society of America, 25*, 1–32.

Sammis, C. G., Rosakis, A. J., & Bhat, H. S. (2009). Effects of off-fault damage on earthquake rupture propagation: Experimental studies. *Pure and Applied Geophysics, 166*, 629–1648. https://doi.org/10.1007/s00024-009-0512-3.

Sonley, E., & Atkinson, G. M. (2005). Empirical relationship between moment magnitude and Nuttli magnitude for small-magnitude earthquakes in Southeastern Canada. *Seismological Research Letters*, *76*(6), 752–755.

Spottiswoode, S. M., & McGarr, A. (1975). Source parameters of tremors in a deep-level gold mine. *Bulletin of the Seismological Society of America*, *65*(1), 93–112.

Taroni, M. (2024). Reintroducing aftershocks in PSHA. *Seismological Research Letters*. https://doi.org/10.1785/0220240391.

Trugman, D. T., Fang, L., Ajo-Franklin, J., Nayak, A., & Li, Z. (2022). Preface to the focus section on big data problems in seismology. *Seismological Research Letters*, *93*(5), 2423–2425.

Tversky, A., & Kahneman, D. (1974). Judgment under uncertainty: Heuristics and biases. *Science*, *185*(4157), 1124–1131.

Uhrhammer, R. A., & Collins, E. R. (1990). Synthesis of Wood-Anderson seismograms from broadband digital records. *Bulletin of the Seismological Society of America*, *80*(3), 702–716.

UNDRO. (1979). Natural disasters and vulnerability analysis. *Report of expert group meeting*. UNDRO - United Nations Disaster Relief Coordinator, Geneva.

van Aswegen, G., & Butler, A. G. (1993). Applications of quantitative seismology in South African gold mines. In R. P. Young (Ed.), *Proceedings 3rd International Symposium on Rockbursts and Seismicity in Mines, Kingston, ON, Canada* (pp. 261–266). Balkema. ISBN 90 5410320 5.

van Aswegen, G., & Mendecki, A. J. (1999). Mine layout, geological features and seismic hazard. *Final report gap 303*. Safety in Mines Research Advisory Committee, South Africa (pp. 1–91).

Vanek, J., Zapotek, A., Karnik, V., Kondorskaya, N. V., Riznichenko, Y. V., Savarensky, E. F., Soloviev, S. L., & Shebalin, N. V. (1962). Standarization of magnitude scales. *Izvestiya, Academy Nauka SSSR, Geophysics*, *2*, 153–158.

Vick, S. G. (2002). *Degrees of belief: Subjective probability and engineering judgment*. American Society of Civil Engineers Press.

Vieira, F. M. C. C., Diering, D. H., & Durrheim, R. J. (2001). Methods to mine the ultra-deep tabular gold-bearing reefs of the Witwatersrand Basin, South Africa. In W. A. Hustrulid & R. L. Bullock, *Underground mining methods: Engineering fundamentals and international case studies* (pp. 691–704). Society for Mining, Metallurgy and Exploration, Inc (SME).

Wadati, K. (1931). Shallow and deep earthquakes. *Geophysical Magazine, Tokyo*, *4*, 231–283.

Wierzchowska, Z., & Dubinski, J. (1973). Methods to calculate energy of seismic events in Upper Silesia (in Polish). Report, Central Mining Institute, Katowice, Poland.

Zhang, M. a. (2022). Loc-flow: An end-to-end machine learning-based high-precision earthquake location workflow. *Seismological Research Letters*, *93*(5), 2426–2438. https://doi.org/10.1785/0220220019.

Open Access This chapter is licensed under the terms of the Creative Commons Attribution 4.0 International License (http://creativecommons.org/licenses/by/4.0/), which permits use, sharing, adaptation, distribution and reproduction in any medium or format, as long as you give appropriate credit to the original author(s) and the source, provide a link to the Creative Commons license and indicate if changes were made.

The images or other third party material in this chapter are included in the chapter's Creative Commons license, unless indicated otherwise in a credit line to the material. If material is not included in the chapter's Creative Commons license and your intended use is not permitted by statutory regulation or exceeds the permitted use, you will need to obtain permission directly from the copyright holder.

Chapter 2
Seismic Sources in Rock

Abstract This chapter starts with the glossary of terms used to describe seismic sources, and a short description of seismic potency, radiated seismic energy, and associated parameters like apparent stress, energy index, and apparent volume. There is a short description of the static solution for a circular crack subjected to a uniform shear strain given by *Eshelby* that explains the basic source scaling relations, and a section on the circular ω^2 source model by *Brune*. While Brune's model fits the spectra of seismic events recorded in deep high stress mines, it does not work as well in many mines with a weaker inhomogeneous rock mass and/or at intermediate depth and in caving mines. Seismic events in such an environment radiate less energy per unit of deformation at source. Therefore Chap. 2 introduces more general point source models suggested by *Beresnev and Atkinson* that allow for "softer spectra", e.g. $\omega^{2.5}$ or ω^3. It also describes the final static deformation and strains induced by the double-couple source, gives the radiation pattern for the near, intermediate, and far fields and derives the expression for the surface of such a source for a given inelastic strain drop.

2.1 Introduction

Mining excavations, whether underground or open cast, induce elastic (reversible) and inelastic (irreversible) strain within the surrounding rock. Elastic strain is defined as a process during which no new micro-defects are nucleated, while all existing micro-defects convect with the mass without growing in size (Krajcinovic & Mastilovic, 1995). The inelastic deformation of brittle rock is mainly due to fracturing and frictional sliding called cataclastic flow. No potential energy, i.e. the energy that could do work, is associated with inelastic strain. The potential energy accumulated during elastic deformation in a given volume of rock may be unloaded due to changes in stress state in this volume, or it may be released gradually due to creep, viscous, or plastic deformation, or it may be released suddenly during the processes of inelastic deformation.

Sudden inelastic deformation associated with fracture and slip radiates seismic waves. The amplitude and frequency of seismic waves radiated from such a source depend, in general, on the strength and state of stress of the rock, the size of the source, and the rate at which the rock is deformed during the fracturing process. The following relations apply here (all other parameters being the same):

1. The amplitude and frequency increase with an increase in rock strength and stress.
2. The amplitude and frequency increase with an increase in co-seismic deformation rate at source.
3. The predominant frequency, i.e. the frequency at which the most energy is radiated, decreases with an increase in source size.[1]

Seismic sensors detect waves caused by inelastic deformation within a certain distance. Since the attenuation of seismic waves increases with their frequency, the higher the amplitude and the lower the frequency, the longer the distance over which one can receive waveforms at a reasonable signal to noise ratio. Quantitative seismological processing of recorded waveforms can routinely provide information on the following parameters pertaining to the source of the seismic radiation:

- Origin time of the event, t_0
- Location, $X = (x, y, z)$
- Seismic potency, P, in m^3, which measures the overall co-seismic inelastic deformation at the source, and its tensor, $P_{i,j}$
- Seismic energy radiated from the source of the seismic event, E, in Joules
- Estimate of stress release at source which scales with the E/P ratio, in Pa, called apparent stress, σ_A

Seismic waveforms do not provide direct information about the absolute stress, but mainly about the strain and stress release at the source. However, sources of seismic events associated with weaker geological features or with a softer or fractured patches of rock yield more slowly under lower driving stress and radiate less seismic energy per unit of inelastic co-seismic deformation, than equivalent sources within strong, solid, and highly stressed rock. Therefore, by comparing the radiated seismic energies of seismic events with similar potencies, one can gain insight into the stresses acting within the part of the rock mass giving rise to these events.

The bulk of seismicity induced by mining originates close to excavations where the rock mass is fractured, and therefore, the E/P ratios of these events are low, in some cases between 10 and 100 Pa. These events are also of a "volumetric" nature, i.e. with a relatively high isotropic component of inelastic deformation at source, and generate lower velocity of ground motion. However, the same mine can have

---

[1] In 1581, Galileo Galilei attended service at the Duomo di Pisa and observed the behaviour of the chandeliers. They were swinging slightly due to the draught. He noticed three things. 1. The period of swinging for chandeliers with the same length was the same regardless of the amplitude of the swing. 2. The chandeliers suspended on longer chains had longer periods of swinging. 3. The period of swinging does not depend on the mass of the chandelier.

very dynamic events that originate away from excavations by rupturing intact rock that deliver E/P ratios between 10^6 and 10^7 Pa and ground motion at source above 1 m/s.

Many mine seismic networks record over 1000, and some over 10000, microseismic events per day, each event being recorded by at least 10, and in some cases more than 50 sites. To get reasonable locations, source parameters and source mechanisms every event should be recorded by preferably 10 or more three-component seismic sensors surrounding the source to ensure adequate sampling of the radiation pattern.

In mines, seismic sensors installed underground are grouted in boreholes away from excavations to avoid site effects. The main reason to install sensors at the skin of an excavation is to measure the amplification of ground motion for support design, and these waveforms are excluded from source inversion. However, over the last 10 years or so, some mines have installed low frequency or strong ground motion sensors on surface to recover the seismic potency of larger events and for ground motion studies.

2.2 Seismic Sources: Glossary of Terms

Seismic Source in Rock A seismic source is a volume of sudden inelastic deformation in rock that radiates perceptible seismic waves. The velocity of that deformation varies from a few tens of cm/s for slow events to a few metres per second for very dynamic events rupturing intact rock. In mines, the average ground motion at source is 0.5 to 1.5 m/s. The volume of inelastic co-seismic deformation, $V = P/\Delta\epsilon$, for strains, $\Delta\epsilon$, greater than 10^{-3} that crack rock, varies from a fraction of a m^3 for events with $\log P \leq -3.0$, to 10^7 m^3 for events with $\log P = 4$. Strain changes between 10^{-3} to 10^{-4} cause minor damage to solid hard rock, and $V (\geq P/10^{-4})$ is correspondingly larger.

Rupture and Slip Rupture is a propagating pulse that precedes slip at a seismic source. Its speed varies from $0.6v_S$ to $0.9v_S$ for sub-shear rupture and can be higher than v_S for super-shear rupture. Rupture speed controls the frequency content of the recorded ground velocity. Rupture may be unilateral, propagating in one direction across the source, bilateral, nucleating at the centre of the source and propagating in both directions or it may be inhomogeneous.

Slip follows rupture, very fast at the tip of the rupture and slowing dramatically past the rupture front. Slip velocity is the velocity of one side of the source with respect to the other. An average slip velocity varies from a few cm/s to a few m/s. Rise time refers to slip duration at a given point in a source.

Directivity It is similar to the Doppler effect. Unilateral rupture directivity will produce earthquake source pulses and source spectra that vary with azimuth (Ben-Menahem, 1961). In the time domain, it produces shorter duration, higher amplitude source time functions in the direction of rupture, and longer duration, lower amplitude source time functions in the opposite direction. However, the area under the source time function is the same, regardless of the azimuth, and is proportional to the seismic potency of the event. In the frequency domain, it produces higher spectral amplitudes at higher frequencies in the direction of rupture and a lack of such high frequency signal in the opposite direction. Low frequency amplitudes of source spectra remain unchanged with azimuth. For a circular crack in a purely elastic media with radius r rupturing outwards from the centre with rupture velocity v_r, the pulse width of ground velocity, or equivalently the rise time of the far-field displacement pulse, is $\tau(\theta) = r/v_r - (\sin\theta) r/v_\pi$, where θ is the take-off angle, i.e. the angle between the normal to the source plane and the ray leaving the source, and v_π is the phase velocity. The full pulse width is the duration between the first break and the second zero crossing, $\Delta T = r/v_r + (\sin\theta) r/v_\pi$. The rise time decreases with increasing θ, and the full pulse width increases with decreasing θ. In a real media, the high frequencies are more attenuated more than low frequencies, resulting in a broadening of the pulse width travel time.

Source Time Function The source time function (STF) defines the deformation $u(t)$, velocity $\dot{u}(t)$, or acceleration $\ddot{u}(t)$ at source versus time. The rise time at a given point at source is the duration of slip at that point, and the average rise time $\langle\tau\rangle = \bar{u}/\langle\dot{u}\rangle \sim 1/f_0$, where f_0 is the corner frequency. In a crack-like model, the slip duration at a given point is comparable to the overall duration of the rupture, i.e. slip at a given point continues until information is received that the rupture has stopped propagating. The stopping of rupture generates strong healing waves that propagate inwards from the rim of the fault. These waves are of three types: P-, S-, and Rayleigh waves. Soon after the passage of the Rayleigh waves, slip rate inside the fault decreases to zero and the fault heals. In some cases, only a portion of the overall rupture surface is undergoing slip at any given point in time, and therefore, there may be larger transient stress changes in the vicinity of the propagating slip pulse (Heaton, 1990). The initial pulse width is the duration between the first break and the first zero crossing of the seismic signal. For wavelengths much larger than source size and for periods much longer than source duration, the source volume is relatively small and can be approximated as a point concentrated in space with finite potency. Shorter, higher frequency waves are sensitive to the finite extent and detailed variation of the slip process at the source, and they require finite source models. The source time function can also be presented as seismic moment or potency and their derivatives in the time domain, e.g. $P(t)$, $\dot{P}(t)$, $\ddot{P}(t)$.

Double Couple A model of seismic source caused by shear slip across an internal surface of zero thickness in an isotropic elastic medium for which the equivalent force system consists of two orthogonal couples with the same moment and opposite sign. The corresponding moment or potency tensors have both zero trace (purely deviatoric) and zero determinant. Physically, this is a representation of a shear dislocation source without any volume changes.

Radiation Pattern A radiation pattern is a geometric description of amplitude and sense of initial motion distributed over the P and S wavefronts on the sphere around the source. If the radius of the sphere is large enough relative to both the size of the source and the dominant wavelength, the radiation pattern represents the far field of a point source. If the radius of the sphere is large compared to the dominant wavelength but comparable with the size of the source, the radiation pattern represents the far field of an extended source. The radiation patterns for the displacement and the velocity are the same since taking the time derivative does not affect the angular distribution. Since radiation from seismic sources reflects the strain distribution near the source, it has a degree of symmetry. The polarity of the initial P-wave pulse leaving the source may be compressional—away from the hypocentre, dilatational—towards the hypocentre, or null at the nodal plane where amplitudes tend to go to zero. For a double-couple point source moving with speed v_r along the OX axis, the angular distribution of displacement is $u_P(\theta) = \sin(2\theta) / [1 - (v_r/v_P)\cos\theta]$, $u_S(\theta) = \cos(2\theta) / [1 - (v_r/v_S)\cos\theta]$, where θ is anticlockwise from OX. It has elongated lobes in the direction of rupture, and more so for faster rupture (Ben-Menahem, 1961; Hirasawa & Stauder, 1965), and it is singular for $v_r \geq v_P$ and $v_r \geq v_S$. For $v_r = 0$, it gives the radiation pattern for a single double-couple non-moving source (Aki & Richards, 2002). A double-couple radiation pattern is symmetric with respect to nodal planes. In mines, tunnels and excavations act as strong scatterers of seismic waves and may modify the radiation patterns of the primary waves. In practice, radiation patterns are observable mainly at lower frequencies.

Source Spectra and Corner Frequency A point source at which stress is released instantaneously would radiate P- and S-wave displacement pulses that propagate as Dirac delta functions in the retarded time $t - R/v_{P,S}$. Their spectra would be flat. Seismic source theory predicts a far-field displacement spectrum that is constant at low frequencies and inversely proportional to some power of the frequency at high frequencies (Aki, 1967). The corner frequency of the displacement spectrum, f_0, is where the high and low frequency trends intersect. In this sense, it is a model parameter. The predominant frequency, f_E, the frequency at which the maximum seismic energy is radiated, i.e. the maximum of the ground motion velocity spectrum, is a more rigorously defined parameter of the seismic spectrum. In the ω^{-2} model, the corner frequency and predominant frequency coincide, $f_0 = f_E$, but in other spectral models they differ.

When the stress is released over a finite time, the radiated pulses are broadened proportionally. There is a corner in both the P-wave and the S-wave displacement spectra at frequencies proportional to the reciprocal of the time for the stress to be released at the source. Similarly, if stress is released instantaneously, but the source size is made finite, the P and S pulses will be broadened, and their spectra will have corner frequencies. The P- and S-waves propagate through an attenuating medium with frequency dependent velocities. This phenomenon is called dispersion, and it leads to a shift in the respective frequencies.

Maximum Frequency, f_{max} While the theoretical acceleration spectrum is flat at high frequencies, the observed spectra are characterised by a trend of exponential decay. Hanks (1979) and Hanks (1982) suggested that the acceleration spectrum is flat above the corner frequency only to a second corner frequency, called f_{max}, above which it decays rapidly. The origin of the rapid decay may be the source or path including the site effect or a combination of the above.

Fault Plane Solution or Focal Mechanism The fault plane solution determines the direction of slip, which is controlled by the orientation of the elastic strain field at the time of rupture, and a possible orientation of the rupture plane. In its simplest form it uses the directions of P-wave motions recorded at a number of stations surrounding the source. Plotting all available directions in the lower hemisphere of a stereographic projection, one can define two orthogonal planes separating compressional and dilatational motion. The axes of maximum shortening and maximum lengthening bisecting the quadrants are known as the P and T axes, respectively. These are the principal strain axes that do not necessarily coincide with the principal stress axes. The P axis lies within the quadrant of dilatational motions, and the T axis lies within the quadrant of compressional motions. Both are orthogonal to the intersection of the two nodal planes where their amplitudes are zero. The axis formed by this intersection is called the B or the null axis. The directions of P-wave motion or moment tensor inversion alone cannot resolve which nodal plane is the rupture plane.

2.3 Seismic Potency and Potency Tensor

A seismic source in rock is a volume of sudden inelastic deformation that radiates perceptible seismic waves. It is always of finite extent in all three dimensions although the thickness of the source is usually much smaller than the other two dimensions. Scalar seismic potency for a physical source is a product of the inelastic strain change and the volume subjected to that strain, $P = \Delta\epsilon V$. Scalar seismic potency modelled as a single dislocation source is the product of an average slip and source area, $P = \bar{u} A$ (Ben-Menahem & Singh, 1981; Ben-Zion & Zhu, 2002). Seismic moment $M = \mu P = \mu \Delta\epsilon V = \Delta\sigma V$, where μ is rigidity.

In hard rock, strain changes $\Delta\epsilon \leq 10^{-6}$ are considered elastic, $\Delta\epsilon \geq 10^{-4}$ damage inhomogeneous rock and $\Delta\epsilon \geq 10^{-3}$ crack intact rock (Jeffreys, 1975). Therefore, a seismic event with $\log P = 2.0$, i.e. $m_{HK} = 2.25$, is not just a $100\,\text{m}^3$ of volume, but such an event creates approximately $V = 10^2/10^{-3} = 10^5\,\text{m}^3$ of rock subjected to cracks, which is not insignificant and frequently leads to damage when its source is close to an excavation.

For small deformations, the total strain at a given point may be written as a sum of elastic ε and inelastic contributions, ϵ. The inelastic strain tensor, ϵ_{ij}, also called transformational strain, represents the inelastic brittle deformation at the seismic source that resets the reference levels of the elastic stress and strain tensors after the event (Eshelby, 1957). The seismic potency tensor, P_{ij}, and seismic moment tensor, M_{ij}, can then be defined as

$$P_{ij} = \int_V \epsilon_{ij} dV \quad \text{and} \quad M_{ij} = \int_V c_{ijkl} \epsilon_{kl} dV, \tag{2.1}$$

where c_{ijkl} is the tensor of elastic moduli of the rock surrounding the source and the product $m_{ij} = c_{ijkl}\epsilon_{kl}$ is also called seismic moment density tensor or stress glut (Backus & Mulcahy, 1976). The scalar potency, $P = \sqrt{2P_{ij}P_{ij}} = \|P_{ij}\|$, and its units are m^3. Assuming zero net torque and zero net rotation, both tensors are symmetric and have six independent components. The advantage of seismic potency is that it makes no assumptions about material properties, c_{ijkl}, outside the source which are poorly constrained.

The seismic potency tensor can be decomposed into an isotropic component, $P_{ISO} = \frac{1}{3}\delta_{ij}\text{tr}(\mathbf{P})$, where $\text{tr}(\mathbf{P}) = \sum_{i=1}^{3} P_{ii}$ and δ_{ij} is the Kronecker delta, and the deviatoric remainder, $P_{DEV} = P_{ij} - P_{ISO}$,

$$P_{ij} = \frac{1}{3}\delta_{ij}\text{tr}(\mathbf{P}) + P'_{ij}. \tag{2.2}$$

A purely isotropic tensor is characterised by three equal eigenvalues. In fluid mechanics, this represents a fluid at rest, and the magnitude of the eigenvalues stands for hydrostatic pressure. Positive eigenvalues signify an ideal explosion and negative signify implosion. To quantify the strength of the isotropic component, Zhu and Ben-Zion (2013) introduced a dimensionless parameter,

$$\xi = \sqrt{\frac{2}{3}} \cdot \frac{\text{tr}(\mathbf{P})}{P}, \tag{2.3}$$

that varies from -1 for implosion to 1 for explosion. Introducing the normalised isotropic tensor, $I_{ij} = (1/\sqrt{3})\delta_{ij}$, and normalised deviatoric tensor, D_{ij}, that satisfies $D_{ii} = 0$ and $D_{ij}D_{ji} = 1$, the potency tensor can be written as

$$P_{ij} = \frac{P}{\sqrt{2}}\left(\xi I_{ij} + \sqrt{1-\xi^2}D_{ij}\right). \tag{2.4}$$

The deviatoric tensor can be decomposed further into a double-couple term, DC, with zero determinant representing shear deformation on a plane, and a remainder non-double-couple term, e.g. a compensated linear vector dipole, CLVD, which is a dipole of magnitude 2 compensated by two unit dipoles (Knopoff & Randall, 1970).

The eigenvalues of the normalised deviatoric tensor are $\lambda_1 \geq \lambda_2 \geq \lambda_3$, where the largest one λ_1 corresponds to the T axis of the deviatoric tensor D_{ij}, the intermediate λ_2 corresponds to the null axis eigenvector N, and λ_3 corresponds to the P axis eigenvector P. When $\lambda_2 = 0$, the deviatoric tensor D_{ij} is a pure double couple. The condition $D_{ii} = 0$ requires that $\lambda_1 + \lambda_2 + \lambda_3 = 0$ and $D_{ij}D_{ij} = 1$ that $\lambda_1^2 + \lambda_2^2 + \lambda_3^2 = 1$. From these three conditions imposed on the eigenvalues, Zhu and Ben-Zion (2013) show that

$$\max(\lambda_2) = \min(\lambda_1) = 1/\sqrt{6} \quad \text{and} \quad \min(\lambda_2) = \max(\lambda_3) = -1/\sqrt{6}. \tag{2.5}$$

While the DC component can be associated with slip on a planar surface, the CLVD component corresponds to compensated uniaxial compression or extension of a volume which requires a more complicated geological setting and may not be well constrained by inversion (Frohlich & Davis, 1999). In addition, the DC-CLVD decomposition is not unique because the CLVD symmetry axis can be aligned with any of the principle axis. Aligning the symmetry CLVD axis with the N axis, Zhu and Ben-Zion (2013) show that

$$D_{ij} = \lambda_1 T_i T_j + \lambda_2 N_i N_j + \lambda_3 P_i P_j = \frac{\lambda_1 - \lambda_3}{\sqrt{2}} D_{ij}^{DC} + \sqrt{\frac{3}{2}} \lambda_2 D_{ij}^{CLVD}, \quad (2.6)$$

where $D_{ij}^{DC} = (T_i T_j - P_i P_j)/\sqrt{2}$ and $D_{ij}^{CLVD} == (2N_i N_j - T_i T_j - P_i P_j)/\sqrt{6}$ are the normalised DC and CLVD tensors, which are orthogonal, and therefore $D_{ij}^{DC} D_{ij}^{CLVD} = 0$. The strength of the CLVD component is given by $\chi = \lambda_2/\sqrt{3/2}$, which varies between $1/2$ and $-1/2$.

Now $D_{ij} = \sqrt{1-\chi^2} D_{ij}^{DC} + \chi D_{ij}^{CLVD}$, and inserting into Eq. (2.4) Zhu and Ben-Zion (2013) obtained

$$P_{ij} = \frac{P}{\sqrt{2}} \left[\xi I_{ij} + \sqrt{1-\xi^2} \left(\sqrt{1-\chi^2} D_{ij}^{DC} + \chi D_{ij}^{CLVD} \right) \right], \quad (2.7)$$

which has six independent parameters, i.e. three amplitude factors: P, ξ, and χ, and three angles that specify the orientation of the principal axes of the deviatoric tensor.

Here the isotropic parameter ξ represents the fractional volumetric source component, and χ represents the relative strength of the CLVD component within the deviatoric tensor. Following the Chapman and Leaney (2012) suggestion to use the squared ratios of the scalar moment of each component to the total scalar moment to represent the relative strengths of the ISO, DC, and CLVD components Zhu and Ben-Zion (2013) obtained,

$$\Lambda^{ISO} = \text{sgn}(\xi)\xi^2 \quad \Lambda^{DC} = \left(1-\xi^2\right)\left(1-\chi^2\right) \quad \Lambda^{CLVD} = \text{sgn}(\chi)\left(1-\xi^2\right)\chi^2, \quad (2.8)$$

so that $\Lambda^{ISO} + \Lambda^{DC} + \Lambda^{CLVD} = 1$.

Figure 2.1 shows the permissible values of the fractional strength parameters. Note that the maximum CLVD strength in this decomposition is 25 per cent, i.e. at $\xi = 0$ and $\chi = \pm 1/2$.

A potency tensor representation is a theoretical relation between the real ground motions at a given site and the potency tensor at the source. To invert for the potency tensor, we need the recorded waveforms and the synthetic waveforms, i.e. Green's functions that describe the impulse response of the rock mass at a given site to a force excitation at the source. Different schemes for moment or potency tensor inversion of mine events are described in Fletcher and McGarr (2005), Malovichko and van Aswegen (2013), Sen et al. (2013), Ma et al. (2018), Willacy et al. (2019).

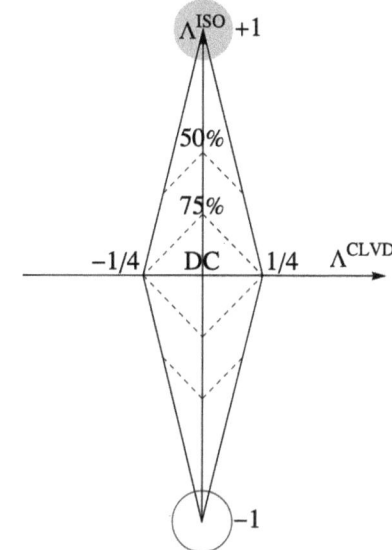

Fig. 2.1 Diagram showing permissible values of the fractional source strengths Λ^{ISO}, Λ^{CLVD}, and Λ^{DC}. The pure explosion and implosion sources are indicated by the grey solid and open circles, respectively. The pure DC source is located at the centre. The contours show DC levels of 75% and 50% (Reproduced from Zhu and Ben-Zion, 2013)

Sources of mid-size and larger events in underground mines interact with excavations and often display a substantial implosive component (McGarr 1992). Seismic moment or potency tensors derived from waveforms of these events processed using elastodynamic Green's functions for an unbounded homogeneous medium may lead to an incorrect interpretation. Malovichko (2020) suggested a correction for conventional expressions for seismic point sources based on the volumetric integral of stress-free strain that includes a term that depends on the displacement at the surface of the excavation. He showed that it affects the type of mechanism and orientations of principal axes.

2.4 Radiated Seismic Energy

The energy release during fracture and frictional sliding is due to the transformation of elastic strain into inelastic strain. This transformation may occur at different rates ranging from slow creep-like events to very fast dynamic seismic events with an average velocity of deformation at the source of up to a few metres per second. Slow type events have a long time duration at the source and thus radiate predominantly lower frequency waves, as opposed to dynamic sources of the same size. Since the excitation of seismic energy can be represented in terms of the temporal derivatives of the source function, one may infer that a slower source process implies less seismic radiation. In terms of fracture mechanics, the slower the rupture velocity, the less energy is radiated; the quasi-static rupture would radiate practically no energy.

To assess the physical sources of radiated energy, let us consider the formula for seismic energy E for the single fracture-type source, expressed in terms of source parameters (Kostrov, 1974; Kostrov & Das, 1988; Rivera & Kanamori, 2005),

$$E = \Delta W - E_F - E_G = \frac{1}{2}\int_A \Delta\sigma_{ij}u_i n_j dA - \int_0^{t_s} dt \int_{A(t)} \dot{\sigma} u_i n_j dA - 2\gamma_{eff} A, \quad (2.9)$$

where γ_{eff}—the effective surface energy, which includes the total loss of mechanical energy, in particular inelastic work, and heat flow from the fracture edge, A—the fracture area with the displacement u_i, $\Delta\sigma_{ij}$—the difference between the final (at the end of the event) and the initial stress, n_j—the unit vector normal to the fracture plane, t_s—the source duration, $\dot{\sigma}_{ij}$—the time derivative of stress or traction rate.

Expression (2.9) allows estimating seismic energy from the stresses and displacements only, and it is not necessary to know the absolute value of stress at source. Therefore, seismic energy depends only on the stress perturbation at source, ΔW, caused by rupture, and is independent of the pre-stress within the medium. However, it is more likely that high pre-stress will drive higher stress change. The term $\Delta\sigma_{ij}u_i n_j$ in Eq. (2.9) cannot be interpreted as the local energy density at a point on the fault plane. It represents the energy released from the tubular volume formed by the integral lines of the vector $F_j \equiv \Delta\sigma_{ij}u_i$ passing through a unit area at a point on the fault plane. Thus, this represents the energy coming from the volume of the medium, rather than the fault plane (Rivera & Kanamori, 2005).

The second term, containing the traction rate, E_F, strongly depends on how the fracture propagates and how it correlates with slip. The energy due to the radiation of high frequency waves during accelerating and decelerating rupture is called radiation friction or radiation loss. If traction rate and slip are uncorrelated, the third term will vanish.

From dimensional analysis, it follows that the first two terms in this equation vary with the fracture area as $A^{3/2}$, whereas the fracture work, $E_G = 2\gamma_{eff}A$, is proportional to A (Kostrov & Das, 1988). Thus, the relative contribution of the fracture work to seismic energy increases with a decrease in the size of the fracture. Consequently, for sufficiently small fractures, the first term may almost cancel the second term, suppressing the acoustic emission so that "silent" fracture would occur.

In the time domain, the radiated seismic energy of the P- or S-wave is proportional to the integral of the radiation pattern corrected far-field velocity pulse squared $\dot{u}^2(t)$ of duration t_s,

$$E_{P,S} = \frac{8}{5}\pi\rho v_{S,P} R^2 \int_0^{t_s} \dot{u}^2_{corr}(t)dt, \quad (2.10)$$

where ρ is rock density, $v_{S,P}$ is S- or P-wave velocity, and R is the distance from the source. Different $u(t)$ may have the same time integral over t_s and thus the same potency P, but their time derivatives $\dot{u}(t)$ may differ, producing different energies E. If $u(t)$ varies very rapidly, the radiated energy can be very large since $E \sim \dot{u}^2(t)$.

In the far field of seismic observations, the P- and S-wave contributions to the total radiated energy are proportional to the integral of the square of the P and S velocity spectrum. For a reasonable signal to noise ratio in the bandwidth of frequencies available on both sides of the corner frequency, the determination of that integral from waveforms recorded by a seismic network is fairly objective. The high frequency component of seismic radiation needs to be recorded by the seismic system if a meaningful insight into the stress regime at the source is to be gained.

2.5 Apparent Stress, Energy Index, and Apparent Volume

Aki (1966) suggested that the ratio of elastic energy released by the source, W, to seismic moment, M, is independent of the average relative displacement at source, $\bar{u}$, and the source area A. Brune (1968) and Wyss and Brune (1971) compared $W = \bar{\sigma} \cdot \bar{u} A = \bar{\sigma} \cdot P$, where $\bar{\sigma}$ is the average of the initial and final stress, and the radiated seismic energy, $E = \xi W$, where ξ is seismic efficiency, and called the product, $\xi \bar{\sigma}$, apparent stress (Wyss & Brune, 1971),

$$E = \xi \bar{\sigma} P \quad \Rightarrow \quad \sigma_A = \xi \bar{\sigma} = E/P. \tag{2.11}$$

Apparent stress is the ratio of the observed radiated seismic energy to seismic potency, and therefore, it is a model independent measure of the dynamic stress release in the source region. Apparent stress is proportional to the integral of the square of the velocity spectrum divided by the amplitude of the low frequency asymptote to the displacement spectrum, or, when the acceleration spectrum is considered, apparent stress depends linearly on the product of the corner frequency and the amplitude of the high frequency asymptote to the acceleration spectrum. It is then a more reliable and less model dependent parameter describing an average stress release at the source than the corner frequency cubed dependent static stress drop.

Figure 2.2 left shows the energy, E vs. $\log P$ for the data set described in Chap. 5 Sect. 1.5, where the green line represents the ordinary least squares fit. The colour here scales with $\log \sigma_A$. Figure 2.2 right shows apparent stress, σ_A, vs. $\log P$ for the same data set where colour indicates the time of the event. In this case, there is an increase in seismic energy with increasing potency.

Although seismic waveforms do not have direct information about absolute stress but merely about the dynamic stress drop at the source, a number of seismological studies and numerous underground observations suggest that a reliable estimate of apparent stress can be used as an indicator of the local level of stress. Gibowicz (1990) considered apparent stress as an independent parameter of stress release in the case when P- and S-wave contributions to the seismic energy are included. (Mendecki, 1993) showed an example where apparent stress associated with seismic events of magnitudes between 1.3 and 1.5 varies from 0.2 to 40 bar, being generally

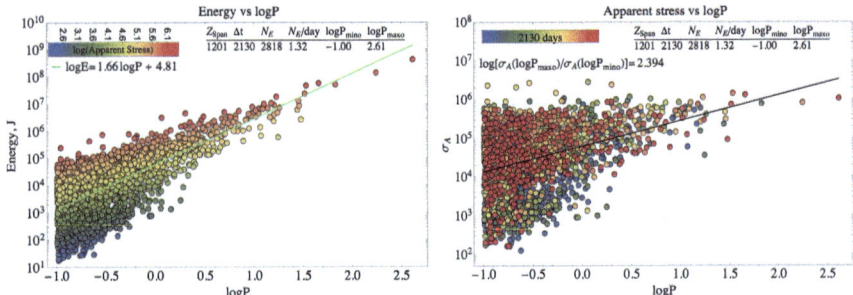

Fig. 2.2 Energy, E, vs. log P plot of events with the least square fit *(left)*. Apparent stress, σ_A, vs. log P for the same data set *(right)*

higher at greater depth and within less faulted rock. In general, the apparent stress expresses the amount of radiated seismic energy per unit volume of inelastic co-seismic deformation.

Let us imagine the source of a seismic event associated with a relatively weak geological feature or with a soft patch in the rock mass. Such a source will yield slowly under lower differential stress producing larger seismic potency and radiating less seismic energy, resulting in a low apparent stress event. The opposite applies to a source associated with a strong geological feature or hard patch in the rock mass. In the case of a so-called complex or multiple event, the rapid deformation process at the initial source can push the stresses in the adjacent volume to a level much higher than could normally be maintained by the rock, producing higher apparent stress sub-event(s) that need not be an indication of a generally high ambient stress prior to the event. Although the estimate of apparent stress does not depend on the rupture complexity (Hanks & Thatcher, 1972), the complexity of the event should be tested before meaningful interpretation in terms of stress can be given.

Seismic sources similar in terms of their potency may differ by up to two orders of magnitude in radiated energies, reflecting stress differences at the place and time of their occurrence (Mendecki, 1993). Since radiated seismic energy broadly increases with increasing seismic potency, to gain insight into the stress regime, one would need to compare the radiated seismic energies associated with seismic events of similar seismic potencies. This notion was translated into a practical tool by van Aswegen and Butler (1993) and called the energy index, EI. The energy index of an event is the ratio of the observed radiated seismic energy of that event E, to the average energy $\bar{E}(P) = 10^{d \log P + c}$ radiated by events of the observed potency P, for a given area of interest,

$$EI = E/\bar{E}(P) = E/10^{d \log P + c} = 10^{-c} E/P^d. \quad (2.12)$$

For $d = 1.0$, the energy index is proportional to apparent stress. In general, it is advisable to use the logarithm of the ratio since it equally affects changes in the

2.5 Apparent Stress, Energy Index, and Apparent Volume

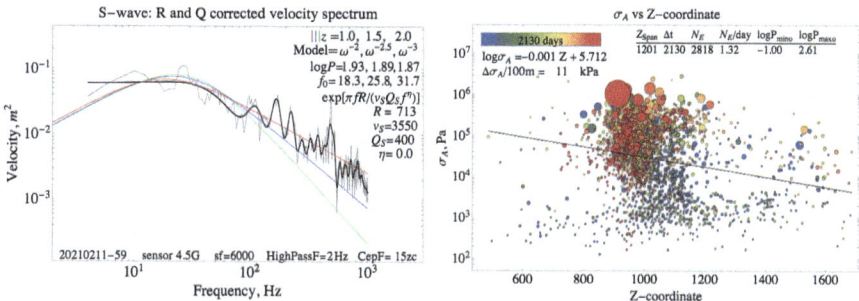

Fig. 2.3 Z-coordinate of events vs. time *(left)* and apparent stress, σ_A, vs. the Z-coordinate of the event for the same data set *(right)*

numerator or denominator. It also solves the problem of lack of symmetry, i.e. if E is greater than $\bar{E}(P)$, the ratio can take theoretically any value greater than 1, but if E is less than $\bar{E}(P)$, the ratio is restricted to the range of 0 to 1. The logged ratio restores the symmetry, i.e. $\log\left(E/\bar{E}(P)\right) = -\log\left(\bar{E}(P)/E\right)$.

Figure 2.3 left shows the Z-coordinate of events vs. time where colour scales with apparent stress, σ_A. Here, the lower the Z the deeper the event. Figure 2.3 right shows apparent stress of events vs. Z-coordinate where colour scales with time. The thin black line shows a linear fit to the data, $\log \sigma_A = -0.001Z + 5.712$, which for this data set indicates that σ_A increases by 11 kPa with every 100 m increase in depth.

The energy index is a relative measure of stress because it is specific to a given data set, i.e. it is a function of the volume and time from which data are selected. Therefore, as more data becomes available, the $\log E$ vs. $\log P$ fit may not be applicable, which makes comparison with the previous data set problematic. However, it is a very useful tool to test for the relative spatial differences in stresses. The energy index of small events is a better test for stress level than larger events, since small ones do not disturb stresses in the area.

Source volume V is the volume of co-seismic inelastic deformation that radiated the recorded seismic waves and can be estimated from $V = M/\Delta\sigma = \mu P/\Delta\sigma$. Since apparent stress scales positively with stress drop and it does not depend on the corner frequency, Mendecki (1993) defined the apparent volume as

$$V_A = \frac{M}{\sigma_A} = \frac{M^2}{\mu E} = \frac{\mu P^2}{E}. \tag{2.13}$$

Equation (2.13) shows that for a given seismic potency source volume scales inversely with seismic energy. Apparent volume, like apparent stress, depends on seismic potency and radiated seismic energy, and, because of its scalar nature, can easily be manipulated in the form of cumulative or contour plots, providing insight into the rate and distribution of co-seismic deformation and/or stress transfer in the rock mass. In a cumulative plot, the apparent volume amplifies the influence

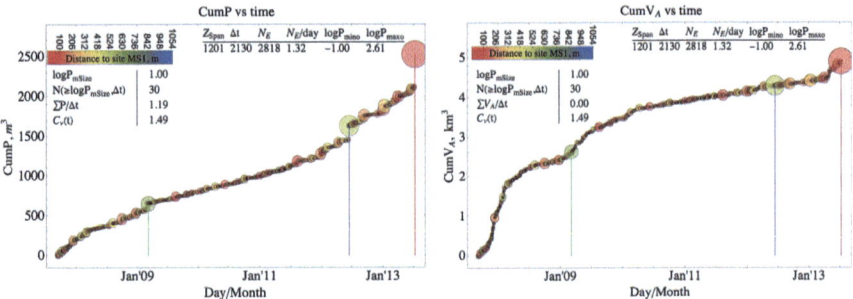

Fig. 2.4 Cumulative potency *(left)* and apparent volume *(right)* for the same data set

of soft events that for a given potency radiate less energy, and reduces the impact of fast dynamic events, and therefore it is useful to observe periods of accelerating deformations.

Figure 2.4 shows the cumulative plots of seismic potency and apparent volume vs. time for the same data set. The way the $CumV_A$ depends on potency and energy makes it more sensitive to softer seismic events and will more likely emphasise the phase of accelerating seismic deformation before larger events, in this case before a $\log P = 2.61$ event, which is not visible on the $CumP$ plot. The main applications of apparent volume are in cumulative and spatial contouring plots.

2.6 Circular Crack: Static Solution by Eshelby

For a circular crack of radius r and a uniform strain drop $\Delta\epsilon$ over the crack surface, $A = \pi r^2$, the displacement profile is given by

$$u(x) = \frac{24}{7\pi} \Delta\epsilon \sqrt{r^2 - x^2}, \qquad (2.14)$$

where x is the radial distance from the centre of the crack and r is the radius of the crack (Eshelby, 1957). The maximum displacement is at the centre, $u_{max} = 24r\Delta\epsilon/(7\pi)$, see Fig. 2.5.

The mean displacement $\bar{u}$ is the integral of $u(x)$ given above divided by the area of the crack, πr^2. Integration in polar coordinates, (x, φ), gives $\int_0^r x\,dx \int_0^{2\pi} d\varphi \sqrt{r^2 - x^2} = (2/3)\pi r^3$, and finally,

$$\bar{u} = \frac{16}{7\pi} \Delta\epsilon r. \qquad (2.15)$$

The ratio of the maximum to the mean displacement $u_{max}/\bar{u} = 1.5$. Equation (2.15) gives seismic potency,

2.6 Circular Crack: Static Solution by Eshelby

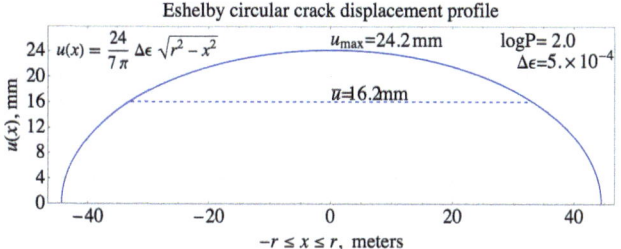

Fig. 2.5 Eshelby (1957) circular crack displacement profile for $\log P = 2.0$ and $\Delta\epsilon = 5 \cdot 10^{-4}$

$$P = \bar{u}\pi r^2 = \frac{16}{7}\Delta\epsilon r^3, \tag{2.16}$$

and the source radius,

$$r = \left(\frac{7P}{16\Delta\epsilon}\right)^{1/3}. \tag{2.17}$$

One can also express $\bar{u}$ as a function of strain drop and potency,

$$\bar{u} = \frac{1}{\pi}\left(\frac{16}{7}\right)^{2/3}\Delta\epsilon^{2/3}P^{1/3} \quad \text{or} \quad P = \pi^3\left(\frac{7}{16}\right)^2 \cdot \bar{u}^3/\Delta\epsilon^2. \tag{2.18}$$

For the average strain change at source $\Delta\epsilon = 5 \cdot 10^{-4}$, the maximum displacement given by Eq. (2.18) is $u_{max} = 1.5 * \bar{u} = 0.0052\sqrt[3]{P}$, which is 13% more than $u_{max} = 0.0046\sqrt[3]{P}$ given by McGarr and Fletcher (2003). Figure 2.6 shows source radius, r, and average displacements, $\bar{u}$, for different $\log P$ events.

The strain energy in an elastically deformed volume is given by $\Delta W = 1/2 \int_V \sigma_{ij}\epsilon_{ij}dV$, where σ_{ij} is the stress tensor, ϵ_{ij} is the strain tensor, and V is the

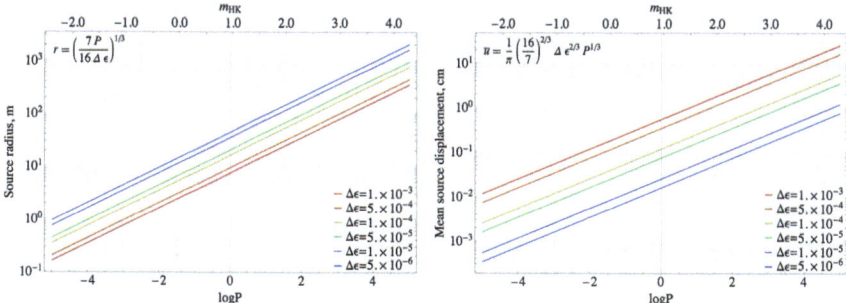

Fig. 2.6 Eshelby source radius (left) and average source displacement (right) vs. $\log P$ and m_{HK} for different strain drops

strained volume. The total change in strain energy ΔW due to a seismic event is $\Delta W = W_0 - W_1$, where W_0 and W_1 are the strain energies before and after the event, respectively. The amount of strain energy released by a seismic event cannot be directly estimated from waveforms since the radiated seismic energy is dependent only on the stress drop and not on the initial stress. This means that two events with the same source displacement can have very different strain energy releases. In the circular crack described above the strain energy, $\Delta W = \mu \Delta \bar{u} A/2 = \Delta\sigma P/2 = 8\mu r^3 \Delta\epsilon^2/7$.

2.7 Circular Crack Model by Brune

Near Field To construct his model, Brune (1970) assumed a shear stress pulse, σ_{eff}, applied to a circular crack with a finite radius r instantaneously at $t = 0$, over the entire interior, i.e. it assumes an infinite rupture velocity, and therefore the crack propagation is neglected. The stress pulse σ_{eff} is the effective stress or the dynamic stress drop, and it represents the difference between the initial stress and the dynamic frictional stress. For a complete stress drop, we can assume that the effective stress is equal to stress drop, $\sigma_{eff} = \Delta\sigma$, or that the effective strain is equal to the dynamic shear strain change (drop), $\epsilon_{eff} = \Delta\epsilon = \sigma_{eff}/\mu$.

The above assumption is equivalent to applying a sudden uniform stress pulse on the interior surface of the crack. Brune also assumed that the crack surface reflects shear waves during rupture. The stress pulse sends a shear pulse propagating orthogonal to the crack, and the initial time function for this pulse follows directly from the boundary conditions,

$$\sigma(x, t) = \sigma_{eff} H(t - x/v_S), \tag{2.19}$$

where x is the distance from the source, v_S is the S-wave velocity, and $H(t) = 0$ for $t < 0$, $H(t) = 1$ for $t \geq 0$ is the Heaviside unit step function. The displacement created by the stress pulse can be obtained by solving the constitutive equation $\mu \partial u/\partial x = -\sigma_{eff}(x, t)$ or $\partial u/\partial x = -\Delta\epsilon(x, t)$,

$$\frac{\partial u(x, t)}{\partial x} = -\frac{\partial(t - x/v_S)}{\partial x} \frac{\partial u}{\partial(t - x/v_S)} = -\Delta\epsilon H(t - x/v_S),$$

and since $\partial(t - x/v_S)/\partial x = -1/v_S$, the differential equation for the initial displacement is

$$\frac{\partial u}{\partial(t - x/v_S)} = v_S \Delta\epsilon H(t - x/v_S),$$

2.7 Circular Crack Model by Brune

that for $t < 0$ gives $u(x = 0, t) = 0$ and for $0 < t < T$

$$u(x = 0, t) = v_S \Delta \epsilon \cdot t, \qquad (2.20)$$

$$\dot{u}(x = 0, t) = v_S \Delta \epsilon, \qquad (2.21)$$

where T is the time required for elastic waves to propagate from the ends of the rupture to the observation point. Therefore, Eqs. (2.23) and (2.22) are for the initial displacement and velocity at a point very close to the centre of the source and neglecting the finite size of the crack. Similar equations were derived by Jeffreys (1962). Assuming $\Delta \epsilon = 3.3 \cdot 10^{-4}$, i.e. $\Delta \sigma = 10$ MPa and $v_S = 3$ km/s Brune, estimated the initial near-field ground velocity at 1 m/s.

The initial ground velocity starts decaying to zero when the effects of the source edges are felt at the observation point, which Brune models by introducing an exponential decay factor,

$$\dot{u}_{NF}(t) = H(t) v_S \Delta \epsilon \exp(-t/\tau), \qquad (2.22)$$

where $\tau \approx r/v_S$, where r is the source radius. For $\Delta \epsilon = 2.5 \cdot 10^{-4}$ at $t = \tau$, it gives $\dot{u}_{NF}(t = \tau) \simeq 0.8$ m/s. Integrating Eq. (2.22) gives

$$u_{NF}(t) = H(t) v_S \Delta \epsilon \cdot \tau \left[1 - \exp(-t/\tau)\right], \qquad (2.23)$$

which for $\Delta \epsilon = 2.5 \cdot 10^{-4}$ at $t = \infty$ gives $u_{NF}(t = \infty) \simeq 0.013$ m. The near-field displacement spectrum is

$$\Omega_{NF}(f) = v_S \Delta \epsilon \left(\frac{1}{2\pi}\right)^2 \cdot \frac{1}{f} \left(f^2 + \tau^{-2}\right)^{-1/2}, \qquad (2.24)$$

which decays as f^{-1} at low and as f^{-2} at high frequencies. Figure 2.7 shows the near-field source time functions for displacement and velocity described by Eqs. (2.23) and (2.22) and the near-field displacement spectra for three different effective strains. The assumptions: $r = 50$ m, $\Delta \epsilon_1 = 2.5 \cdot 10^{-4}$, $\Delta \epsilon_2 = 5 \cdot 10^{-5}$, $\Delta \epsilon_3 = 1 \cdot 10^{-5}$, $v_S = 3250$ m/s, $f_{0S} = 0.373 v_S/r = 24.2$ Hz, and $\tau = 1/(2\pi f_{0S}) = 0.0066$ seconds.

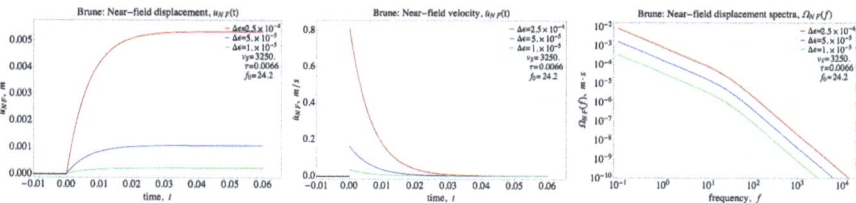

Fig. 2.7 Brune near-field displacement and velocity source time functions and displacement spectra described by Eqs. (2.23), (2.22), and (2.24) for selected effective strains

Far Field The near field is essentially determined by the motion on one side of the source, while the far field represents the contributions from both sides of the source, i.e. the elastic waves radiated by the opposing side of the source diffract around the source surface and differentiate the far-field spectrum modifying its low frequency part (Keilis-Borok et al., 1960). Brune approximated this effect by multiplying the displacement function by an exponential with decay time of the order of r/v_S and by a factor r/R to correct for spherical spreading,

$$u_{FF}(t) = \Lambda_S v_S \Delta\epsilon \, (r/R) \cdot t \cdot \exp(-t/\tau), \qquad (2.25)$$

$$\dot{u}_{FF}(t) = \Lambda_S v_S \Delta\epsilon \, (r/R) \cdot (1 - t/\tau) \exp(-t/\tau), \qquad (2.26)$$

$$\ddot{u}_{FF}(t) = \Lambda_S v_S \Delta\epsilon \, (r/R) \, (1/\tau) \cdot (-2 + t/\tau) \cdot \exp(-t/\tau), \qquad (2.27)$$

where R is the distance, and $\Lambda_S = \sqrt{2/5} = 0.632$ is the root mean square radiation pattern of the S-wave, i.e. the mean energy radiation computed by integrating the squared radiation pattern over the surface of the unit sphere (Aki & Richards, 2002). Here, the displacement, $u_{FF}(t)$, continues indefinitely, although $\lim_{t\to\infty}(t \cdot \exp(-t/\tau)) = 0$. Velocity and acceleration have jumps at $t = 0$. The Fourier amplitude spectra of the S-wave far-field displacement, velocity, and acceleration are

$$\Omega_{FFS}(f) = \Lambda_S v_S \Delta\epsilon \left(\frac{r}{R}\right) \cdot \frac{1}{(2\pi f_0)^2} \left[1 + (f/f_0)^2\right]^{-1}, \qquad (2.28)$$

$$\dot{\Omega}_{FFS}(f) = \Lambda_S v_S \Delta\epsilon \left(\frac{r}{R}\right) \cdot \frac{f}{2\pi f_0^2} \left[1 + (f/f_0)^2\right]^{-1}, \qquad (2.29)$$

$$\ddot{\Omega}_{FFS}(f) = \Lambda_S v_S \Delta\epsilon \left(\frac{r}{R}\right) \cdot \left(\frac{f}{f_0}\right)^2 \left[1 + (f/f_0)^2\right]^{-1}, \qquad (2.30)$$

where f_0 is the corner frequency that represents the transition between the flat and sloping parts of the displacement spectrum. It is used to estimate source radius, although this relation is poorly constrained. The displacement spectra decay with f^{-2} past the corner frequency; therefore, it is a so-called ω^{-2} model. The acceleration spectrum predicts that ground acceleration is flat for arbitrarily high frequencies. The acceleration spectrum usually decays after a high frequency corner identified as f_{max}.

Since the seismic energy scales with velocity squared, the maximum energy is radiated at the predominant frequency, i.e. the frequency where the velocity spectrum is at maximum. In this case, i.e. the ω^{-2} model, the corner frequency and predominant frequency are the same, which is not the case for other models. Figure 2.8 shows the distance corrected far-field S-wave displacement, velocity, and acceleration in time and their spectra. The assumptions are: $r = 50$ m, $R = 500$ m, $\Delta\epsilon_1 = 2.5 \cdot 10^{-4}$, $\Delta\epsilon_2 = 5 \cdot 10^{-5}$, $\Delta\epsilon_3 = 1 \cdot 10^{-5}$, $v_S = 3250$ m/s,

2.7 Circular Crack Model by Brune

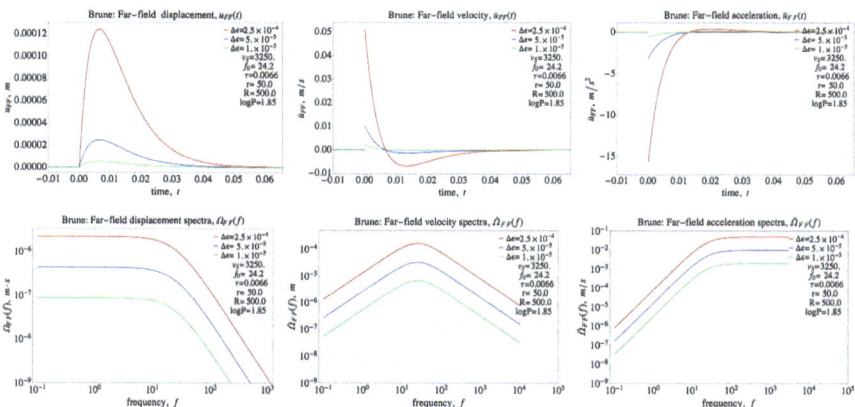

Fig. 2.8 Far-field distance corrected S-wave displacement, velocity, and acceleration in time *(top)* and spectra *(bottom)* and for selected effective strains

$f_{0S} = 0.373 v_S/r = 24.2$ Hz, $\tau = 1/(2\pi f_{0S}) = 0.0066$ seconds, and potency $P = (16/7) \Delta \epsilon r^3$ which for $\Delta \epsilon_1 = 2.5 \cdot 10^{-4}$ gives $\log P = 1.85$. No Q correction was applied at this stage. Note that the theoretical acceleration spectrum extends to infinity, which is not physical; however, the observed acceleration spectra decay above a certain frequency called f_{max}, (Hanks, 1982).

Corner Frequency Brune estimated spherical average corner frequency by comparing the zero frequency limit of the shear wave displacement spectrum given by Eq. (2.28) with Keylis-Borok (1959) equation for the far-field S-wave spectrum at zero frequency, $\Omega_{FFS}(f = 0) = \Omega_{0S}$,

$$\Omega_{FFS}(f = 0) = \Omega_{0S} = \Lambda_S \mu P / \left(4\pi \rho R v_S^3\right)$$

$$= \Lambda_S v_S \Delta\epsilon \left(\frac{r}{R}\right) \left[1 + (f/f_0)^2\right]^{-1} / (2\pi f_0)^2,$$

and taking $P = (16/7) \Delta \epsilon r^3$ (see Eq. 2.16) and $\mu = \rho v_S^2$, he obtained,

$$f_{0S} = k \cdot v_S/r = 0.373 v_S/r \quad \text{or} \quad r = k \cdot v_S/f_{0S}, \tag{2.31}$$

where $k = \sqrt{7/(16\pi)} = 0.373$ is the corner frequency normalised by the wave velocity and source radius, referred as the normalised corner frequency when comparing the results of different source scenarios. If we assume that source radius can also be recovered from the P-wave radiation, then, taking $r = k v_P/f_{0P}$, the ratio of P- to S-wave corner frequency is

$$f_{0S}/f_{0P} = v_S/v_P. \tag{2.32}$$

Corner frequency is not a physical parameter of a seismic source, and there are considerable variations when computing f_0 using different source models. Madariaga (1976) developed a dynamic model, with a stress singularity at the rupture front, and assuming rupture velocity $v_r = 0.9 v_S$ gave an average $k_S = 0.21$ for the S-wave and $k_P = 0.32$ for P-wave. Brune et al. (1979) considered different source models with different assumptions and suggested an average $k = 1/3$. Kaneko and Shearer (2014) studied a quasi-dynamic circular crack model with a cohesive zone that prevents a stress singularity at the rupture front and, for $v_r = 0.9 v_S$, gave $k_S = 0.26$ and $k_P = 0.38$.

Keylis-Borok (1959) related the average static stress drop on a circular source to the average slip, $\Delta \sigma = 7 \pi \mu \bar{u} / (16 r)$, that for the S-wave can be written as

$$\Delta \sigma = \frac{7}{16} \mu P \cdot \frac{1}{r^3} = \frac{7}{16} \mu P \left(\frac{f_{0S}}{k v_S} \right)^3. \tag{2.33}$$

Here, any uncertainty in the corner frequency f_0 is cubed when computing static stress $\Delta \sigma$. Aki (1966) and Wyss and Brune (1971) proposed to use the two independent source parameters, seismic potency and radiated seismic energy, to estimate stress release at source, $\sigma_A = E/P$, that does not require an estimate of the source dimension.

If we assume that seismic potency can also be estimated from the zero frequency limit of the P-wave displacement spectrum, $\Omega_{0P} = \Lambda_P \mu P / (4 \pi \rho R v_P^3)$, then the ratio

$$\frac{\Omega_{0S}}{\Omega_{0P}} = \frac{\Lambda_S}{\Lambda_P} \cdot \left(\frac{v_P}{v_S} \right)^3, \tag{2.34}$$

where $\Lambda_P = \sqrt{4/15} = 0.516$. For a Poissonian solid, i.e. $v_P = \sqrt{3} \cdot v_S$, the ratio is $(3/2) \sqrt{27} \simeq 7.8$. Seismic potency derived from the low frequency limit of the displacement spectra of P- and S-waves should be the same,

$$P_P = 4 \pi R v_P^3 \Omega_{0P} / \left(\Lambda_P v_S^2 \right) = P_S = 4 \pi R v_S \Omega_{0S} / \Lambda_S.$$

2.8 More General Point Source Models

The simplest representation of a seismic source is the far-field point source model, i.e. a one moment tensor source with a single source time function radiating from a point. In contrast, an extended seismic source is rupturing and slipping at all points behind the propagating rupture front, and it can be treated as a set of point-like sub-sources.

2.8 More General Point Source Models

There are crack-like rupture models, discussed in this section, where the duration of deformation at a given point at source is comparable to the overall duration of rupture, and pulse-like rupture models where the slipping portion is constrained to the vicinity of the propagating rupture tip and the duration of deformation at a point is short relative to the rupture duration (Heaton, 1990). Pulse-like models deliver a spectrum with two corner frequencies: the lower one that scales inversely with the overall duration of rupture, while the higher one is inversely proportional to pulse duration.

The rupture of a fault and the subsequent slip are fast non-linear physical processes involving very large irrecoverable deformations and violation of the continuity of the material. The effect of a seismic source on the rock mass in the immediate vicinity of the fault is to induce motions which cannot be described within the framework of linear elasticity. It is only at a certain distance from the source, where the local displacements are below the plasticity threshold, that one can consider the local wave field as a superposition of the waves radiated by all parts of the source. It is the applicability of the superposition principle which separates the domain of non-linear material behaviour around the source from the rest of the rock mass which resembles the laws of linear elasticity. The terms near field, intermediate field, and far field apply only to the rock mass outside the broken zone at the source. In that sense, a seismic source is always volumetric even when the rupture can be assumed to take place on a 2D surface.

Ideally, seismic source time functions that describe the displacement, velocity, and acceleration of ground motion at source should have the following properties:

- The motion must start at a given moment of time, say at $t = 0$, with no motion before that.
- The motion must stop after a finite period of time has elapsed.
- There must be no change in the direction of the velocity during the motion.
- There must be no intermittent motion, i.e. no stop and restart.
- There must be no discontinuities or infinite values in the displacement and velocity.
- The acceleration time function cannot become infinite but can exhibit discontinuities, i.e. sudden changes in the driving force.
- The far-field displacement source pulse of a simple dislocation source is one-sided, and the Fourier transform of such a pulse is always flat at low frequencies and falls off at high frequencies.
- If the far-field displacement is a simple one-sided pulse, then the velocity pulse is two-sided, and the acceleration pulse is three-sided with zero area, i.e. zero velocity at end of record.

The far-field displacement due to a point source is controlled by the slip rate, $u_{FF}(t) = C \cdot \dot{u}(t - r/v_{P,S})$. This statement dictates the following relation between the Fourier transform of the far-field displacement and the final static displacement at the source, $u_{FF}(\omega) = C \int_{-\infty}^{\infty} \dot{u}(t) \exp(-i\omega t)\, dt$, where ω is angular frequency and C a constant.

Therefore, the low frequency limit of the far-field displacement spectrum is

$$\lim_{\omega \to 0} u_{FF}(\omega) = C \int_{-\infty}^{\infty} \dot{u}(t)\,dt = C \cdot u_{src}(t \to \infty),$$

since $u_{NF}(t = -\infty) = 0$ and is related to the scalar seismic potency.

In the frequency domain, the Fourier spectrum of the displacement in the far field should be the same, up to a frequency independent factor, as the spectrum of the velocity at the source. The spectrum of the velocity in the far field should be related to the spectrum of the acceleration at the source, and the spectrum of the acceleration in the far field can be expressed in terms of the spectrum of the acceleration rate at the source. This means that it will be enough to obtain the spectra of the displacement, velocity, acceleration, and acceleration rate at source in order to have all spectra in the near field as well as in the intermediate and far field.

Source Time Functions In the near field, there is no separation between the P- and S-wave because the arrivals of these waves are delayed by less time than the total seismic source duration. They arrive very close together and interfere to produce a static displacement field; therefore, the near-field term refers to ground motion unrelated to a specific type of wave. The far-field approximation is valid at distances R much larger than the dominant wavelength λ, i.e. for $R/\lambda \gg 1$, and $\lambda = v/f$, where v is the wave velocity and f the predominant frequency. Seismic waveforms contain a range of frequencies. The high frequency part of the spectrum reflects the details of the far-field ground motion, while the low frequency is determined by the properties of the motion at the source and in the near field. Since in the far field there should be no residual static deformation and because the maximum peak ground velocities, PGV, are usually associated with the lower frequencies of S-waves, one can consider the far field to be at distances where the recorded seismic strains, $PGV/v_S < 10^{-6}$, which are assumed elastic for a hard rock. Practically, sites in hard rock that record the maximum ground velocity from the S-wave of less than 3 mm/s can be considered far field. If such low ground velocities are recorded at distances that give a short S-P window, then spectra should be calculated from the S-wave only.

2.8.1 Source Time Functions and Spectra

Beresnev (2019) proposed a simple expression for the displacement source time function that caters for different far-field spectral models,

$$u(t, z) = H(t) u_\infty \left[1 - \frac{\Gamma(z+1, t/\tau)}{\Gamma(z+1)} \right], \quad (2.35)$$

2.8 More General Point Source Models

where $H(t)$ is the Heaviside step function, $\Gamma(z+1, t/\tau) = \int_{t/\tau}^{\infty} x^z \exp(-x)\,dx$ is the upper incomplete gamma function, and, for a non-negative integer $\Gamma(z+1) = z!$, u_∞ is the final displacement at $t = \infty$ and τ is related to the rise time. Parameter $z = 1$ gives the most frequently used omega-square model, ω^{-2}, $z = 2$ gives the omega-cube model, ω^{-3}, and $z = 1.5$ gives an $\omega^{-2.5}$ model. For real positive values of z, the slip velocity and acceleration at source are

$$\dot{u}(t, z) = H(t) \frac{u_\infty}{\Gamma(z+1)} \frac{1}{\tau} (t/\tau)^z \exp(-t/\tau). \tag{2.36}$$

$$\ddot{u}(t, z) = H(t) \frac{u_\infty}{\Gamma(z+1)} \frac{1}{\tau^2} \left[z(t/\tau)^{z-1} - (t/\tau)^z \right] \exp(-t/\tau). \tag{2.37}$$

Note that the acceleration time function for the ω^{-2} model, i.e. for $z = 1$, exhibits a finite jump at $t = 0$, which means that the acceleration is undefined there. For all values of $z \geq 1$, the velocity is continuous and equal to zero for $t = 0$. The velocity is maximum at $t = z \cdot \tau$, where its value is

$$\dot{u}_{max} = \frac{u_\infty}{\tau} \cdot \frac{z^z}{z! \exp(z)} \quad \Rightarrow \quad \tau = \frac{u_\infty}{\dot{u}_{max}} \cdot \frac{z^z}{z! \exp(z)}. \tag{2.38}$$

The velocity spectrum is controlled by two physical parameters: the final displacement and the maximum slip velocity defined by the parameters z and τ.

Figure 2.9 shows the displacement, velocity, and acceleration source time functions given by Eqs. (2.35), (2.36), and (2.37) for $z = 1$ that defines the ω^2 model, $z = 1.5$ for the $\omega^{2.5}$ model, and $z = 2$, for the ω^3 model. Assumptions: $\Delta\epsilon = 2.5 \cdot 10^{-4}$, $r = 50$ m, give $P = (16/7)\,\Delta\epsilon r^3 = 71.43$ and $\log P = 1.85$, the final max displacement $u_\infty = 24r\Delta\epsilon/(7\pi) = 0.0136$ m, and $v_S = 3250$ m/s gives $f_{0S} = 0.21 v_S/r = 13.65$ Hz and $\tau = 1/(2\pi f_{0S}) = 0.01166$ seconds. The maximum slip velocities are $\dot{u}_1 = 0.43$ m/s at $t_{z1} = 0.01166$ for the ω^{-2} model, $\dot{u}_{1.5} = 0.36$ m/s at $t_{z1.5} = 0.0175$ for the $\omega^{-2.5}$ model, and $\dot{u}_2 = 0.32$ m/s at $t_{z2} = 0.02332$ seconds for the ω^{-3} model.

The acceleration rate time function at source is

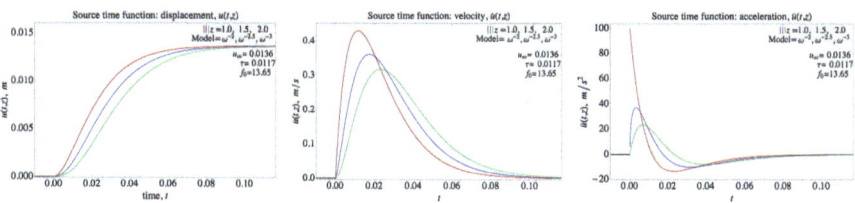

Fig. 2.9 Displacement, velocity, and acceleration source time functions given by Eqs. (2.35), (2.36), and (2.37) for the ω^2 model ($z = 1$) in red, the $\omega^{2.5}$ model ($z = 1.5$) in blue, and the ω^3 model ($z = 2$) in green

$$\dddot{u}(t,z) = \frac{u_\infty}{\Gamma(z+1)} \left(\frac{1}{\tau}\right)^2 \left[z\left(\frac{t}{\tau}\right)^{z-1} - \left(\frac{t}{\tau}\right)^z\right] \delta(t) \exp\left(-\frac{t}{\tau}\right)$$

$$+ \frac{u_\infty}{\Gamma(z+1)} \left(\frac{1}{\tau}\right)^3 \left[z(z-1)\left(\frac{t}{\tau}\right)^{z-2} - 2z\left(\frac{t}{\tau}\right)^{z-1} + \left(\frac{t}{\tau}\right)^z\right]$$

$$\exp\left(-\frac{t}{\tau}\right) H(t),$$

which is not zero for $t = 0$. However, for any $z > 1$, regardless how small, the Dirac delta term in the acceleration rate at $t = 0$ will be zero

$$\dddot{u}(t,z) = \frac{u_\infty}{\Gamma(z+1)} \left(\frac{1}{\tau}\right)^3 \left[z(z-1)\left(\frac{t}{\tau}\right)^{z-2} - 2z\left(\frac{t}{\tau}\right)^{z-1} + \left(\frac{t}{\tau}\right)^z\right]$$

$$\exp\left(-\frac{t}{\tau}\right) H(t),$$

where $\tau = 1/(2\pi f_0)$.

Source Spectra The amplitude spectra of the displacement, Velocity, and acceleration source time functions are

$$\Omega(f,z) = u_\infty \cdot \frac{1}{2\pi f} \left[1 + (f/f_0)^2\right]^{-(z+1)/2}, \quad (2.39)$$

$$\dot{\Omega}(f,z) = u_\infty \cdot \left[1 + (f/f_0)^2\right]^{-(z+1)/2}, \quad (2.40)$$

$$\ddot{\Omega}(f,z) = u_\infty \cdot 2\pi f \left[1 + (f/f_0)^2\right]^{-(z+1)/2}. \quad (2.41)$$

At low frequencies, the displacement spectrum behaves as f^{-1}, the velocity spectrum is flat, and acceleration spectrum is proportional to f. This is true for all three cases irrespective of z. Figure 2.10 shows the displacement, velocity, and acceleration source spectra given by Eqs. (2.39), (2.40), and (2.41) for $z = 1$ that defines the ω^2 model, $z = 1.5$ being the $\omega^{2.5}$ model, and $z = 2$, for the ω^3 model.

2.8.2 Far Field in Time Domain and Spectra

Far Field in Time The far-field P- and S-wave displacements radiated by a shear dislocation at an individual source patch area element A at distance R can be expressed as

2.8 More General Point Source Models

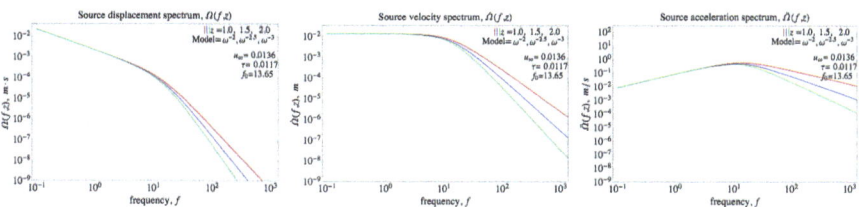

Fig. 2.10 Displacement, velocity, and acceleration source spectra given by Eqs. (2.39), (2.40), and (2.41), for the ω^2 model ($z = 1$) in red, the $\omega^{2.5}$ model ($z = 1.5$) in blue, and the ω^3 model ($z = 2$) in green

$$u_{FF}(R,t) = \frac{\Lambda_P v_S^2}{4\pi v_P^3} \frac{1}{R} \cdot A\dot{u}(t_P') + \frac{\Lambda_S}{4\pi v_S} \frac{1}{R} \cdot A\dot{u}(t_S'), \tag{2.42}$$

where $\dot{u}(t_P')$ and $\dot{u}(t_S')$ are the displacement rate functions at source for the respective retarded times $t_P' = t - R/v_P$ and $t_S' = t - R/v_S$ for P- and S-wave (Aki & Richards, 2002). The term $1/R$ stands for geometrical spreading with distance for body waves in a homogeneous elastic medium. Taking potency $P = u_\infty A$, the far-field displacement time function for the P- and S-wave is

$$u_{FFP,S}(R,t,z) = H(t_{P,S}') \Omega_{P,S} \cdot \frac{1}{\Gamma(z+1)} \frac{1}{\tau} \left(\frac{t_{P,S}'}{\tau}\right)^z \exp\left(-\frac{t_{P,S}'}{\tau}\right), \tag{2.43}$$

where $\Omega_{P,S} = C_{FFP,S} \cdot P/R$, $C_{FFP} = \Lambda_P v_S^2 / (4\pi v_P^3)$, $C_{FFS} = \Lambda_S / (4\pi v_S)$, and $\Lambda_P = \sqrt{4/15} = 0.516$ and $\Lambda_S = \sqrt{2/5} = 0.632$ are the root mean square radiation patterns of P- and S-waves, respectively, and P is seismic potency.

The displacement time function reaches its maximum for

$$t\left[\max(u_{FFP,S})\right] = z\tau + R/v_{P,S} \qquad u_{maxFFP,S} = C_{FFP,S}(1/\tau)(z)^z \exp(-z).$$

The velocity function in the far field for the P- and S-wave is

$$\dot{u}_{FFP,S}(R,t,z) = H(t_{P,S}') \Omega_{P,S} \cdot \frac{1}{\tau^2} \left[z\left(\frac{t_{P,S}'}{\tau}\right)^{z-1} - \left(\frac{t_{P,S}'}{\tau}\right)^z\right] \exp\left(-\frac{t_{P,S}'}{\tau}\right). \tag{2.44}$$

For $z = 1$, it is not zero at the onset of motion, $t_{P,S}' = 0$. This indicates a kinematic inconsistency of the ω^{-2} model since a discontinuity in the slip velocity, i.e. a jump from zero to a finite value at the beginning of the motion, would require an infinite driving force. The function behaves properly for any $z > 1$, regardless of how small.

As expected, the slip velocity time function for $z = 1$ is $\dot{u}_{FFP,S}(R, t \to 0, z = 1) = \Omega_{P,S}(1/\tau^2) \neq 0$, while for $z = 1 + \epsilon$, $\epsilon > 0$, one has

$$\ddot{u}_{FFP,S}(R, t \to 0, z = 1+\epsilon) = \Omega_{P,S} \cdot \frac{1}{\tau^2}\left[(1+\epsilon)\left(\frac{t'_{P,S}}{\tau}\right)^{\epsilon} - \left(\frac{t'_{P,S}}{\tau}\right)^{1+\epsilon}\right] = 0.$$

The acceleration time function in the far field for the P- and S-wave is

$$\ddot{u}_{FFP,S}(R, t, z)$$

$$= H(t'_{P,S})\Omega_{P,S}\left(\frac{1}{\tau^3}\right)\left[z(z-1)\left(\frac{t'_{P,S}}{\tau}\right)^{z-2} - 2z\left(\frac{t'_{P,S}}{\tau}\right)^{z-1} + \left(\frac{t'_{P,S}}{\tau}\right)^{z}\right]$$

$$\exp\left(-\frac{t'_{P,S}}{\tau}\right)$$

$$+ \delta(t'_{P,S})\Omega_{P,S}\left(\frac{1}{\tau^2}\right)\left[z\left(\frac{t'_{P,S}}{\tau}\right)^{z-1} - \left(\frac{t'_{P,S}}{\tau}\right)^{z}\right]\exp\left(-\frac{t'_{P,S}}{\tau}\right). \quad (2.45)$$

The Dirac delta singularity at $t'_{P,S} = 0$ is cancelled for $z > 1$ but survives for $z = 1$, i.e. for the ω^{-2} model, which is the effect from the discontinuity in the slip velocity for that value of the parameter z.

Figure 2.11 shows the distance corrected far-field time domain displacements, velocities, and accelerations for $z = 1, 1.5,$ and 2.0, with the same assumptions as in Fig. 2.9. For $z < 2$, the power of the term $\left(t'_{P,S}/\tau\right)^{z-2}$ in Eq. (2.45) is negative and goes to infinity for $t = 0$, see Fig. 2.11 right.

In general, the modelled time functions can have kinks and singularities which can reflect on their behaviour and on the behaviour of their derivatives. Nevertheless, such functions can produce regular and smooth Fourier transforms as long as the original goes sufficiently fast to zero at infinity and provided that its singularities are integrable. In this case, the Fourier transforms of the far-field time functions are smooth.

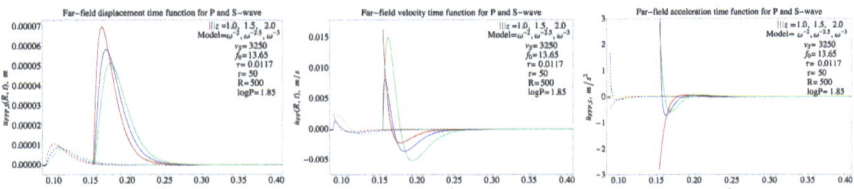

Fig. 2.11 Displacement, velocity, and acceleration far-field time functions given by Eqs. (2.43), (2.44), and (2.45) for the ω^2 model ($z = 1$) in red, the $\omega^{2.5}$ model ($z = 1.5$) in blue, and the ω^3 model ($z = 2$) in green. Dashed lines for P-wave and solid for S-wave

2.8 More General Point Source Models

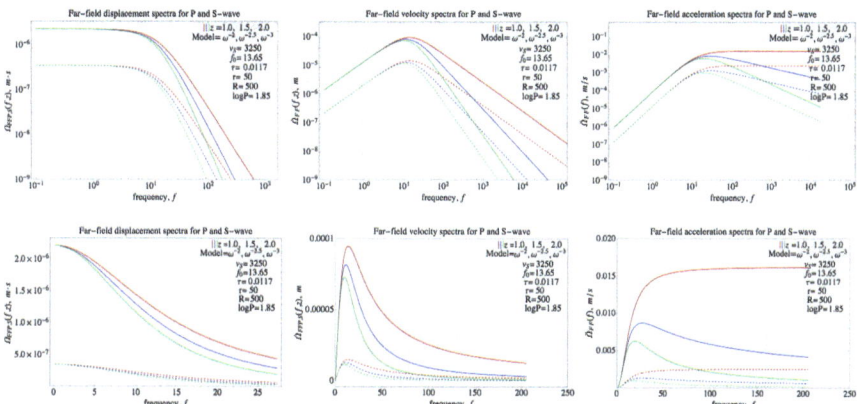

Fig. 2.12 *Top row.* Log-log plots of the far-field displacement, velocity, and acceleration spectra given by Eqs. (2.46), (2.47), and (2.50) for the ω^2 model ($z = 1$) in red, the $\omega^{2.5}$ model ($z = 1.5$) in blue, and the ω^3 model ($z = 2$) in green. Dashed lines for the P-wave and solid for the S-wave. *Bottom row.* The same as top row but in a linear scale to demonstrate their asymmetry with respect to the corner frequency

Far Field Spectra The expression for the slip velocity at the source defines, up to a constant factor, the displacement time function in the far field, i.e. the Fourier transform of the ground displacement in the far field will be proportional to the complex-valued function,

$$F(f) = \int_0^\infty \left[(1/\tau) \, (t/\tau)^z \exp(-t/\tau) \right] \exp(-i2\pi f t) dt = \Gamma(z+1) \, (1 + i f/f_0)^{-z-1},$$

where $f_0 = 1/(2\pi\tau)$.

The modulus of $F(f) = \sqrt{[\text{Re}(F)]^2 + [\text{Im}(F)]^2}$ defines the amplitude spectra of the far-field displacement, velocity, and acceleration for a given z. Figure 2.12 top row shows the log-log plots and bottom row log-linear plots of the far-field displacement, velocity, and acceleration spectra.

The displacement spectra for P- and S-wave are given by

$$\Omega_{FFP,S}(R, f, z) = \Omega_{P,S} \cdot \left[1 + (f/f_0)^2 \right]^{-(z+1)/2}, \qquad (2.46)$$

where $\Omega_{P,S} = C_{FFP,S} \cdot P/R$. For $f = 0$, it gives the Keylis-Borok equation for the zero frequency displacement spectral level of S-wave, $\Omega_{FFS}(f = 0, z) = \Omega_{0S} = \Lambda_S P/(4\pi R v_S)$. The velocity spectrum is

$$\dot{\Omega}_{FFP,S}(R, f, z) = \Omega_{P,S} \cdot (2\pi f) \left[1 + (f/f_0)^2 \right]^{-(z+1)/2}. \qquad (2.47)$$

The maxima of the P- and S-wave far-field velocity spectra are at frequency,

$$f\left[\max\left(\dot{\Omega}_{maxFFP,S}\right)\right] = f_0/\sqrt{z}, \qquad (2.48)$$

and their magnitudes are

$$\dot{\Omega}_{maxFFP,S} = \Omega_{P,S} \cdot 2\pi f_0 \, (z)^{z/2} \, [1+z]^{-(z+1)/2} \, . \qquad (2.49)$$

The P- and S-wave acceleration spectrum in the far field is

$$\ddot{\Omega}_{FFP,S}(R, f, z) = \Omega_{P,S} \cdot (2\pi f)^2 \left[1 + (f/f_0)^2\right]^{-(z+1)/2}. \qquad (2.50)$$

2.8.3 Q Corrections and Site Effects

Q Corrections For a single or a narrow band frequency, the amplitude at source, A_0, of a plain P- or S-wave decays with distance as

$$A(R) = A_0 \exp\left[-\pi f R/\left(v_{P,S} Q_{P,S} f^\eta\right)\right], \qquad (2.51)$$

where η is the exponent in the power law frequency dependence of Q, and Q^{-1} is the fractional loss of energy per cycle of oscillation due to intrinsic absorption and due to scattering caused by energy redistribution. For $\eta = 0$, the amplitude decay increases exponentially with frequency. Larger Q implies less attenuation, and high frequency waves will attenuate faster than low frequency waves. For $\eta = 1$, the amplitude decay is frequency independent.

For a relatively solid hard rock mass, Q is well above 500, but for a softer and/or fractured rock mass, it can be as low as 25. For strains between 10^{-3} and 10^{-6}, attenuation is strain dependent, and therefore, amplitudes decay more rapidly in the intermediate field of seismic radiation, where strains are larger than 10^{-3} and where Q may be as low as 10. In general, Q is frequency dependent (e.g. Fedotov & Boldyrev, 1969; Rautian & Khalturin, 1978; Aki, 1980), and therefore, it is larger for P-waves than for S-waves.

To correct for amplitude decay due to attenuation and scattering, one should multiply the observed far-field spectra by $\exp\left[\pi f R/\left(v_{P,S} Q_{P,S} f^\eta\right)\right]$, where $R/v_{P,S}$ is the travel time from source to the recording site.

Solid lines in Fig. 2.13 left show the shapes of function (2.51) for $Q_S = 200$ and $\eta = 0$, $\eta = 0.1$ and 0.2. Dashed lines show the same but for $Q_S = 25$, i.e. for a strongly attenuating rock mass. Figure 2.13 centre and right shows a shift in the predominant frequency and decay in spectral amplitudes of the far-field velocity model spectra due to attenuation over distance. Solid lines represent P- and S-wave velocity spectra with no Q attenuation and dashed lines with attenuation for $Q_P =$

2.8 More General Point Source Models

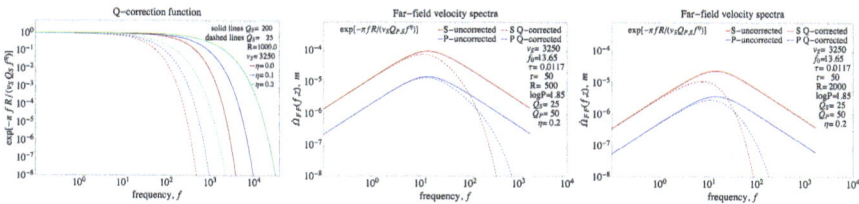

Fig. 2.13 The shape of function 2.51 *(left)*. P- and S-wave of the ω^{-2}, i.e. $z = 1$ model, far-field unattenuated (solid lines) and Q attenuated (dashed lines) velocity spectra over distances of 500 *(centre)* and 2000 m *(right)*.

50 and $Q_S = 25$ at distances 500 m (centre) and 2000 m (right). The relatively low values of Q were selected to illustrate the point.

An alternative model for spectral decay at high frequencies is $a(f) = A_0 \exp(-\pi f \kappa)$, where $\kappa = R/(vQ)$, and its frequency dependent modification described below under f_{max} (Anderson & Hough, 1984 and Hendel et al., 2020).

Site Effect Site effect may be defined as the modification of the amplitude, frequency, and duration of the incoming wave field due to the specific mechanical properties and geometrical features of the ground surrounding the sensor site. In most cases, the ground motions are amplified at certain frequencies in fractured rock relative to the motion in solid rock. Seismic systems in mines are designed to locate events and to estimate their source parameters. For this reason, sensors are installed at 6 to 10 m in boreholes that penetrate the stress-induced fractures surrounding the excavations to avoid the surface noise and the very site effects that amplify ground motion at the skin of excavations. The further away is the sensor from the skin of an excavation the lower the site effect. Site effect at sites 10 m away from an excavation is negligible.

One way to reduce site effect is by averaging the corrected spectra over the sites accepted for source inversion and then estimating source parameters from the averaged spectrum. This is effective if site effects at different recording sites are at different frequencies and not too strong, which is mostly the case for sensors installed underground in boreholes. Surface sites are more prone to site effect, and to remove it, one needs to estimate the site transfer function and then deconvolve by division in the frequency domain. However, if the site response is coherent, it could easily be enhanced rather than reduced by the conventional averaging of spectra over a number of stations. An effective way to correct spectra for such propagation effects is to employ homomorphic deconvolution in the cepstral domain, e.g. Dysart et al. (1988) and Mendecki (1993). An advantage of this technique is that one needs no prior knowledge about the site effect one is trying to remove. This filtering procedure starts with the displacement amplitude spectrum. A complex function in the frequency domain is then constructed of which the real part is the logarithm of the amplitude spectrum and the imaginary part is zero. Taking the logarithm transforms the product to a sum so that the cepstrum contains the superposition of the source and the spike train. The inverse Fourier transform of this zero phase spectrum is taken, and the resulting cepstrum is filtered by excising peaks or troughs and/or by zeroing portions of the trace after the first or second zero crossings. A forward Fourier transform is then performed on the filtered cepstrum, and the filtered amplitude spectrum is constructed from the antilog of the transformed function.

Empirical Green's Functions The removal of path effect can also be done by deconvolving seismograms of the event by seismograms of a suitable small event of similar mechanism located close, preferably within one source dimension, to the hypocentre of the target event e.g. Aki (1967), Bakun and Bufe (1975), Hartzell (1978), Viegas et al. (2010), Abercrombie (2015), and recently by Ross and Ben-Zion (2016) for large data sets. The small event then acts as an empirical Green's function (EGF), i.e. simulating the impulse response of the medium. Therefore, the source properties of the EGF event cannot be imprinted in the frequency range used to deconvolve. It is a very effective method if such a suitable small event is available. One of the conflicting requirements is that the size, i.e. the source duration, of the candidate EGF event should be considerably smaller than that of the larger event to simulate an impulse response, while the signal to noise ratio should be above the background noise in the frequency range of interest at all recording sites used in the source inversion. That limits the useful frequency bandwidth to only that with signal from both large and small events. Source parameters are estimated by comparing the observed spectral ratio between the target $\Omega_1(f)$ and EGF event $\Omega_2(f)$ and a given spectral model,

$$\frac{\Omega_T(f)}{\Omega_{EGF}(f)} = \frac{P_T}{P_{EGF}} \left[\frac{1+(f/f_{0T})^2}{1+(f/f_{0EGF})^2} \right]^{(z+1)/2}, \qquad (2.52)$$

where P_T, f_{0T} and P_{EGF}, f_{0EGF} are the potencies and corner frequencies of the target and the EGF event, respectively (Abercrombie & Rice, 2005). If z is fixed and the P_{EGF} and f_{0EGF} are known, there are only two parameters to be inverted, P_T/P_{EGF} ratio and f_{0T}. The same can be done to the velocity spectrum to estimate energy. A comprehensive review of the applications of EGFs can be found in Hutchings and Viegas (2012).

Maximum Frequency, f_{max} While the theoretical acceleration spectrum for the ω^{-2} model is flat at high frequencies, the observed acceleration spectra are characterised by a trend of exponential decay. Hanks (1979) and Hanks (1982) suggested that the acceleration spectrum is flat above the corner frequency only to a second corner frequency, called f_{max}, above which it decays rapidly. The origin of the rapid decay may be the source or path including the site effect or the combination of the above. Cases with $f_{max} < f_0$ would limit the resolution of corner frequency, f_0, and therefore impose an artificial minimum source dimension on seismic events.

There are two types of parametric models widely used to shape the observed high frequency acceleration spectra. Boore (1983a) proposed the power law frequency decay model,

2.8 More General Point Source Models

$$P(f) = \left[1 + (f/f_{max})^{2s}\right]^{-1/2}, \tag{2.53}$$

where s controls the decay rate at high frequencies, and Anderson and Hough (1984) the exponential model,

$$a(f) = A_0 \exp(-\pi f \kappa), \tag{2.54}$$

where $\kappa = R/(vQ)$ is the spectral decay parameter and can be derived from $\partial \ln a(f)/\partial f = -\pi \kappa$, for frequencies above f_{max}. Some studies suggest that that the slope of the acceleration spectrum in log-linear space is not constant but rather curved, so the estimated κ does depend on the chosen frequency band of analysis. Hendel et al. (2020) modified the kappa model to account for the non-linear spectral decay at high frequencies by incorporating a power law frequency dependent Q which they call the zeta model, $a(f) = \exp\left[-\pi \zeta f^{1-\eta}\right]$, where $\zeta = Rf_0/(vQ)$ and η is a parameter. The $P(f)$, κ and the zeta models are to be applied to the ω^{-2} source model in which the acceleration spectrum is flat above the corner frequency. The $\omega^{-1.5}$ and ω^{-3} spectra are not flat above corner frequency, see Fig. 2.12 right column, and provide a natural high-cut filtering to be modelled as a source effect.

Figure 2.14 shows S-wave far-field acceleration spectra for the ω^{-2} model ($z = 1$) in red, the $\omega^{-2.5}$ model ($z = 1.5$) in blue, and the ω^{-3} model ($z = 2$) in green all attenuated for constant Q (left) and the frequency dependent $Q(f)$ (centre). The right figure shows the ω^{-2} model attenuated for constant Q and for f_{max}, and $\omega^{-2.5}$ and ω^{-3} models corrected only for a constant Q. It is more difficult to resolve the corner frequency when $f_0/f_{max} = 1$ because then source spectra are strongly attenuated for high frequencies past the f_{max}.

Figure 2.15 shows acceleration waveforms (left column), integrated velocity waveforms (centre column), and the smoothed FFT of the S-wave window (right column) of two events recorded at the same site by three-component accelerometers installed in a borehole in an underground hard rock mine.

The first event with $\log P = 1.05$ located 17 km away from the recording site (top row), and the second event with $\log P = 0.87$ located 243 m away (bottom row). The smoothed FFT of noise is marked by dashed lines. The

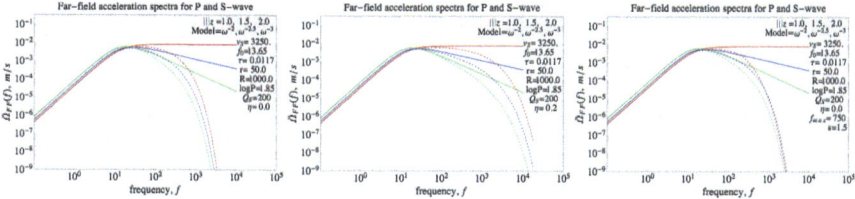

Fig. 2.14 S-wave far-field acceleration spectra for the ω^{-2} model ($z = 1$) in red, the $\omega^{-2.5}$ model ($z = 1.5$) in blue, and the ω^{-3} model ($z = 2$) in green all corrected for frequency dependent $Q(f)$ *(left)* and for constant Q *(centre)*. The right figure shows the ω^{-2} model corrected for constant Q and for f_{max} and $\omega^{-2.5}$ and ω^{-3} models corrected only for a constant Q

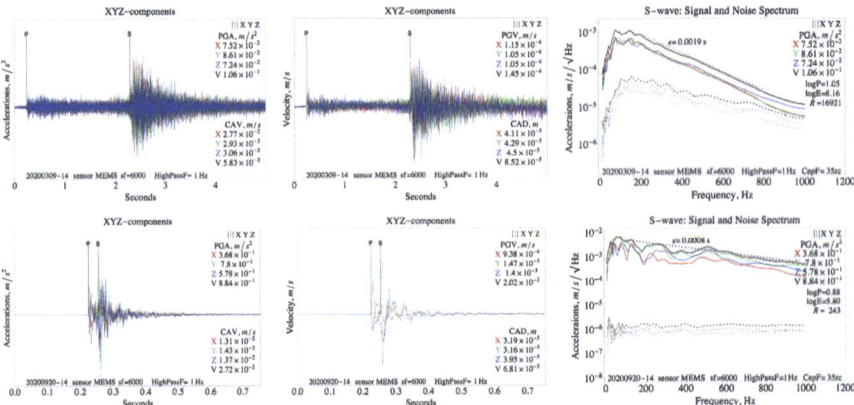

Fig. 2.15 Acceleration waveforms *(left)*, integrated velocity waveforms *(centre)*, and smoothed FFT of accelerations *(right)* of a $\log P = 1.05$ event recorded 17 km away *(top row)* and a $\log P = 0.87$ event recorded 243 m away *(bottom row)*

spectrum was smoothed with the Konno and Ohmachi function, $w(f, f_c) = \{\text{sinc}\,[b \log (f/f_c)]\}^4$, where $b = 40$ is the coefficient for bandwidth and f_c is the centre frequency (Konno & Ohmachi, 1998). The distant event shows a considerable depletion of high frequencies over the frequency range 1 to 1000 Hz and maintains a constant slope of the acceleration spectrum in log-linear domain up to 800 Hz. The casual fit gives $\kappa = 0.0019$ second and would imply a constant Q over that frequency range.

The second, close event shows very limited depletion in high frequency over the same frequency band, and the spectrum is scalloping up to 500 Hz but overall reasonably linear up to 1000 Hz with $\kappa = 0.0008$ second. Note that although the second event has 8.3 times higher PGA and 14 times higher PGV, its cumulative absolute velocity $CAV = \int_0^{t_d} |a(t)|\,dt$ is 2.1 times lower, and the cumulative absolute displacement, $CAD = \int_0^{t_d} |v(t)|\,dt$, 1.25 times lower than that of the larger distant event. The reason is the duration of ground motion, t_d, which for the distant event is 8 times longer. This illustrates the importance of duration. The first event with lower PGA and PGV but longer duration will consume more deformation capacity of support than the second event.

2.9 Frequency Range $\log P$ and $\log E$

Potency Recovery The potency and energy range of recorded seismic events is limited by the frequency range of the monitoring system, (f_1, f_2), which is mainly determined by the capabilities of the seismic sensors. The following ratio of the far-field displacement spectra quantifies the potency recovery as a function of the ratio

2.9 Frequency Range log P and log E

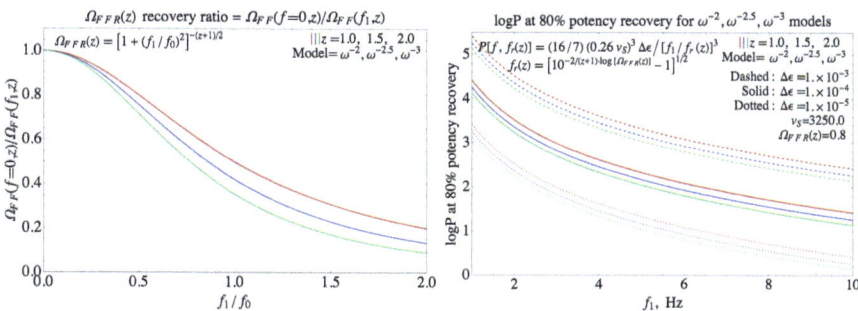

Fig. 2.16 Recovery of seismic potency as a function of f_1/f_0 *(left)* and the S-wave log P at 80% potency recovery as a function of f_1 for the ω^{-2}, $\omega^{-2.5}$, and ω^{-3} models and for three selected strain drops *(right)*

of available frequencies at the lower end of the spectrum to corner frequency f_1/f_0, for different spectral models,

$$\Omega_{FF}(R, f_1/f_0, z) / \Omega_{FF}(R, f_1/f_0 = 0, z) = \Omega_{FFR}(z) = \left[1 + (f_1/f_0)^2\right]^{-(z+1)/2}, \quad (2.55)$$

which applies to the P-wave if $f_0 = f_{0P}$ and to the S-wave if $f_0 = f_{0S}$. Figure 2.16 (left) shows that the ω^{-2} model has the least under-recovery followed by $\omega^{-2.5}$ and ω^{-3}. It shows that with f_1 at the corner frequency the ω^{-2} model recovers 50% of seismic potency, $\omega^{-1.5}$ recovers 42%, and ω^{-3} 35%. The ω^{-2} model recovers 80% of seismic potency at $f_1/f_0 = 0.5$, $\omega^{-2.5}$ at $f_1/f_0 = 0.44$, and ω^{-3} at $f_1/f_0 = 0.4$. The 80% recovery underestimates log P by $\log(0.8P) = \log P - 0.0969$, and the m_{HK} by 0.0646, and at 50% recovery, i.e. $f_1/f_0 = 1.0$ the log P is underestimated by 0.3 units and the m_{HK} by 0.2.

The rate of deformation at seismic sources in softer or in fractured rock is slower than in stronger solid rock, and such sources radiate the predominant portion of their energy at lower frequencies, resulting in a lower corner frequency. Therefore, the recovery of seismic potency will be lower in low strain drop conditions. Potency recovery as a function of the lowest frequency available to calculate spectra, f_1, taking into account strain drop, can be derived from $P = (16/7) \Delta\epsilon r^3$ (Eshelby, 1957) and, for the S-wave, $r = 0.26v_S/f_0$ (Kaneko & Shearer, 2014),

$$P[f, f_r(z)] = (16/7)(0.26v_S)^3 \Delta\epsilon / [f_1/f_r(z)]^3, \quad (2.56)$$

where $f_0 = f_1/f_r(z)$ and $f_r(z)$ is the (f_1/f_0) ratio derived from Eq. (2.55) for the required potency recovery $\Omega_{FFR}(z)$,

$$f_r(z) = \left[10^{-2/(z+1)\cdot\log[\Omega_{FFR}(z)]} - 1\right]^{1/2}. \quad (2.57)$$

Figure 2.16 (right) shows $\Omega_{FFR}(z) = 0.8$, i.e. the 80% potency recovery as a function of f_1 for selected strain drops (Eqs. 2.56 and 2.57). Clearly, the recovery of seismic potency increases in harder rock, i.e. higher $\Delta\epsilon$.

Energy Recovery The predominant frequency f_E, i.e. the frequency at which the maximum energy is radiated, is at the maximum of the velocity power spectrum. Solving the equation $\partial \dot{\Omega}_{FF}(f, z)/\partial f = 0$,

$$f_E f_0^2 \left[1 + (f_E/f_0)^2\right]^{-1-z} + f_E^3 (-z - 1)\left[1 + (f_E/f_0)^2\right]^{-2-z} = 0,$$

gives

$$f_E(z) = f_0/\sqrt{z}. \tag{2.58}$$

Therefore, for the ω^{-2} model, i.e. $z = 1$, the predominant frequency is at the corner frequency, $f_E(z = 1) = f_0$. However, for the $\omega^{2.5}$ model, i.e. $z = 1.5$, the $f_E(z = 1.5) = 0.816 f_0$, and for ω^3 model, i.e. $z = 2$, the $f_E(z = 2) = 0.707 f_0$.

Seismic radiated energy is proportional to the integral of the velocity power spectrum,

$$S_{V2}(z, 0, \infty) = 2 \int_0^\infty \left[\dot{\Omega}_{FF}(R, f, z)\right]^2 df = 2\pi^{5/2} \frac{\Gamma(z - 1/2)}{\Gamma(z + 1)} \Omega_{FF}^2 f_0^3, \tag{2.59}$$

which applies to the P-wave if $\dot{\Omega}_{FF} = \dot{\Omega}_{FFP}$ and to the S-wave if $\dot{\Omega}_{FF} = \dot{\Omega}_{FFS}$.

The ratio $S_{V2}(z = 1.5, 0, \infty)/S_{V2}(z = 1, 0, \infty) \simeq 0.85$ shows that the $\omega^{-2.5}$ model produces 85% of the seismic energy of the ω^{-2} model and the ratio $S_{V2}(z = 2, 0, \infty)/S_{V2}(z = 1, 0, \infty) = 0.5$ that the ω^{-3} model produces 50% of the seismic energy of the ω^{-2} model.

The model independent seismic energy can be calculated by $\tilde{E}_{P,S} = 4\pi \rho v_{P,S} \tilde{S}_{V2P,S}$, where $\tilde{S}_{V2P,S}$ is the observed, instrument, distance, and Q corrected velocity power spectrum for the P- or S-wave. However, the observed spectrum is limited by the frequency range of the system, and we can recover only the (f_1, f_2) part of it. The theoretical energy recovery as a function of z can be defined as

$$ER(z) = S_{V2}(z, f_1, f_2)/S_{V2}(z, 0, \infty). \tag{2.60}$$

The integral $2 \int_{f_1}^{f_2} \left[\dot{\Omega}_{FF}(f, z)\right]^2 df$ does not have a simple analytical solution for all z, but it does if z is either integer or half integer. For $z = 1$, $z = 1.5$, and $z = 2$, the respective integrals give

2.9 Frequency Range $\log P$ and $\log E$

$$S_{V2}(z=1, f_1, f_2) = 4\pi^2 \Omega_{FF}^2 f_0^3 \left[(f_1/f_0)/c_1 - (f_2/f_0)/c_2 + A\right],$$

$$S_{V2}(z=1.5, f_1, f_2) = \frac{8}{3}\pi^2 \Omega_{FF}^2 f_0^3 \left[-(f_1/f_0)^3/c_1^{3/2} + (f_2/f_0)^3/c_2^{3/2}\right], \quad (2.61)$$

$$S_{V2}(z=2, f_1, f_2) = \pi^2 \Omega_{FF}^2 f_0^3 \left[2(f_1/f_0)/c_1^2 - (f_1/f_0)/c_1 - 2(f_2/f_0)/c_2^2 \right.$$
$$\left. + (f_2/f_0)/c_2 + A\right],$$

where $A = \arctan(f_2/f_0) - \arctan(f_1/f_0)$, $c_1 = 1 + (f_1/f_0)^2$, $c_2 = 1 + (f_2/f_0)^2$. The following equations quantify the portion of the velocity power spectrum lost in the frequency range $(0, f_1)$ and (f_2, ∞) for $z = 1$, $z = 1.5$, and $z = 2$.

$$S_{V2}(z=1, 0, f_1) = 4\pi^2 \Omega_{FF}^2 f_0^3 \left[\arctan(f_1/f_0) - (f_1/f_0)/c_1\right],$$

$$S_{V2}(z=1, f_2, \infty) = 4\pi^2 \Omega_{FF}^2 f_0^3 \left[\frac{\pi}{2} - \arctan(f_2/f_0) + (f_2/f_0)/c_2\right],$$

$$S_{V2}(z=1.5, 0, f_1) = \frac{8}{3}\pi^2 \Omega_{FF}^2 f_0^3 \left[(f_1/f_0)^3/c_1^{3/2}\right], \quad (2.62)$$

$$S_{V2}(z=1.5, f_2, \infty) = \frac{8}{3}\pi^2 \Omega_{FF}^2 f_0^3 \left\{1 - (f_2/f_0)^3/c_2^{3/2}\right\},$$

$$S_{V2}(z=2, 0, f_1) = \pi^2 \Omega_{FF}^2 f_0^3 \left[(f_1/f_0)/c_1 - 2(f_1/f_0)/c_1^2 + \arctan(f_1/f_0)\right],$$

$$S_{V2}(z=2, f_2, \infty) = \pi^2 \Omega_{FF}^2 f_0^3 \left[\frac{\pi}{2} - (f_2/f_0)/c_2 + 2(f_2/f_0)/c_2^2 \right.$$
$$\left. - \arctan(f_2/f_0)\right].$$

The ratio $S_{V2}(z, f_0, \infty)/S_{V2}(z, 0, \infty)$ shows that the ω^{-2} model gives 82% of the seismic energy above the corner frequency, while the $\omega^{-2.5}$ model produces 65% and the ω^{-3} model only 50%. The theoretical energy recovery due to bandwidth limitation is $S_{V2}(z, f_1, f_2)/S_{V2}(z, 0, \infty)$. Figure 2.17 shows the recovery of seismic energy given by Eq. (2.60) as a function of f_2/f_0 for $f_1/f_0 = 0.2$.

Corner Frequency and Potency The far-field displacement power spectrum is

$$S_{D2}(z, 0, \infty) = 2\int_0^\infty [\Omega_{FF}(R, f, z)]^2 df = \sqrt{\pi}\frac{\Gamma(z+1/2)}{\Gamma(z+1)} \Omega_{FF}^2 f_0, \quad (2.63)$$

that for the ω^{-2} model, i.e. $z = 1$ gives $S_{D2} = (\pi/2)\Omega_{FF}^2 f_0$, for the $\omega^{-2.5}$ model, i.e. $z = 1.5$ $S_{D2} = (4/3)\Omega_{FF}^2 f_0$, and for the ω^{-3} model, i.e. $z = 2$ $S_{D2} = (3\pi/8)\Omega_{FF}^2 f_0$.

Corner frequency can be derived from the ratio of the velocity to displacement power spectra

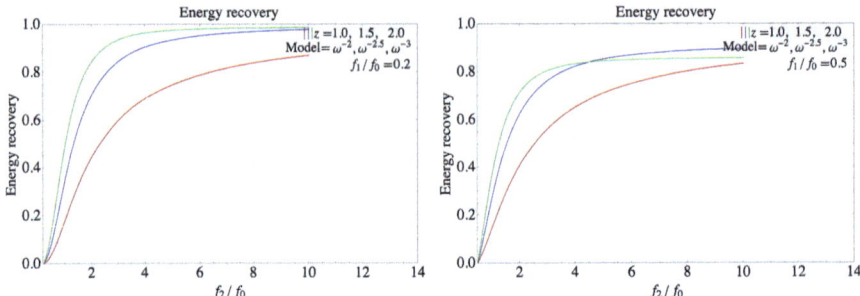

Fig. 2.17 Recovery of seismic energy as a function of f_2/f_0 when $f_2/f_0 = 0.2$ *(left)* and $f_2/f_0 = 0.5$ *(right)*

$$\frac{S_{V2}(z,0,\infty)}{S_{D2}(z,0,\infty)} = 2\pi^2 \frac{\Gamma(z-1/2)}{\Gamma(z+1/2)} f_0^2 = \frac{2\pi^2}{z-1/2} f_0^2$$

$$\Rightarrow f_0 = \frac{1}{2\pi}\sqrt{2z-1}\sqrt{\frac{S_{V2}(z,0,\infty)}{S_{D2}(z,0,\infty)}}, \quad (2.64)$$

which for the ω^{-2} model, i.e. $z = 1$ gives $f_0 = \sqrt{S_{V2}/S_{D2}}/(2\pi)$, see also Boore (1983b) and Andrews (1986)). For the $\omega^{-2.5}$ model, i.e. $z = 1.5$ $f_0 = \sqrt{S_{V2}/S_{D2}}/\left(\pi\sqrt{2}\right)$, and for the ω^{-3} model, i.e. $z = 2$ $f_0 = 3\sqrt{S_{V2}/S_{D2}}/(2\pi)$. Inserting $f_0(z)$ into Eq. (2.63) gives the low frequency displacement spectral plateau, $\Omega_{PS}(z)$, as a function of z,

$$\Omega_{FF}^2 = \sqrt{2\pi}\frac{\Gamma(z+1)}{\Gamma(z+1/2)}\sqrt{\frac{\Gamma(z-1/2)}{\Gamma(z+1/2)}}\cdot\frac{S_{D2}^{3/2}}{S_{V2}^{1/2}} = \frac{2\sqrt{\pi}}{\sqrt{2z-1}}\frac{\Gamma(z+1)}{\Gamma(z+1/2)}\cdot\frac{S_{D2}^{3/2}}{S_{V2}^{1/2}}, \quad (2.65)$$

and P- and S-wave potency, $P_P = 4\pi R v_P^3 \Omega_P/(\Lambda_P v_S^2)$ and $P_S = 4\pi R v_S \Omega_S/\Lambda_S$. The following equations quantify the portion of the displacement power spectrum in the frequency range (f_1, f_2) for $z = 1$, $z = 1.5$ and $z = 2$,

$$S_{D2}(z=1, f_1, f_2) = \Omega_{FF}^2 f_0 [(f_2/f_0)/c_2 - (f_1/f_0)/c_1 + A], \quad (2.66)$$

$$S_{D2}(z=1.5, f_1, f_2) = \frac{2}{3}\Omega_{FF}^2 f_0 \left[3(f_2/f_0)/c_2^{1/2}\right.$$
$$\left. - (f_2/f_0)^3/c_2^{1/2} - 3(f_1/f_0)/c_1^{1/2} + (f_1/f_0)^3/c_1^{1/2}\right],$$

$$S_{D2}(z=2, f_1, f_2) = \frac{1}{4}\Omega_{FF}^2 f_0 \left[2(f_2/f_1)/c_2^2 + 3(f_2/f_1)/c_2\right.$$
$$\left. - 2(f_1/f_0)/c_1^2 - 3(f_1/f_0)/c_1 + 3A\right].$$

2.9 Frequency Range $\log P$ and $\log E$

The following equations quantify the portion of the displacement power spectrum lost in the frequency range $(0, f_1)$ and (f_2, ∞) for $z = 1$, $z = 1.5$ and $z = 2$.

$$S_{D2}(z = 1, 0, f_1) = \Omega_{FF}^2 f_0 \left[(f_1/f_0)/c_1 + \arctan(f_1/f_0) \right],$$

$$S_{D2}(z = 1, f_2, \infty) = \Omega_{FF}^2 f_0 \left[\frac{\pi}{2} - (f_2/f_0)/c_2 - \arctan(f_2/f_0) \right],$$

$$S_{D2}(z = 1.5, 0, f_1) = \frac{2}{3}\Omega_{FF}^2 f_0 \left[3(f_1/f_0)/c_1^{1/2} - (f_1/f_0)^3/c_1^{1/2} \right], \quad (2.67)$$

$$S_{D2}(z = 1.5, f_2, \infty) = \frac{2}{3}\Omega_{FF}^2 f_0 \left[2 - 3(f_2/f_0)/c_2^{1/2} + (f_2/f_0)^3/c_2^{1/2} \right],$$

$$S_{D2}(z = 2, 0, f_1) = \frac{1}{4}\Omega_{FF}^2 f_0 \left[2(f_1/f_0)/c_1^2 + 3(f_1/f_0)/c_1 \right.$$
$$\left. + 3\arctan(f_1/f_0) \right],$$

$$S_{D2}(z = 2, f_2, \infty) = \frac{1}{4}\Omega_{FF}^2 f_0 \left[\frac{3\pi}{2} - 2(f_2/f_0)/c_2^2 - 3(f_2/f_0)/c_2 \right.$$
$$\left. - 3\arctan(f_2/f_0) \right].$$

In practice we estimate the power spectra in the frequency band (f_1, f_2), and therefore we underestimate seismic energy and introduce a bias into the corner frequency and other derived source parameters. The corner frequency, f_0, corrected for bandwidth limitation as a function of z is

$$f_0(z) = \tilde{f}_0 \left[\frac{S_{V2}(z, 0, \infty)}{S_{D2}(z, 0, \infty)} \cdot \frac{S_{D2}(z, f_1, f_2)}{S_{V2}(z, f_1, f_2)} \right]^{1/2}, \quad (2.68)$$

where $\tilde{f}_0$ is the corner frequency estimated within the frequency range (f_1, f_2). The following equations give the corrected f_0 for $z = 1$, $z = 1.5$, and $z = 2$.

$$f_0(z = 1) = \tilde{f}_0 \left[\frac{\tilde{A} + \tilde{B}}{\tilde{A} - \tilde{B}} \right]^{1/2}, \quad (2.69)$$

$$f_0(z = 1.5) = \tilde{f}_0 \left[\frac{\tilde{F}}{3\tilde{E} - \tilde{F}} \right]^{1/2} \quad (2.70)$$

$$f_0(z = 2) = \tilde{f}_0 \left[\frac{\tilde{A} + \tilde{B} - 2\tilde{D}}{3\tilde{A} + 3\tilde{B} + 2\tilde{D}} \right]^{1/2}, \quad (2.71)$$

where $\tilde{f}_0$ is calculated using Eq. (2.64), but with bandwidth limited by given sensors to (f_1, f_2) and

$$\tilde{A} = \arctan\left(f_2/\tilde{f}_0\right) - \arctan\left(f_1/\tilde{f}_0\right), \tilde{c}_1 = 1 + \left(f_1/\tilde{f}_0\right)^2, \tilde{c}_2 = 1 + \left(f_2/\tilde{f}_0\right)^2,$$
$$\tilde{B} = \left(f_2/\tilde{f}_0\right)/\tilde{c}_2 - \left(f_1/\tilde{f}_0\right)/\tilde{c}_1,$$
$$\tilde{D} = \left(f_2/\tilde{f}_0\right)/\tilde{c}_2^2 - \left(f_1/\tilde{f}_0\right)/\tilde{c}_1^2, \tilde{E} = \left(f_2/\tilde{f}_0\right)/\tilde{c}_2^{1/2} - \left(f_1/\tilde{f}_0\right)/\tilde{c}_1^{1/2}, \tilde{F} = \left(f_2/\tilde{f}_0\right)^3/\tilde{c}_2^{3/2} - \left(f_1/\tilde{f}_0\right)^3/\tilde{c}_1^{3/2}.$$

For $z = 1$, i.e. the ω^{-2} model, Eq. (2.69) was derived by Di Bona and Rovelli (1988).

Bandwidth corrections for seismic energy can be carried out by dividing the observed energy, $\tilde{E} = 4\pi\rho v_{P,S}\tilde{S}_{V2P,S}$, estimated for a given frequency range (f_1, f_2) by the energy recovery, $E(z) = \tilde{E}/\widetilde{ER}(z)$, given by Eq. (2.60) with Eq. (2.61), for $z = 1$, $z = 1.5$ and $z = 2$, respectively, which gives

$$E(z = 1) = \tilde{E}/\left[(2/\pi)\left(\tilde{A} - \tilde{B}\right)\right], \quad (2.72)$$

$$E(z = 1.5) = \tilde{E}/\left(-\tilde{F}\right), \quad (2.73)$$

$$E(z = 2) = \tilde{E}/\left[(2/\pi)\left(2\tilde{D} + \tilde{B} + \tilde{A}\right)\right], \quad (2.74)$$

where $\tilde{S}_{V2P,S}$, is the observed, instrument, distance, and Q corrected velocity spectrum. Since the recovery of Ω_{FF}, and therefore seismic potency, is limited by f_1 only, the bandwidth correction can be done by dividing the observed $\tilde{\Omega}_{FF}$ by the potency recovery given by Eq. (2.55),

$$\Omega_{FF}(z) = \tilde{\Omega}_{FF}/\left[1 + \left(f_1/\tilde{f}_0\right)^2\right]^{-(z+1)/2}. \quad (2.75)$$

2.10 Source Parameters from Spectra: Examples

First estimates of spectral source parameters of mine events were conducted by Smith et al. (1974), Spottiswoode and McGarr (1975), Gibowicz et al. (1977), and Cichowicz (1981), all using digitised analog waveforms, so it was not conducive to routine application. The introduction of digital seismic systems to mines in the late 80s facilitated a real time quantitative seismology, whereby apart from its timing and location each seismic event is quantified by seismic potency, or seismic moment, their tensors, by radiated seismic energy and the associated predominant frequency (Mendecki, 1993). With the recent introduction of machine learning that automates the phase picking and event classification (e.g. Gal et al., 2021; Zhu et al., 2022), we estimate there are approximately 10^6 mine seismic events processed and quantified that way per day, bulk of them using the ω^{-2} model.

2.10 Source Parameters from Spectra: Examples

However, the ω^{-2} model fits spectra of seismic sources originating in a hard, relatively homogeneous rock well, the harder and more homogeneous the rock the better the fit. In mines with a softer and/or less homogeneous rock mass, the ω^{-2} model tends to overestimate higher the frequencies. The difficulties are that in some mines some seismic events are well described by the ω^{-2} model, while in another part of the same mine they tend to conform to the $\omega^{-2.5}$ or even ω^{-3} model, therefore a need for adaptive processing. Since there is a trade-off between Q and the exponent of the spectral decay, it is not recommended to invert Q together with Ω_0 and corner frequency f_0. In mines, most attenuation and scatter occur close to seismic sources due to the mine workings and fractured rock around them, while in crustal earthquakes most inelastic attenuations and site effects are close to surface (Cichowicz et al., 1988 and Cichowicz & Green, 1989).

In the frequency domain, the recorded waveform, $W(f)$, is a convolution of the instrument response $I(f)$, the source function, $\Omega(f)$, the path effect modelled by the attenuation and scattering function $Q(f)$, and the site effect $S(f)$. To get source, we need to deconvolve the instrument response, path, and site effect, $\Omega(f) = W(f)/[I(f)Q(f)S(f)]$. The instrument response for a 4.5 Hz geophone with damping $b = 0.72$ is given by $I(f) = \|f^2/[20.25 + (0 + 6.48i)f - f^2]\|$, and for 14 Hz omnidirectional geophone with the same damping used in longer boreholes, $I(f) = \|f^2/[196 + (0 + 20.16i)f - f^2]\|$, see Mountfort and Mendecki (1997). Frequently the default damping is between 0.5 and 0.7.

Spectral parameters, namely the zero frequency asymptote Ω_0, and corner frequency f_0, are derived from the instrument, distance, and Q corrected spectrum of each recorded waveform and then averaged. A more efficient method is to correct the individual spectra and then average, or stack, each spectral component over all the sites involved. This method tends to average out incoherent site effects, source directivity, and random fluctuations of the high frequency spectrum.

In many cases, most frequently in smaller mines, the distances from seismic sources to the nearest sensors are too short to meet the requirement of being in the far field. Since in the near field the displacement spectra at low frequencies decay as f^{-1} and at high frequencies as f^{-2}, it may introduce bias to Ω_0 and to seismic energy, see Eq. (2.24) and Fig. 2.7. If we assume that the intermediate field is where co-seismic strains are of the order 10^{-5}, then for hard rock with $v_S = 3000$ m/s seismic sites that record $PGV \geq 3$ cm/s would be affected. In addition a short distance from source to sensor makes the P-wave spectral window too short for spectral analysis and it should be rejected. Other complications in mines are frequently used single-component sensors, mains electrical noise, and, in some cases, long cables that attenuate the signal before A/D conversion. Below are three examples of events with similar energy but with spectra that fit different models.

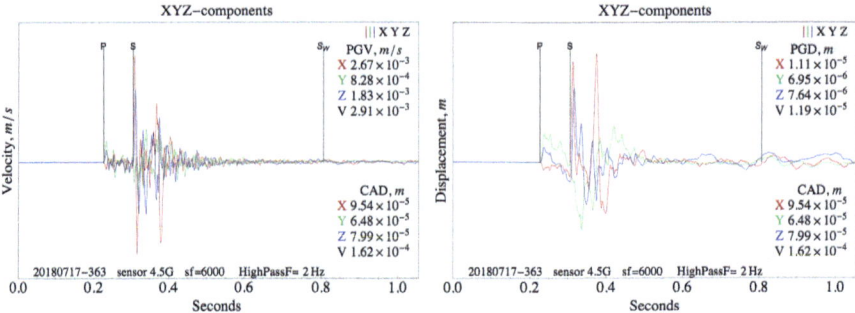

Fig. 2.18 Velocity waveforms of a mid-size seismic event recorded by 4.5 Hz geophones in an 8 m borehole away from the excavation in an u/g mine (left), and integrated displacement waveforms (right)

Example 1, $\log E \simeq 7.1$, 4.5 Hz Sensor, $\omega^{-2.5}$ Model Figure 2.18 shows a 1.05 second snap of the 4.4 second buffer waveforms recorded in an underground hard rock caving mine 1.2 km below surface by a three-component 4.5 Hz geophone grouted at the end of an 8 m borehole away from the skin of the excavation.

The event located 607 m from the site. The vector $PGV = 2.91 \cdot 10^{-3}$ m/s, $PGD = 1.19 \cdot 10^{-5}$ m, and the cumulative absolute displacement, defined as the integral of the absolute value of a velocity time series, $CAD = \int_0^{td} |v(t)| dt$, calculated for the full 4.4 s buffer, $CAD = 1.62 \cdot 10^{-4}$ m. The P-wave window for spectral analysis is taken from just before the P-arrival to the S-arrival, and the S-wave window from just before the S-arrival to S_w. Sampling frequency, $s_f = 6$ kHz, and after the DC offset removal waveforms are high passed by a 2nd order 1 Hz Butterworth filter run both ways. No site effect deconvolution was needed for this event since it was recorded by sensors embedded in solid rock.

Figure 2.19 shows the velocity and displacement cepstra filtered at the 15th zero crossing and the resulting velocity and displacement spectra.

Figure 2.20 shows the distance, Q corrected, and cepstral filtered S-wave velocity and displacement spectra fitted with ω^{-2}, $\omega^{-2.5}$, and ω^{-3} models. The observed spectra deviate from the ω^{-2} model, shown here in red, and fit $\omega^{-2.5}$ better shown in blue. Note that seismic energy estimates from one site can be strongly affected by the radiation pattern.

Example 2, $\log E \simeq 7.1$, 4.5 Hz Sensor, ω^{-3} Model Figure 2.21 top row shows a 1.2 second snap of the 5.0 second buffer waveforms recorded in an underground hard rock caving mine 850 m below surface by a three-component 4.5 Hz geophone grouted at the end of an 8 m borehole away from the skin of the excavation. The event located 370 m from the site. The vectors $PGV = 4.43 \cdot 10^{-3}$ m/s, $PGD = 3.91 \cdot 10^{-5}$ m, and the $CAD = 2.64 \cdot 10^{-4}$ m, calculated for the full 5.0 second buffer. Sampling frequency, $s_f = 6$ kHz, and after DC offset removal, the waveforms are high passed by a 2nd order 1 Hz Butterworth filter run both ways.

2.10 Source Parameters from Spectra: Examples

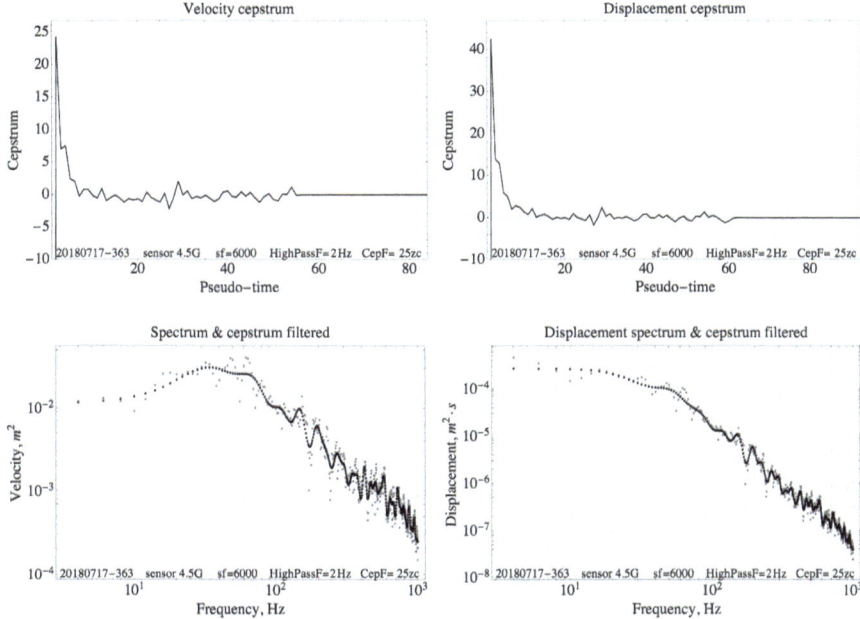

Fig. 2.19 Velocity and displacement cepstra filtered at 15 zero crossing *(top row)* and velocity and displacement spectra before, in grey, and after, in black, cepstral filtering

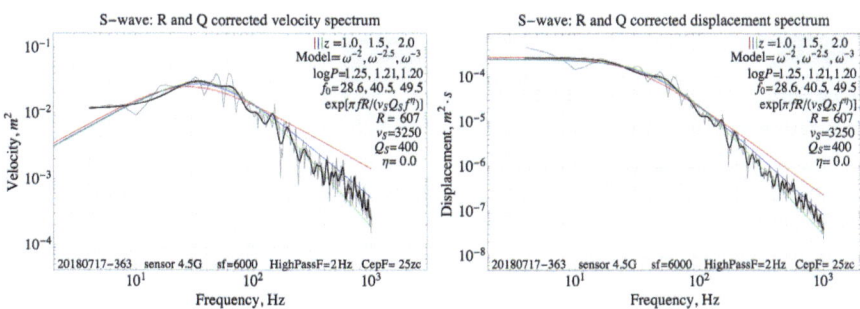

Fig. 2.20 Distance and Q corrected and cepstral filtered S-wave velocity *(left)* and displacement *(right)* spectra for ω^{-2}, $\omega^{-2.5}$, and ω^{-3} models

Figure 2.21 bottom row shows the distance, Q corrected, and cepstral filtered S-wave velocity and displacement spectra fitted with ω^{-2}, $\omega^{-2.5}$, and ω^{-3} models. In this case, the observed spectra fit the ω^{-3} model shown in green. It indicates a lower stress environment at the site of the event due to the shallow depth and a more inhomogeneous and weaker rock mass than in the first example.

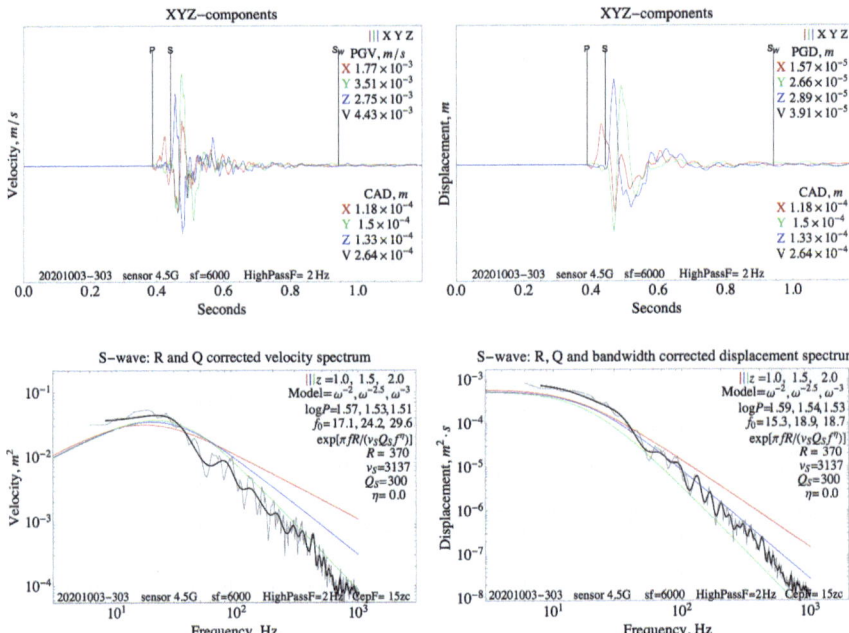

Fig. 2.21 Velocity and displacement waveforms of a mid-size seismic event recorded by 4.5 Hz geophones in an 8 m borehole away from the excavation in an u/g mine *(top row)*. The distance, Q corrected, and cepstral filtered S-wave velocity and displacement spectra fitted with ω^{-2}, $\omega^{-2.5}$, and ω^{-3} models *(bottom row)*

Example 3, $\log E \simeq 7.8$, **4.5 Hz Sensor,** ω^{-2} **Model** Figure 2.22 top row shows 1.2 second snap of the 5.0 second buffer waveforms recorded in an underground tabular hard rock gold mine 2000 m below surface by a three-component 4.5 Hz geophone grouted at the end of an 8 m borehole away from the skin of the excavation.

The event located 713 m from the site on a known normal fault. The vectors $PGV = 5.31 \cdot 10^{-3}$ m/s, $PGD = 3.59 \cdot 10^{-5}$ M, and the $CAD = 4.4 \cdot 10^{-4}$ m, calculated for the full 5.0 second buffer. Sampling frequency, $s_f = 6$ kHz, and after the DC offset removal waveforms are high passed filtered by a 2nd order 1 Hz Butterworth filter run both ways.

Figure 2.23 bottom row shows the distance, Q corrected, and cepstral filtered S-wave velocity and displacement spectra fitted with ω^{-2}, $\omega^{-2.5}$, and ω^{-3} models. The observed spectra fit the ω^{-2} model, shown in red, which is expected for events associated with fault slip in high stress environment.

2.11 Final Static Deformation for Double-Couple Source

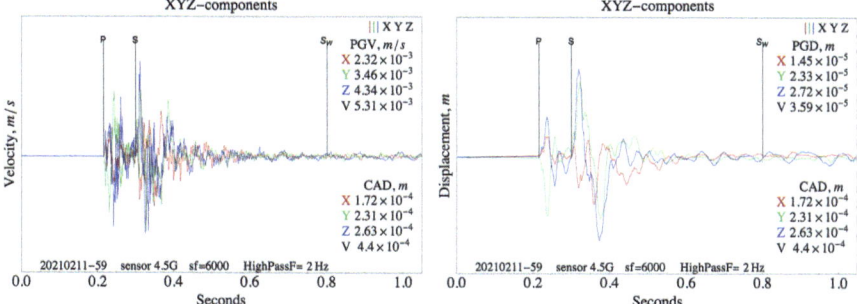

Fig. 2.22 Velocity waveforms of a mid-size seismic event recorded by 4.5 Hz geophones in an 8 m borehole away from the excavation in an u/g mine (left), and integrated displacement waveforms (right)

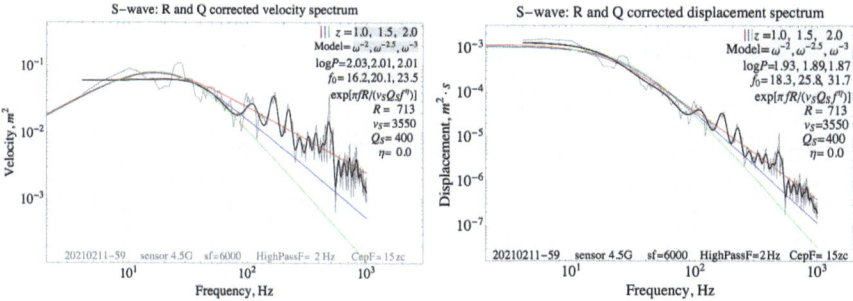

Fig. 2.23 The distance, Q corrected, and cepstral filtered S-wave velocity *(left)* and displacement *(right)* spectra fitted with ω^{-2}, $\omega^{-2.5}$, and ω^{-3} models

2.11 Final Static Deformation for Double-Couple Source

2.11.1 Radiation Patterns

Aki and Richards (2002) gave a relatively simple formula for the displacement vector due to a moment or potency tensor corresponding to a point shear dislocation. This equation shows the following:

1. The far-field displacements attenuate as R^{-1} and are proportional to particle velocity at the source.
2. The far-field and the near-field radiation patterns are similar.
3. There is a final static displacement that attenuates as R^{-2}.

Two coordinate systems are assumed: One is Cartesian, with the source at the origin and the slip in the XY-plane, and the other is spherical polar with the same origin and the polar axis along the Cartesian OZ. The azimuth ϕ is measured from the OX axis of the Cartesian system.

The famous Aki and Richards's Eq. (4.32) can be expressed in terms of the scalar seismic potency time function $P(t)$,

$$u(R, \theta, \phi; t) = \frac{\Lambda_N v_S^2}{4\pi} \frac{1}{R^4} \int_{R/v_P}^{R/v_S} \zeta P(t - \zeta) d\zeta$$

$$+ \frac{\Lambda_{IP} v_S^2}{4\pi v_P^2} \frac{1}{R^2} P(t_P') + \frac{\Lambda_{IS}}{4\pi} \frac{1}{R^2} P(t_S')$$

$$+ \frac{\Lambda_{FP} v_S^2}{4\pi v_P^3} \frac{1}{R} \dot{P}(t_P') + \frac{\Lambda_{FS}}{4\pi v_S} \frac{1}{R} \dot{P}(t_S'), \qquad (2.76)$$

where (R, θ, ϕ) are the spherical polar co-ordinates of the receiver, the integration variable ζ has units of time, $t_P' = t - R/v_P$ and $t_S' = t - R/v_S$ are the respective retarded times, and the radiation patterns, Λ_N, Λ_{IP}, Λ_{IS}, Λ_{FP}, and Λ_{FS} are vectors as given by Aki and Richards (2002) in Eq. (4.33) as

$$\Lambda_N = 9a\hat{r} - 6b\hat{\theta} + 6c\hat{\phi}, \quad \Lambda_{IP} = 4a\hat{r} - 2b\hat{\theta} + 2c\hat{\phi}, \quad \Lambda_{FP} = a\hat{r}$$

$$\Lambda_{IS} = -3a\hat{r} + 3b\hat{\theta} - 3c\hat{\phi}, \quad \Lambda_{FS} = b\hat{\theta} - c\hat{\phi} \qquad (2.77)$$

$$\Lambda_{FSV} = b\hat{\theta} + c\hat{\phi}, \qquad \Lambda_{FSH} = -c\hat{\phi},$$

where Λ_{FSV} and Λ_{FSH} are the two polarisations of the far-field S-wave. The scalar coefficients a, b, and c and the Cartesian components of the unit vectors $\hat{r}$, $\hat{\theta}$ and $\hat{\phi}$ are

$$a = \sin(2\theta)\cos(\phi) \qquad \hat{r} = [\sin(\theta)\cos(\phi), \sin(\theta)\sin(\phi), \cos(\theta)]^T$$

$$b = \cos(2\theta)\cos(\phi) \qquad \hat{\theta} = [\cos(\theta)\cos(\phi), \cos(\theta)\sin(\phi), -\sin(\theta)]^T \qquad (2.78)$$

$$c = \cos(\theta)\sin(\phi) \qquad \hat{\phi} = [-\sin(\phi), \cos(\phi), 0]^T.$$

The first term with $1/R^4$ in Eq. (2.76) is named near field as it is negligible at distances far away from the source. Madariaga et al. (2019) wrote the first term of Eq. (2.76) in a slightly different form to stress that the near and intermediate terms decay as $1/R^2$ and cannot be separated in the study of seismic radiation. Indeed, applying the mean value theorem of calculus, one can write

$$\frac{1}{R^4} \int_{R/v_P}^{R/v_S} \zeta P(t-\zeta) d\zeta \approx \frac{1}{R^4} \left(\frac{R}{v_P} - \frac{R}{v_S} \right) \cdot \frac{1}{2} \left[\frac{R}{v_P} + \frac{R}{v_S} \right] P(\xi) \text{ for } \frac{R}{v_P} < \xi < \frac{R}{v_S},$$

which is of order $1/R^2$. The P- and S-waves cannot be separated at very short distances from source, which does not mean that the two types of seismic waves do

2.11 Final Static Deformation for Double-Couple Source

not exist there. The reason that we cannot separate them is that they arrive almost simultaneously and that their amplitudes are not significantly different.

The next two terms in 2.76 are said to belong to the intermediate field of the seismic radiation. Both the near field and the intermediate field terms are fully determined by the potency time function at the source. The last two terms are dominant in the far field and are determined by the potency rate. They decrease with the distance to the source as $1/R$ and go to zero when the time goes to infinity, unlike the near-field term and the two intermediate field terms.

A common feature of the individual terms in 2.76 is that in each of them the angular dependence is factored out that makes it possible to visualise those terms as 3D surfaces in polar coordinates,

$$\|\Re_\Xi\| = \sqrt{\Lambda_{\Xi R}^2 a^2 + \Lambda_{\Xi\theta}^2 b^2 + \Lambda_{\Xi\phi}^2 c^2}, \qquad (2.79)$$

where the subscript Ξ stands for N, IP, IS, FP or FS and the components $\Lambda_{\Xi R}$, $\Lambda_{\Xi\theta}$, and $\Lambda_{\Xi\phi}$ are given in Fig. 2.24 bottom right, see also Eq. (2.77).

The RMS radiation pattern is obtained by averaging $\|\Re_\Xi\|^2$ over the unit sphere around the source and then taking the square root of the result,

$$\Lambda_{\Xi,\text{RMS}} = \sqrt{\frac{1}{4\pi} \int_0^{2\pi} d\phi \int_0^\pi \sin(\theta) \left[\Lambda_{\Xi R}^2 a^2 + \Lambda_{\Xi\theta}^2 b^2 + \Lambda_{\Xi\phi}^2 c^2 \right] d\theta}. \qquad (2.80)$$

Figure 2.24 illustrates the spatial distribution or the radiation from a double-couple source in the near, intermediate, and far fields. Note that the vector $\Lambda_{IP} = 4a\hat{r} - 2b\hat{\theta} + 2c\hat{\phi}$ is neither orthogonal to the wavefront at the point of the receiver to qualify it as a purely P- or pressure wave of longitudinal polarisation, nor it is tangential to the wavefront at the point of the receiver, to qualify it as a purely S- or shear wave of transverse polarisation. The same applies to $\Lambda_{IS} = -3a\hat{r} + 3b\hat{\theta} - 3c\hat{\phi}$. Therefore, the intermediate field terms in the classical expression do not describe seismic waves of different polarisation but rather do so for groups of waves which arrive at the receiver almost at the same time. The same can be said for the near-field term.

2.11.2 Final Static Displacement and Induced Strain

Nearly all seismic related damage in underground mines is observed in excavations relatively close to the seismic sources where rock is subjected to large inelastic deformations. Dynamic strains of the order of 10^{-4} or greater are associated with ground velocities over 30 cm/s, which are frequently associated with localised

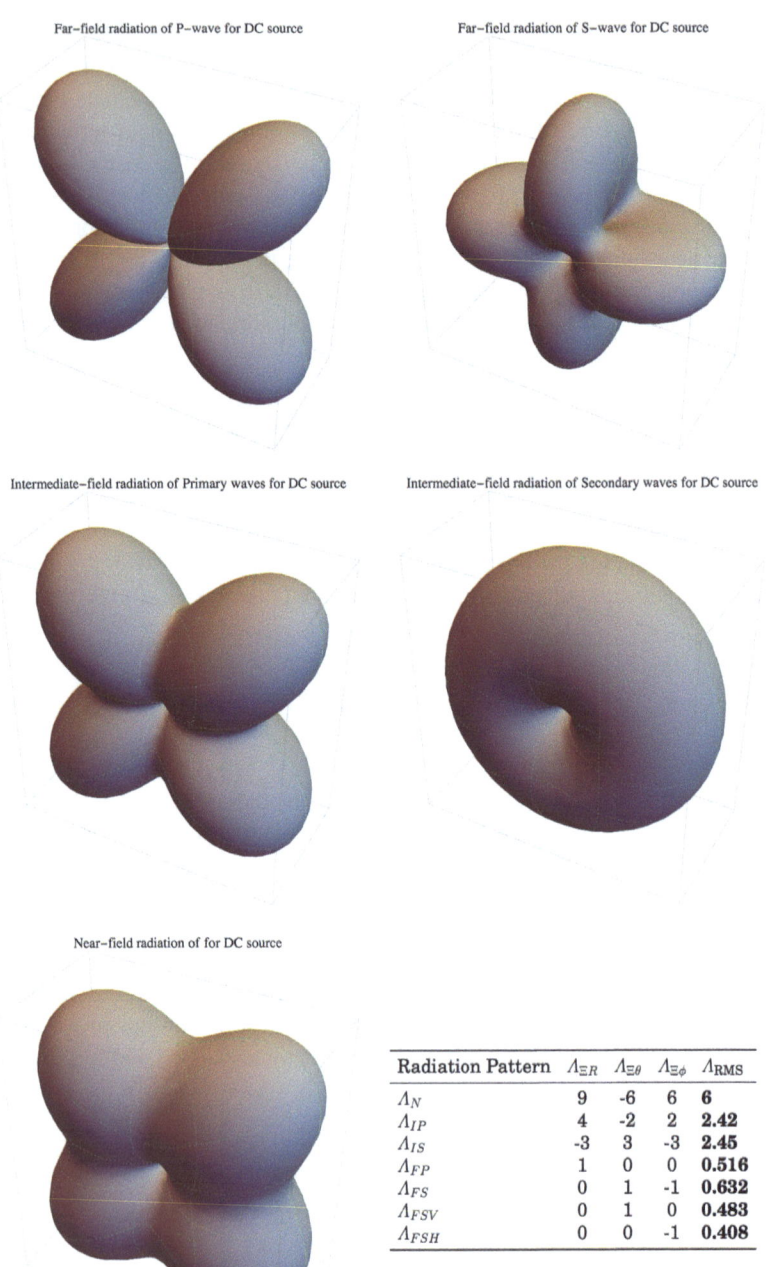

Fig. 2.24 Spatial distribution of the radiation of seismic waves for a double-couple source. Table bottom right gives parameters $\Lambda_{\Xi R}$, $\Lambda_{\Xi\theta}$, $\Lambda_{\Xi\phi}$ and the RMS values

2.11 Final Static Deformation for Double-Couple Source

damage to underground infrastructure. Larger strains crack intact rock. It is therefore important to gain insight into the extent of that deformation caused by seismic sources.

Taking $t \to \infty$ in Eq. (2.76), i.e. $u(R, \theta, \phi; t \to \infty)$, gives the final static displacement field for a shear dislocation by potency P that does not depend on the details of the source time function. This permanent deformation is due to the near-field and the intermediate field terms since in the far field $\lim_{t \to \infty} \dot{P}(t) = 0$. The convolution integral, i.e. the near-field term, can be taken exactly in the limit $t \to \infty$,

$$\lim_{t \to \infty} \int_{R/v_P}^{R/v_S} \zeta P(t - \zeta) d\zeta = P \frac{1}{2} \left[\left(\frac{R}{v_S} \right)^2 - \left(\frac{R}{v_P} \right)^2 \right] = PR^2 \frac{(v_P^2 - v_S^2)}{2 v_P^2 v_S^2},$$

and the final expression for the static displacement at (R, θ, ϕ) is

$$u(R, \theta, \phi, t \to \infty) = \frac{1}{4\pi} \frac{P}{R^2} \left[\frac{1}{2} \left(1 - \frac{v_S^2}{v_P^2} \right) A_N + \frac{v_S^2}{v_P^2} A_{IP} + A_{IS} \right], \quad (2.81)$$

which decays along direction (θ, ϕ) as $1/R^2$, see Aki and Richards (2002) equation (4.34). The maximum static displacement is obtained for $\theta = \pi/4$ and $\phi = 0$, which for a Poissonian solid gives

$$u_{max} = 1.333 \frac{P}{4\pi R^2}. \quad (2.82)$$

The components of the strain tensor in spherical polar coordinates can be expressed as partial derivatives with respect to R, θ and ϕ of the displacement vector components relative to the spherical polar system,

$$\begin{aligned}
\epsilon_{RR} &= \frac{\partial}{\partial R} u_R, \\
\epsilon_{\theta\theta} &= \frac{1}{R} \left[\frac{\partial}{\partial \theta} u_\theta + u_R \right], \\
\epsilon_{\phi\phi} &= \frac{1}{R \sin(\theta)} \left[\frac{\partial}{\partial \phi} u_\phi + u_R \sin(\theta) + u_\theta \cos(\theta) \right], \\
\epsilon_{R\theta} &= \frac{1}{2} \left[\frac{1}{R} \frac{\partial}{\partial \theta} u_R + \frac{\partial}{\partial R} u_\theta - \frac{1}{R} u_\theta \right], \\
\epsilon_{R\phi} &= \frac{1}{2} \left[\frac{1}{R \sin(\theta)} \frac{\partial}{\partial \phi} u_R + \frac{\partial}{\partial R} u_\phi - \frac{1}{R} u_\phi \right], \\
\epsilon_{\theta\phi} &= \frac{1}{2R} \left[\frac{1}{\sin(\theta)} \frac{\partial}{\partial \phi} u_\theta + \frac{\partial}{\partial \theta} u_\phi - u_\phi \frac{\cos(\theta)}{\sin(\theta)} \right].
\end{aligned} \quad (2.83)$$

The spherical polar components of the displacement vector in Eq. (2.75) are obtained by decomposing Λ_N, Λ_{IP}, and Λ_{IS} as combinations of the spherical polar base vectors $\hat{r}$, $\hat{\theta}$ and $\hat{\phi}$ and then grouping the terms. The spherical polar components of the static co-seismic displacement are

$$u_R = \frac{P}{4\pi R^2} \left[\frac{3}{2} \left(1 - \frac{1}{3}\frac{v_S^2}{v_P^2} \right) \sin(2\theta) \cos(\phi) \right],$$

$$u_\theta = \frac{P}{4\pi R^2} \left[\frac{v_S^2}{v_P^2} \cos(2\theta) \cos(\phi) \right], \qquad (2.84)$$

$$u_\phi = \frac{P}{4\pi R^2} \left[-\frac{v_S^2}{v_P^2} \cos(\theta) \sin(\phi) \right].$$

The spherical polar components of the induced static strain tensor ϵ_{ij} are

$$\epsilon_{RR} = \frac{P}{4\pi R^3} \left[-\left(3 - \frac{v_S^2}{v_P^2} \right) \sin(2\theta) \cos(\phi) \right],$$

$$\epsilon_{\theta\theta} = \frac{P}{4\pi R^3} \left[\frac{1}{2} \left(3 - 5\frac{v_S^2}{v_P^2} \right) \sin(2\theta) \cos(\phi) \right],$$

$$\epsilon_{\phi\phi} = \frac{P}{4\pi R^3} \left[\frac{3}{2} \left(1 - \frac{v_S^2}{v_P^2} \right) \sin(2\theta) \cos(\phi) \right],$$

$$\epsilon_{R\theta} = \frac{P}{4\pi R^3} \left[\left(\frac{3}{2} - 2\frac{v_S^2}{v_P^2} \right) \cos(2\theta) \cos(\phi) \right], \qquad (2.85)$$

$$\epsilon_{R\phi} = \frac{P}{4\pi R^3} \left[-\left(\frac{3}{2} - 2\frac{v_S^2}{v_P^2} \right) \cos(\theta) \sin(\phi) \right],$$

$$\epsilon_{\theta\phi} = \frac{P}{4\pi R^3} \left[\frac{v_S^2}{v_P^2} \sin(\theta) \sin(\phi) \right],$$

which can be written $\epsilon_{ij}(P, v) = P/(4\pi R^3)\, e_{ij}$, where the components of the new tensor $e_{ij}(\theta, \phi, v)$ are equal to the respective expressions in the square brackets in Eq. (2.85) and depend on θ, ϕ and the Poisson ratio, v, since

$$\frac{v_S^2}{v_P^2} = (1 - 2v)/(2 - 2v) \quad \text{or} \quad v = \left[1 - 2\left(v_S^2/v_P^2\right) \right] / \left[2 - 2\left(v_S^2/v_P^2\right) \right].$$

2.11 Final Static Deformation for Double-Couple Source

The second order invariant of the strain tensor, $I_2 = \mathrm{Tr}\,(\epsilon \cdot \epsilon)$, can be used to define a scalar measure for the level of inelastic deformation ϵ_p,

$$\epsilon_p = \sqrt{\mathrm{Tr}\,(\epsilon \cdot \epsilon)} = \frac{P}{4\pi R^3}\sqrt{\mathrm{Tr}\,(e \cdot e)}, \tag{2.86}$$

where

$$\mathrm{Tr}(\epsilon \cdot \epsilon) = (\epsilon_{RR})^2 + (\epsilon_{\theta\theta})^2 + (\epsilon_{\phi\phi})^2 + 2\left[(\epsilon_{R\theta})^2 + (\epsilon_{R\phi})^2 + (\epsilon_{\theta\phi})^2\right]$$

$$= \left(\frac{P}{4\pi R^3}\right)^2 \mathrm{Tr}\,(e \cdot e).$$

Equation (2.86) can be solved for R to get a surface of the source volume with $\epsilon \geq \epsilon_p$,

$$R(P, \epsilon_P, \nu; \theta, \phi) = \left[\frac{P}{4\pi\epsilon_P}\sqrt{\mathrm{Tr}\,(e \cdot e)}\right]^{\frac{1}{3}}, \quad \text{for} \quad 0 < \theta < \pi, \quad 0 < \phi < 2\pi. \tag{2.87}$$

The volume of the double-couple source with $\epsilon \geq \epsilon_p$ can be computed by taking the integral,

$$V(P, \epsilon_P, \nu) = \int_0^{2\pi} d\phi \int_0^{\pi} d\theta \cdot \sin(\theta) \int_0^{R(,\theta,\phi)} r^2 dr = \frac{1}{3}\int_0^{2\pi} d\phi \int_0^{\pi} d\theta \cdot \sin(\theta) \cdot R^3(\theta, \phi),$$

that gives

$$V(P, \epsilon_P, \nu) = \frac{P}{12\pi\epsilon_P}\int_0^{2\pi} d\phi \int_0^{\pi} d\theta \cdot \sin(\theta) \cdot \sqrt{\mathrm{Tr}\,(e \cdot e)}.$$

Figure 2.25 shows the shape of the source volume for an event with $\log P = 2.0$ for $\epsilon_p \geq 10^{-6}$ in grey and $\epsilon_p \geq 10^{-4}$ in red, as given by Eq. (2.87) assuming $v_S^2/v_P^2 = 1/3$, i.e. Poisson ratio $\nu = 0.25$.

Figure 2.26 left shows the source volume with $\epsilon_p \geq 10^{-4}$ for a seismic event with $\log P = 2.0$ as a function of Poisson ratio, and Fig. 2.26 right shows the Poisson ratio as a function of v_S/v_P, for reference.

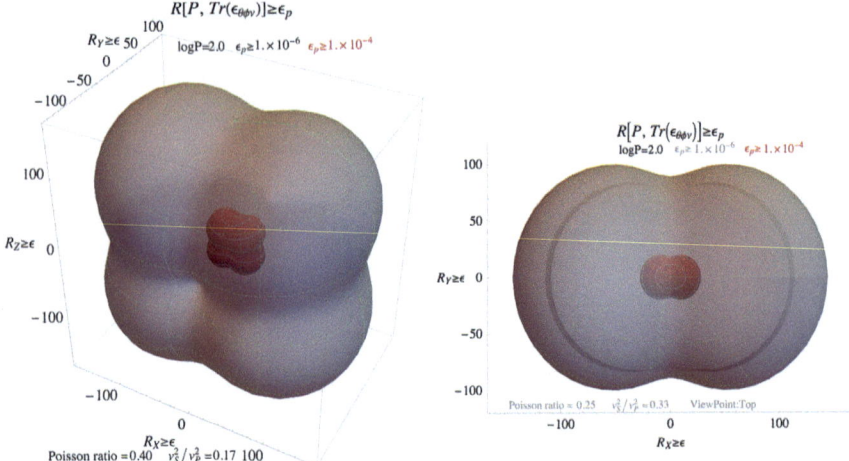

Fig. 2.25 Shape of the source volume with $\epsilon_p \geq 10^{-6}$ in grey and $\epsilon_p \geq 10^{-4}$ in red given by Eq. (2.87) in a 3D view *(left)* and top view *(right)*

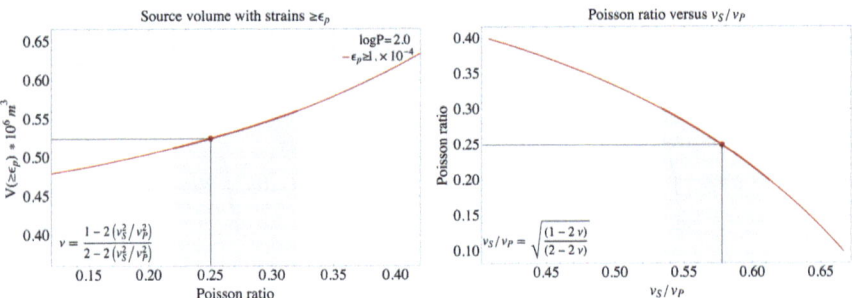

Fig. 2.26 Volume of the source region with strains $\epsilon_p \geq 10^{-4}$ as a function of Poisson ratio *(left)* and Poisson ratio as a function of v_S/v_P for reference *(right)*

References

Abercrombie, R. E. (2015). Investigating uncertainties in empirical Green's function analysis of earthquake source parameters. *Journal of Geophysical Research: Solid Earth, 120*, 4263–4277. https://doi.org/10.1002/2015JB011984.

Abercrombie, R. E., & Rice, J. R. (2005). Can observations of earthquake scaling constrain slip weakening? *Geophysical Journal International, 162*, 406–424. https://doi.org/10.1111/j.1365-246X.2005.02579.x.

Aki, K. (1966). Generation and propagation of G waves from the Niigata earthquake of June 16, 1964. Part 2: Estimation of earthquake moment, released energy, and stress strain drop from the G-wave spectrum. *Bulletin Earthquake Research Institute Tokyo University, 44*, 73–88.

Aki, K. (1967). Scaling law of seismic spectrum. *Journal of Geophysical Research, 72*, 1217–1231.

Aki, K. (1980). Scattering and attenuation of shear waves in the lithosphere. *Journal of Geophysical Research, 85*(B11), 6496–6504.

Aki, K., & Richards, P. G. (2002). *Quantitative seismology* (2nd ed.). University Science Books.

References

Anderson, J. G., & Hough, S. E. (1984). A model for the shape of the Fourier amplitude spectrum of acceleration at high frequencies. *Bulletin of the Seismological Society of America, 74*(5), 1969–1993.

Andrews, D. J. (1986). Objective determination of source parameters and similarity of earthquakes of different size. In S. Das, J. Boatwright, & C. H. Scholz (Eds.), *Earthquake source mechanics* (Vol. 6, pp. 259–267). American Geophysical Monograph 37.

Backus, G. E., & Mulcahy, M. (1976). Moment tensor and other phenomenological descriptions of seismic sources 1. Continous displacements. *Geophysical Journal of the Royal Astronomical Society, 46*(2), 341–361.

Bakun, W. H., & Bufe, C. G. (1975). Shear-wave attenuation along the San Andreas fault zone in central California. *Bulletin Seismological Society of America, 65*, 439–460; *65*(2), 439–459.

Ben-Menahem, A. (1961). Radiation of seismic surface-waves from finite moving sources. *Bulletin of the Seismological Society of America, 51*(3), 401–435.

Ben-Menahem, A., & Singh, S. J. (1981). *Seismic waves and sources.* Springer-Verlag.

Ben-Zion, Y., & Zhu, L. (2002). Potency-magnitude scaling relation for southern California earthquakes with 1.0 <M <7.0. *Geophysical Journal International, 148*, F1–F5.

Beresnev, I. A. (2019). Interpretation of kappa and fmax filters as source effect. *Bulletin of the Seismological Society of America, 109*(2), 822–826.

Boore, D. M. (1983a). Stochastic simulation of high-frequency ground motions based on seismological models of the radiated spectra. *Bulletin of the Seismological Society of America, 73*(6), 1865–1894.

Boore, D. M. (1983b). Stochastic simulation of high frequency ground motions based on seismological models of the radiated spectra. *Bulletin of the Seismological Society of America, 73*(6), 1865–1894.

Brune, J. N. (1968), Seismic moment, seismicity and rate of slip along major fault zones. *Journal of Geophysical Research, 73*(2), 777–784.

Brune, J. N. (1970), Tectonic stress and the spectra of seismic shear waves from earthquakes. *Journal of Geophysical Research, 75*(26), 4997–5009.

Brune, J. N., Archuleta, R. J., & Hartzell, S. (1979). Far-field S-wave spectra, corner frequencies, and pulse shapes. *Journal of Geophysical Research, 84*, May.

Chapman, C. H., & Leaney, W. S. (2012). A new moment-tensor decomposition for seismic events in anisotropic media. *Geophysical Journal International, 188*, 343–370.

Cichowicz, A. (1981). *Determination of source parameters from seismograms of mining tremors and the inverse problem for a seismic source* (Vol. 147). Publications of The Institute of Geophysics Polish, Series A-11.

Cichowicz, A., & Green, R. W. E. (1989). Changes in the early part of the seismic coda due to localized scatterers: The estimation of q in a stope environment. *Pure and Applied Geophysics, 129*(3/4), 497–511.

Cichowicz, A., Green, R. W. E., & van Zyl Brink, A. (1988). Coda polarisation properties of high-frequency microseismic events. *Bulletin of the Seismological Society of America, 78*(3), 1297–1318.

Di Bona, M., & Rovelli, A. (1988). Effects of the bandwidth limitation of stress drops estimated from integrals of the ground motion. *Bulletin of the Seismological Society of America, 78*(5), 1818–1825.

Dysart, P. S., Snoke, J. A., & Sacks, I. S. (1988). Source parameters and scaling relations for small earthquakes in the Matsushiro region, southwest Honshu, Japan. *Bulletin of the Seismological Society of America, 78*(2), 571–589.

Eshelby, J. D. (1957), The determination of the elastic field of an ellipsoidal inclusion and related problems. *Proceedings of the Royal Society of London, Series A, Mathematical and Physical Sciences, 241*(1226), 376–396.

Fedotov, S. A., & Boldyrev, S. A. (1969). On dependence of body wave absorption on the frequency in the crust and upper mantle of the Kurile Island arc. *Izvestiya Rossiiskoi Akademii Nauk, Seriya Fizika Zemli, 9*, 17–33.

Fletcher, J. B., & McGarr, A. (2005). Moment tensor inversion of ground motion from mining-Induced earthquakes, Trail Mountain, Utah. *Bulletin of the Seismological Society of America*, 95(1), 48–57. https://doi.org/10.1785/0120040047.

Frohlich, C., & Davis, S. D. (1999). How well constrained are well-constrained t, b and p axes in moment tensor catalogs? *Journal of Geophysical Research*, 104, 4901–4910.

Gal, M., Lotter, E., Olivier, G., Green, M., Meyer, S., Dales, P., & Reading, A. M. (2021). CCLoc—an improved interferometric seismic event location algorithm applied to induced seismicity. *Seismological Research Letters*, 92, 3492–3503, https://doi.org/10.1785/0220210068.

Gibowicz, S. J. (1990). Seismicity induced by mining. *Advances in Geophysics*, 32, 1–74.

Gibowicz, S. J., Cichowicz, A., & Dybel, T. (1977). Seismic moment and source size of mining tremors in Upper Silesia. *Acta Geophysica Polonica*, 25, 201–218.

Hanks, T. C. (1979). b value and omega seismic source models: Implications for tectonic stress variations along active crustal fault zones and the estimation of high-frequency strong ground motion. *Journal of Geophysical Research*, 84(B5), 2235–2242.

Hanks, T. C. (1982). fmax. *Bulletin of the Seismological Society of America*, 72(6), 1867–1879.

Hanks, T. C., & Thatcher, W. (1972). A graphical representation of seismic source parameters. *Journal of Geophysical Research*, 77(23), 4393–4405.

Hartzell, S. H. (1978). Earthquake aftershocks as Green's functions. *Geophysical Research Letters*, 5(1), 1–4. https://doi.org/10.1029/GL005i001p00001.

Heaton, T. H. (1990). Evidence for and implications of self-healing pulses of slip in earthquake rupture. *Physics of the Earth and Planetary Interiors*, 64, 1–20.

Hendel, A., Anderson, J. G., Pilz, M., & Cotton, F. (2020). A frequency-dependent model for the shape of the Fourier amplitude spectrum of acceleration at high frequencies. *Bulletin of the Seismological Society of America*, 1–12. https://doi.org/10.1785/0120200118.

Hirasawa, T., & Stauder, W. (1965). On the seismic body waves from a finite moving source. *Bulletin of the Seismological Society of America*, 55(2), 237–262.

Hutchings, L., & Viegas, G. (2012). Application of empirical Green's functions in earthquake source, wave propagation and strong ground motion studies, earthquake research and analysis - new frontiers in seismology (pp. 87–140). http://www.intechopen.com/books/earthquake-research-and-analysis-new-frontiers-in-seismology/application-of-empirical-green-s-functions-in-earthquake-source-wave-propagation-and-strong-ground-m.

Jeffreys, H. (1962). *The Earth*. Cambridge University Press.

Jeffreys, H. (1975). The importance of damping in geophysics. *Geophysical Journal of the Royal Astronomical Society*, 40(1), 23–27.

Kaneko, Y., & Shearer, P. M. (2014). Variability of seismic source spectra, estimated stress drop, and radiated energy, derived from cohesive-zone models of symmetrical and asymmetrical circular and elliptical ruptures. *Journal of Geophysical Research: Solid Earth*. https://doi.org/10.1002/2014JB011642.

Keilis-Borok, V. I., Bessanova, E. N., Gotsadze, O. D., Kirilova, I. V., Kogan, S. D., Kikhtikova, T. I., Malinovskaya, C. N., Pavola, G. I., & Sarskii, A. A. (1960). Investigation of the mechanism of earthquakes (English translation). *American Geophysical Union, Consultants Bureau, New York*.

Keylis-Borok, V. I. (1959). On estimation of the displacement in an earthquake source and of source dimensions. *Annali di Geofisica*, 12(2), 205–214.

Knopoff, L., & Randall, M. J. (1970). The compensated linear-vector dipole: A possible mechanism for deep earthquakes. *Journal of Geophysical Research*, 75(26), 4957–4963.

Konno, K., & Ohmachi, T. (1998). Ground motion characteristics estimated from spectral ratio between horizontal and vertical components of microtremors. *Bulletin of the Seismological Society of America*, 88(1), 228–241.

Kostrov, B. V. (1974). Seismic moment and energy of earthquakes and seismic flow of rock. *Izvestiya, Physics of the Solid Earth*, 13, 13–21.

Kostrov, B. V., & Das, S. (1988). *Principles of earthquake source mechanics*, Cambridge University Press.

Krajcinovic, D., & Mastilovic, S. (1995). Some fundamental issues of damage mechanics. *Mechanics of Materials*, 21, 217–230.

References

Ma, J., Dineva, S., Ceska, S., & Heimann, S. (2018). Moment tensor inversion with three-dimensional sensor configuration of mining induced seismicity (Kiruna Mine, Sweden). *Geophysical Journal International*, *213*, 2147–2160. https://doi.org/10.1093/gji/ggy115.

Madariaga, R. (1976). Dynamics of an expanding circular fault. *Bulletin of the Seismological Society of America*, *66*(3), 639–666.

Madariaga, R., Ruiz, S., Reivera, E., Leyton, F., & Baez, J. R. (2019). Near-field spectra of large earthquakes. *Pure and Applied Geophysics*, *176*, 983–1001.

Malovichko, D. (2020). Description of seismic sources in underground mines: Theory. *Bulletin of the Seismological Society of America*, *110*(5), 2124–2137.

Malovichko, D., & van Aswegen, G. (2013). Testing of the source processes of mine related seismic events. In *Proceedings 8th International Symposium on Rockbursts and Seismicity in Mines*.

McGarr, A. (1992). Moment tensors of ten witwatersrand mine tremors. *Pure and Applied Geophysics*, *139*(3–4), 781–800.

McGarr, A., & Fletcher, J. B. (2003). Maximum slip in earthquake fault zones, apparent stress, and stick-slip friction. *Bulletin of the Seismological Society of America*, *93*(6), 2355–2362.

Mendecki, A. J. (1993). Real time quantitative seismology in mines: Keynote Address. In R. P. Young (Ed.), *Proceedings 3rd International Symposium on Rockbursts and Seismicity in Mines, Kingston, ON, Canada* (pp. 287–295). Balkema.

Mountfort, P., & Mendecki, A. J. (1997). Seismic transducers. In A. J. Mendecki (Ed.), *Seismic monitoring in mines* (1 ed., chap. 1, pp. 1–20). Chapman and Hall.

Rautian, T. G., & Khalturin, V. I. (1978). The use of the coda for determination of the earthquake source spectrum. *Bulletin of the Seismological Society of America*, *68*(4), 923–948.

Rivera, L., & Kanamori, H. (2005). Representations of the radiated energy in earthquakes. *Geophysical Journal International*, *162*(1), 148–155.

Ross, Z. E., & Ben-Zion, Y. (2016). Towards reliable automated estimates of earthquake source properties from body wave spectra. *Journal of Geophysical Research*, *121*, 4390–4407. https://doi.org/10.1002/2016JB01300.

Sen, A. T., Cesca, S., Bischoff, M., Meier, T., & Dahm, T. (2013). Automated full moment tensor inversion of coal mining-induced seismicity. *Geophysical Journal International*, *195*, 1267–1281. https://doi.org/10.1093/gji/ggt300.

Smith, R. B., Winkler, P. L., Anderson, J. G., & Scholz, C. H. (1974). Source mechanisms of microearthquakes associated with underground mines in eastern Utah. *Bulletin of the Seismological Society of America*, *64*(4), 1295–1317.

Spottiswoode, S. M., & McGarr, A. (1975). Source parameters of tremors in a deep-level gold mine. *Bulletin of the Seismological Society of America*, *65*(1), 93–112.

van Aswegen, G., & Butler, A. G. (1993). Applications of quantitative seismology in South African gold mines. In R. P. Young (Ed.), *Proceedings 3rd International Symposium on Rockbursts and Seismicity in Mines, Kingston, ON, Canada* (pp. 261–266). Balkema. ISBN 9054103205.

Viegas, G. M., Abercrombie, R. E., & Kim, W. Y. (2010). The 2002 M5 Au Sable Forks, NY, earthquake sequence: Source scaling relationships and energy budget. *Journal of Geophysical Research: Solid Earth*, *115*(B07310). https://doi.org/10.1029/2009JB006799.

Willacy, C., van Dedem, E., Minisini, S., Li, J., Blokland, J.-W., Das, I., & Droujinine, A. (2019). Full-waveform event location and moment tensor inversion for induced seismicity. *Geophysics*, *84*(2), KS39–KS57. https://doi.org/10.1190/GEO2018-0212.1.

Wyss, M., & Brune, J. N. (1971). Regional variations of source properties in southern California estimated from the ratio of short- to long-period amplitudes. *Bulletin of the Seismological Society of America*, *61*(5), 1153–1167.

Zhu, L., & Ben-Zion, Y. (2013). Parameterization of general seismic potency and moment tensors for source inversion of seismic waveform data. *Geophysical Journal International*, *194*, 839–843.

Zhu, W., Tai, K. S., Mousavi, S. M., Bailis, P., & Beroza, G. (2022). An end-to-end earthquake detection method for joint phase picking and association using deep learning. *JGR Solid Earth*, *127*(3), e2021JB023,283. https://doi.org/10.1029/2021JB023283.

Open Access This chapter is licensed under the terms of the Creative Commons Attribution 4.0 International License (http://creativecommons.org/licenses/by/4.0/), which permits use, sharing, adaptation, distribution and reproduction in any medium or format, as long as you give appropriate credit to the original author(s) and the source, provide a link to the Creative Commons license and indicate if changes were made.

The images or other third party material in this chapter are included in the chapter's Creative Commons license, unless indicated otherwise in a credit line to the material. If material is not included in the chapter's Creative Commons license and your intended use is not permitted by statutory regulation or exceeds the permitted use, you will need to obtain permission directly from the copyright holder.

Chapter 3
Monitoring Rock Mass Stability

Abstract This chapter is a continuation of the seismic stability analysis suggested by Mendecki (1993; Real time quantitative seismology in mines: Keynote Address. In R. P. Young (Ed.), *Proceedings 3rd International Symposium on Rockbursts and Seismicity in Mines, Kingston, Ontario, Canada* (pp. 287–295). Balkema, Rotterdam.). The assumption here is that an inhomogeneous rock mass subjected to loading displays certain seismic symptoms when approaching instability. (1) An overall softening, measured by a decrease in the average value of the apparent stress. (2) Increased rate or accelerating deformation, measured by an increase in the activity rate and/or apparent volume. (3) An associated increase in correlation length where one would expect to observe an increase in the spatial distribution of seismic activity, measured, for example, by seismic diffusivity. (4) A decrease in the dimensionality, or localisation, of seismic activity as can be measured by the shape factor. The objective of seismic stability analysis is not to predict instability or to manage seismic exposure in the short term, but to guide control measures to mitigate seismic hazard.

3.1 Note on Time, Order, Disorder, and Stability

In 1708, Gottfried Wilhelm von Leibniz said "Time and space are not things, but order of things", (Wiener, 1951). Or, as Albert Einstein put it "Time and space are modes by which we think, and not condition in which we live" (Forsee, 1967). Therefore, time is nature's way to keep everything from happening all at once, (Atiyah, 1988). And John Wheeler (1911–2008) quipped that "Space is what prevents everything happening to me". In essence, anything is possible if it happens too fast to be detected.

Time, as incorporated in the basic laws of physics, e.g. by Newton, Maxwell, or Einstein, does not distinguish between past and future, all processes are time reversible, meaning that they can proceed backwards as well as forwards through time. In his excellent book, Eddington (1928) introduced the concept of the arrow of time. He writes that "cause and effect are closely bound up with arrow of time, i.e. the cause must precede the effect. Thus in primary physics, which knows nothing of time's arrow, there is no discrimination of cause and effect, but events are connected by a symmetrical causal relation which is the same viewed from either end". He then states that the future is associated with more randomness and the past with less randomness.[1] The subject of arrow of time and irreversibility was thoroughly explored by the Nobel prize winner Ilya Prigogine in Prigogine (1980) and Prigogine (1997), where he states that irreversibility can no longer be identified with a mere appearance that would disappear if we had perfect knowledge. He used the concept of entropy to distinguish between reversible and irreversible processes. Only irreversible processes contribute to entropy production, e.g. chemical reactions, heat conduction, diffusion. The second law of thermodynamics then states that irreversible processes lead to a kind of "one-sidedness" of time. The positive time direction is associated with the increase of entropy. The law of entropy increase is simply a law of increasing disorganisation. In this process, the initial conditions are forgotten. As Prigogine (1997) speculates: "The big bang was an event associated with an instability within the medium that produced our universe. It marked the start of our universe but not the start of time. Although our universe has an age, the medium that produced our universe has none. Time has no beginning, and probably no end".

On a macroscopic scale, all processes in nature are dissipative. Natural systems consist of a large number of elements which, at any given time, are not in the same state. Therefore, in order to accommodate the differences, a macroscopic system spontaneously generates local flow of energy and momentum in the form of localised small-scale events, in addition to those imposed by the external conditions. Close to equilibrium, the distribution of fluctuations is more or less random, the correlation time and the correlation length are short, and nonlinearities are mostly hidden. In other words, near equilibrium fluctuations are harmless. Away from equilibrium, the system is more susceptible to the action of intermittent intrinsic instabilities that, due to their nonlinear nature, are agents of spatial and temporal correlations. Here the distribution of fluctuations is broader than Gaussian, with slower, power law like, decays or with additional peaks, facilitating rare but larger events that dominate the behaviour of the system. The finite values of spatial and temporal correlations measure the distance from equilibrium, and as this distance grows, the influence of nonlinearities increases. When the spatial range of fluctuations increases, elements of the different parts of the system interact, and the system can generate and maintain a reproducible relationship among its distant parts. In this process of self-organisation, the system creates spatial and temporal patterns that are not directly imposed by external forces. When the range of correlations becomes comparable with the system size, the resulting coherence, or

---

[1] Without any mystic appeal to consciousness, it is possible to find a direction of time on the four-dimensional map by a study of organisation. Let us draw an arrow arbitrarily. If as we follow the arrow, we find more and more of the random element in the state of the world, then the arrow is pointed towards the future; if the random element decreases, the arrow points towards the past. That is the only distinction known to physics. This follows at once if our fundamental contention is admitted that the introduction of randomness is the only thing which cannot be undone.

3.1 Note on Time, Order, Disorder, and Stability

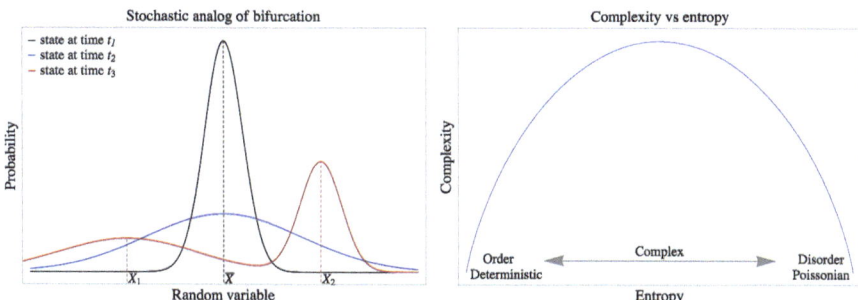

Fig. 3.1 Illustrations of stochastic analogue of bifurcation *(left)* and complexity vs. entropy *(right)*

order, may influence its behaviour qualitatively, and the system may become critical and undergo bifurcation, i.e. transition to another state (Nicolis & Prigogine, 1977). The divergence of correlation length indicates that the details of the system are irrelevant to its critical behaviour.

In essence, the traditional classical science evolved around stability and certitudes, and probability was associated with ignorance, while in reality we experience fluctuations, instabilities, multiple choices, and limited predictability. Once instability is included, the meaning of "law of nature" changes, and they no longer express certitudes but rather possibilities and need to be formulated on the statistical level (Prigogine, 1996) or, as Wallerstein (1999) said, "Probability is the only scientific truth there is".

Figure 3.1 left illustrates the stochastic analogue of bifurcation described by Nicolis and Prigogine (1989).

The system at time t_1 is described by a sharp unimodal peak around its mean value $\langle X \rangle$. In other words, the system is firmly in the current state, and the internal fluctuations and/or the external influences are weak. At time t_2 due to increased strength of fluctuations and/or changes in boundary conditions, the range of spatial correlation increased, and the system has a flat distribution, exploring regions of phase space that are far from the most likely value, preparing for bifurcation, i.e. a qualitative change of its behaviour due to a small smooth change in the parameter values. At this stage, one would expect seismic activity to develop spatial correlations. For example, the seismic response to blasting would have far wider spatial range than at the time t_1. This is a state of marginal stability. Such correlations can only arise in systems in which the regime of Poissonian fluctuation, or disorder, has been overcome, i.e. there is an increased coherence and associated lower dimensionality of the system. In a Poissonian regime, the inter-event times are distributed with the same probability density function, i.e. the process is memoryless, and the vanishing of the covariance in the joint probability distribution in two sub-volumes implies the complete absence of spatial correlations, i.e. no interactions between events. Such a state can therefore be considered as the prototype of disorder. At time t_3, the system transitioned into a multi-hump regime beyond bifurcation and is more likely to be at $\langle X_2 \rangle$ than $\langle X_1 \rangle$.

In general, the approach to the critical point and the nature of the instability depend on the complexity of the system that depends, in turn, on the degree of order or disorder in the system. A highly entropic state is random and not so complex, and a low entropic state is not complex because of its regularities (Crutchfield & Young, 1989; Kaneko & Tsuda, 2001), see Fig. 3.1 right. Therefore one would expect the state of marginal stability, state at time t_2, to represent high complexity where the system itself hesitates what to do next. For complexity to feature, the system needs to be open to interaction with the environment where there is a net flow of energy. Closed systems would proceed towards an ordered or disordered equilibrium state, which is not complex.

Mining is an open system where transitions between ordered and disordered states are driven by stress and strain gradients imposed by rock extraction and by resulting seismic and aseismic deformation. Moreover, mining is not a spontaneous process. Rock extraction takes place at a particular place, at a particular time, and at a particular rate which are all highly variable compared to the tectonic regime. This interferes with the process of self-organisation. The average rate of deformation induced by mining is at least two orders of magnitude greater than the average slip rate of tectonic plates. The bulk of seismic activity in mines starts with rock extraction, increases with the extraction ratio of the ore body, and stops rather quickly with cessation of mining. Larger events tend to occur after the extraction ratio, or the depth of mining, or both reach a certain level. The complexity of the rock mass response to mining is driven mainly by two competing spatio-temporal processes of excitation and relaxation. In terms of the stress-strain behaviour of a given volume of rock, it would be the mode switching and the balance between strain hardening and softening that drives complexity. Since softer, heterogeneous systems spend more time after the peak stress and dissipate more, they generate more entropy at the costs of complexity. Thus, it may be that the relative drop in complexity, due to increase in entropy, gives rise to lower seismic hazard and more visible alerts. Increase in disorder leads to slower transitions and to diffused instabilities, while highly homogeneous systems may crack in one go with little warning or precursory behaviour.

An important agent in the development of spatial and temporal correlations in highly stressed rock is seismic activity itself. By breaking numerous asperities, seismic events smooth the system, allowing transfer of stresses over larger distances and thus paving the way for even larger events. Disorder, on the other hand, plays a stabilising role and, to a degree, can be engineered into the system by scattered layout and/or by a scattered sequence of mining or blasting. Disordered directions of local stresses and a slower loading rate, i.e. slower rate of mining, may also play a role, since it promotes healing and thus stress roughening.

Assuming that large seismic events are those small ones that were not stopped soon enough, the one way to manage seismic hazard is to engineer a mine layout that, together with the natural structures, is "rough" enough to limit the extent of ruptures. It has been postulated that mining scenarios that introduce spatial heterogeneity, or roughness, may de-correlate the system and be less likely to develop larger dynamic instabilities (e.g. van Aswegen & Mendecki, 1999; Handley et al., 2000; Mendecki, 2001, Figure 8; Mendecki, 2005).

However, there is a degree of confusion regarding the role disorder plays in rock mass stability. Since order is prerequisite of human survival, the impulse to produce orderly arrangements is inbred by evolution. In common speech, order describes organisation, structural regularity, and is associated with stability. We can achieve more if we act in organised manner, and therefore, there is a tendency to design mining a layout with simple geometrical shapes with straight lines. When in the late 80s I argued for a disordered or scattered mining layout in South African gold mines, a senior mine executive replied "a long-wall is a long-wall is a long-wall", which in an essence promoted straight lines. It took a few years and countless rockbursts to change that perception.

But rock subjected to excessive stress also ruptures along straight lines. Yes, in aseismic mines, an orderly design may facilitate better financial outcome, but in deep hard rock mines it is the order or the smoothness of the mine layout and geological structures that promote more frequent and more destructive rockbursts. Put simply, the entropy or disorder measures the degree to which energy is mixed up inside a system, and therefore, it scales positively with the quantity of energy no longer available to do physical work.

This introduction explains why this chapter is about the degree of rock mass stability rather than prediction or forecasting. Keilis-Borok (1994) described the earthquake generating part of the solid Earth as a hierarchical nonlinear dissipative system. The same can be said about a rock mass subjected to mining. It consists of a hierarchy of geological structures under load, random heterogeneities, patches of rock resisting deformation where stresses are increasing, patches of fractured rock where stresses are decreasing, and passive volumes of rock not influenced by loading. Such a system shows partial self-similarity, fractality, and self-organisation, and its parts may remain in a sub-critical state even after larger events.

3.2 Stability of Deformation and Stability of a System

The surface which bounds all stress states that correspond to elastic deformation in six-dimensional stress space is called the yield or damage surface. The evolution of the yield surface with continued deformation specifies the manner in which the material hardens or softens during the deformation. The direction of the inelastic strain increment beyond the yield surface is determined by the inelastic potential. The inelastic potential surface is always normal to the inelastic strain rate vector $\dot{\epsilon}_{in}$. The associated flow or normality rule applies to the material where yield and potential surfaces coincide, that is when the inelastic strain rate vector is always normal to the yield surface. Under the associated flow with a smooth yield surface, one would expect a strain softening behaviour before the instability. This could be described by the classical instability criterion formulated by Hadamard in 1903 (Truesdell & Noll, 2004), and rediscovered by Pearson (1956) and Hill (1958),

$$\int_{\delta V} \dot{\sigma} \cdot \dot{\epsilon}_e \cdot dV + \int_{\delta V} \dot{\sigma} \cdot \dot{\epsilon}_{in} \cdot dV \begin{matrix} > 0 \text{ stable} \\ < 0 \text{ unstable} \end{matrix}, \qquad (3.1)$$

where $\dot{\epsilon}_e$ and $\dot{\epsilon}_{in}$ are elastic and inelastic strain rates, $\dot{\sigma}$ is the rate of change of stress, and dV the volume of interest. The second inequality in Eq. (3.1) states that for unstable deformation of dV the inner product of the next increment of stress with the next increment of strain should be negative,

$$d\sigma d\epsilon = d\sigma d\epsilon_e + d\sigma d\epsilon_{in} < 0.$$

For the elastic strain increments $d\sigma d\epsilon_e > 0$, therefore the deformation will be unstable only when the inelastic term balances the elastic term, so that there is a net strain softening. Near the end of the hardening regime, however, almost all further strain increments are inelastic. As stress and strain are tensors, there are many different components that may be influential in causing the instability.

At some stage, the system may cease to deform in a homogeneous mode and undergoes bifurcation. In general, bifurcation refers to a qualitative change of the object under study due to a change of parameters on which the object depends. In this case, it is the non-uniqueness of the deformation path. A bifurcation mode may or may not be amplified by continued deformation past the bifurcation point. After the displacement corresponding to the bifurcation point, there may be more than one incremental displacement field that satisfies the equilibrium and boundary conditions. If $\dot{\sigma}_1$ and $\dot{\sigma}_2$ are the stress rate fields corresponding to two such solutions, with $\dot{\epsilon}_1$ and $\dot{\epsilon}_2$ the respective strain rates, then the necessary condition for bifurcation to occur is that there exists an incremental displacement field such that $\int_V \Delta\dot{\sigma} \Delta\dot{\epsilon} = 0$, where $\Delta\dot{\sigma} = \dot{\sigma}_1 - \dot{\sigma}_2$ and $\Delta\dot{\epsilon} = \dot{\epsilon}_1 - \dot{\epsilon}_2$.

If the bifurcation grows, it may be symmetrical or asymmetrical. A symmetrical bifurcation (e.g. barrelling of the specimen) is stable, so $d\sigma d\epsilon > 0$, i.e. the deformation is stable when it is easier to start deformation elsewhere rather than to pursue it where it has begun. Asymmetrical bifurcation, characteristic of materials that show frictional and dilatational responses to stress, results in strain localisation in single, conjugate, or quasi-periodic multiple shear bands (Rudnicki & Rice, 1975; Hobbs et al., 1990; Muhlhaus et al., 1992). The possibility of localisation arises when one or more stress components within a volume are able to decrease with increasing strain (Cundall, 1990). Here the deformation may be unstable. Alternatively, strain localisation can be caused by relative strain softening, where material within a localised strain zone hardens at a lower rate than the adjacent body of rock (Hobbs et al., 1990; Jessell & Lister, 1991). The localisation is driven by the release of strain energy from an unloading region outside the localisation zone. Strain localisation may occur under both strain softening and strain hardening material response and is promoted by non-associated flow and/or by the presence of vertices on the yield surface (Hobbs et al., 1990; Olson, 1992).

Fredrich et al. (1989) conducted the first laboratory study to quantify all the constitutive parameters for pressure-sensitive dilatant materials in both the brittle

3.2 Stability of Deformation and Stability of a System

and semi-brittle regime. They measured an internal friction coefficient, a dilatancy factor, being the ratio between the increments of inelastic volume strain and the inelastic shear strain, and a hardening modulus for Carrera marble deformed at room temperature and at different confining pressures. At a confining pressure of 5 MPa, they showed a brittle failure mode, and at 40–190 MPa, semi-brittle failure was observed. In the brittle region, shear localisation was not evident until the sample was deformed well into the post-failure region. In the semi-brittle regime, with increasing confining pressure, the hardening modulus increased, whereas the friction coefficient, dilatancy factor, and Poisson ratio decreased. Thus, the bifurcation model predicts that shear localisation is inhibited in the semi-brittle regime as long as the rock continues to strain harden.

Wawersik et al. (1990) investigated localisation of deformation theoretically and in the laboratory. They found that deviation from normality, i.e. associated flow, promotes localisation before peak stress for plane-strain compression. Triaxial deformation was characterised by a broad peak followed by gradual strain softening, with breakdown instability occurring only well beyond the peak stress.

In the case where the nucleation zone is localised and is developing far from the free surface, the shear band(s) are constrained by both elastic and, to a limited extent, inelastic deformation in the less deformed rock surrounding the shear zone. In such a case, hardening of the system undergoing deformation is to be expected even if the material softens within the nucleation zone. The critical states of stability of a three-dimensional body can generally occur only if the compression stress is of the same order of magnitude as the shear modulus of the material (Bazant & Cedolin, 1991).

Whether or when the process of strain softening and/or strain localisation becomes unstable depends on the energy release and the dissipation inside the softening zone and on the exchange of mechanical energy between the softening zone and the ambient rock. Stress decrease associated with softening within the nucleation zone induces some elastic un-straining and redistribution of stresses in the ambient rock. Energy released by elastic rebound of the surrounding rock is fed into the softening region and accelerates its deformation. The amount of energy fed into the nucleation zone correlates positively with its size and with the strain rate or, rather, rate of un-straining in the ambient rock. As soon as an input of elastic energy exceeds the energy dissipation due to the inelastic processes in the growing nucleation zone, the system becomes unstable.

One should distinguish between the stability of deformation defined by the stress-strain response of the material and the stability of the system being deformed. For a given loading, the necessary condition for instability of a system, e.g. a tunnel, stope, mined out fault, or dyke, is that the volume of net softening needs to reach a critical size, which introduces a size scale into the instability criterion. The necessary condition for the dynamic instability is that this critical volume needs to be overloaded suddenly, which introduces a critical time scale.

There is no universal rule how to determine the critical length scale; however, one can try for a proxy to gauge the proximity of the system to the critical point. Some of the proxies are based on the assumption that the stability of the rock mass subjected to mining can be related to its stiffness, i.e. its ability to resist deformation with increasing stress. While the overall stiffness of the rock mass is being maintained, the seismic response to mining, measured by the cumulative inelastic co-seismic deformation, e.g., seismic potency P, is expected to be proportional to the effective volume mined, $\sum P \sim V_{meff}$. As mining progresses, the overall stiffness of the rock mass is being degraded and the rate of potency release may increase. With further degradation in stiffness, the response may become nonlinear with accelerating potency release, frequently associated with an increase in activity rate, signifying potential for larger instability. The dynamics of such instability depends on the ratio of the stiffness of the potentially unstable volume of rock to the stiffness of the surrounding rock mass. The higher this ratio the more energy will be released per unit of inelastic deformation at the source.

Another guiding idea in the interpretation of seismic activity is the concept of self-organisation into critical state, i.e. a state at which the correlation length becomes comparable with the system size. Intermittently, the correlations length may reach or even exceed the system size creating conditions conducive for larger instabilities. Here one would expect an increase in the mean distance between consecutive events $\langle d \rangle$. It is assumed that the growth of long-range correlations within the rock mass allows for progressively larger events to be generated. Zoller et al. (2001) tested the critical state concept for earthquakes in California in terms of the spatial correlation range and found a scaling relation $\log R \sim 0.7$ m between the main shock magnitude m and the critical region R. However, in mines, seismic activity follows rock extraction, and a scattered mining operation may obscure spatial correlation and influence the distribution of distances between events.

3.3 Seismic Softening and Accelerating Deformation

Seismic Softening The growth of the deformation processes up to the point of instability is called nucleation. Breakdown instability will only take place once a quasi-static and/or quasi-dynamic inelastic deformation has occurred within the critical volume of rock. The entire nucleation process prior to overall instability includes aseismic creep (e.g. Dieterich, 1992; Ohnaka, 1992), sub-critical crack growth, and dynamic instabilities of local to small scale. Experimental data on subcritical crack growth in synthetic quartz crystals indicates that this process starts only above certain stress levels, which could be of the order of 50% of the rupture stress (Darot & Gueguen, 1986). During sub-critical crack growth, crack advance occurs by discrete, individually dynamic events, but with slow average rupture velocity. Though individual microseismic events may be dynamic, their evolution can be modelled as a continuous, quasi-static process because they release only small amounts of stress or seismic potency compared to the breakdown instability. The larger the event the stronger its influence on the stress and strain environment in the area and the more likely it will affect both the time and the size of the next seismic event.

3.3 Seismic Softening and Accelerating Deformation

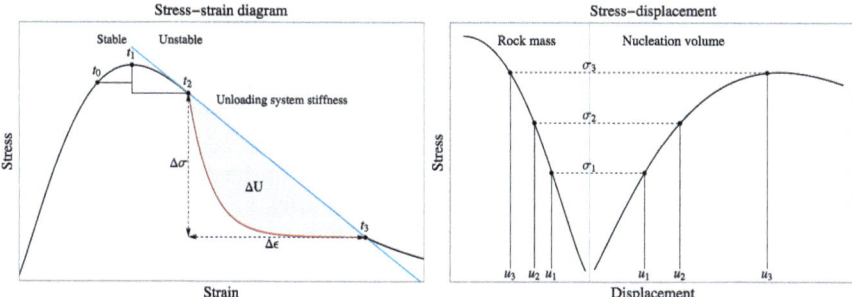

Fig. 3.2 Interaction of inelastic behaviour of the nucleation volume with the stiffness of the surrounding rock mass *(left)*. Typical stress-strain diagram of a rock sample: stable sequence from t_0 to t_2 and an instability from t_2 to t_3 *(right)*

Since the development of the nucleation zone is associated with overall strain softening, sources of seismic events located within the nucleation zone would be, on average, of a slower or softer nature. This effect would be magnified in cases where the nucleation zone interfaces through the fractured zone with the opening. In the laboratory experiments, crack branching and distributed damage associated with the fracture are observed only if the strength variation in the source region is greater than the stress concentration (Labuz et al., 1985; Labuz et al., 1985; Cox & Paterson, 1990). Outside, or at the interface of the nucleation zone, one would expect seismic events of a harder or faster nature characterised by higher apparent stress. The overall size of the nucleation zone increases with the size of the potential instability (Scholz, 1990) and with the degree of inhomogeneity and decreases with the increase in the strain rate (Kato et al., 1992).

One can consider an interpretation of instability in terms of the relative stiffness of the components of the system—an instability will occur when the stiffness of the nucleation volume is equal to or exceeds the unloading stiffness of the surrounding rock, see Fig. 3.2 left. Stuart (1981) stated that, for earthquakes, the proximity to instability can be measured by the ratio of the stiffness of the fault zone to that of the elastic surroundings—stability decreases as this ratio approaches unity.

The stress and the strain jumps at the point of instability, $\Delta\sigma$ and $\Delta\epsilon$, depend on the properties of the rock within and, to a certain extent, outside the nucleation volume. The stronger and more homogeneous the rock mass the higher the stress drop and the higher the ratio of the stress drop over strain drop associated with instability. In particular, instability is predicted much nearer the peak load for localised nucleation volumes and for states of deviatoric simple shear than for states of axisymmetric compression (Rudnicki, 1977).

Accelerating Deformation It is the nature of rock fracture and friction that breakdown instability does not occur without some preceding phase of accelerating deformation (e.g. Hirata et al., 1987; (Rudnicki, 1988); Scholz, 1990 Mendecki, 1993; Dieterich, 1994). In some cases, this phase is detected by the seismic monitoring system and in other cases not.

An increase in the rate of co-seismic deformation before instability can be reflected by an increase in the rate of seismic events, an increase in the number of mid-size events or, for the same rate and sizes of events, by the softer nature of individual events, occurring within the nucleation volume. A small but statistically significant decrease in seismic activity (quiescence) has been observed at the beginning of the strain softening stage, followed by an increase just before the instability (Brady, 1977b; Brady, 1977a; Main et al., 1992).

Figure 3.2 right illustrates the acceleration of deformation preceding instability. It is assumed that the surrounding rock mass is stiffer than the nucleation zone and undergoes strain hardening at the time when the nucleation zone softens. As a result, for the same increments of loading stress, $\sigma_2 - \sigma_1$, and $\sigma_3 - \sigma_2$, the nucleation volume experiences considerably larger increments of displacement, u_1, u_2, u_3, than the surrounding rock (see Rudnicki, 1988). The longer the surrounding rock maintains its stiffness the later and less dynamic is the instability.

3.4 Statistical Parameters of Co-seismic Deformation

Seismic events can routinely be quantified by the following four independent parameters derived from recorded waveforms: the time of the event, t, location, $X = x, y, z$, seismic potency, P, and the radiated seismic energy, E. From seismic potency and energy, one can derive apparent stress, $\sigma_A = E/P$, energy index, EI, i.e. the ratio of the observed radiated seismic energy of that event E, to the average energy $\bar{E}(P) = 10^{d \log P + c}$ radiated by events of the observed potency P, for a given area of interest ((van Aswegen & Butler, 1993)), and apparent volume, $V_A = \mu P^2 / E$, (Mendecki, 1993).

From the four independent quantities derived from waveforms, we can derive a number of seismicity parameters related to co-seismic deformation and associated changes in the strain rate, stress, and rheology of the process.

3.4.1 Seismic, Strain, Strain Rate, Stress, and Stiffness

Brune (1968) calculated average earthquake slip rates for major fault zones by $\langle u \rangle / \Delta t = \sum M / (\mu A_T)$ or $\langle u \rangle / \Delta t = \sum P / (A_T)$, where M is seismic moment, P potency, A_T is the total area of the shear zone involved in seismic slip, and μ is the assumed rigidity. Kostrov (1974) generalised Brune's formula to get an average strain and strain rate produced by the n events randomly distributed within a volume ΔV rather than along an individual fault, over time Δt,

$$\epsilon_s(\Delta V, \Delta t) = \frac{\sum P}{2 \Delta V} \quad \text{and} \quad \dot{\epsilon}_s(\Delta V, \Delta t) = \frac{\sum P}{2 \Delta V \Delta t}. \tag{3.2}$$

Then taking $E = \sigma_s \dot{\epsilon}_s \Delta V \Delta t$, he defined the average seismic stress, σ_s,

$$\sigma_s(\Delta V, \Delta t) = \frac{1}{\dot{\epsilon}_s \Delta V \Delta t} \sum E = \frac{2 \sum E}{\sum P}. \tag{3.3}$$

Seismic stiffness, K_s, then can be defined as the ratio of seismic stress to seismic strain,

$$K_s(\Delta V, \Delta t) = \frac{\sigma_s}{\epsilon_s} = \frac{4 \Delta V \sum E}{\left(\sum P\right)^2}. \tag{3.4}$$

Seismic stiffness measures the rock mass ability to resist deformation with increasing stress.

3.4.2 Seismic Viscosity, Relaxation Time, and Deborah Number

Rock mass resistance to seismic deformation can also be measured by seismic viscosity, η_s, in Pa·s, defined as the ratio of seismic stress to seismic strain rate (Kostrov & Das, 1988; Mendecki, 1997),

$$\eta_s(\Delta V, \Delta t) = \frac{\sigma_s}{\dot{\epsilon}_s} = \frac{4 \Delta V \Delta t \sum E}{\left(\sum P\right)^2}. \tag{3.5}$$

The concept of seismic viscosity is similar to the fluid mechanics concept of turbulent or eddy viscosity. Unlike ordinary or molecular viscosity, eddy viscosity does not describe the physical properties of the medium but characterises the statistical properties of the flow. Therefore, it does not have to be constant but varies in time and space. For a fixed ΔV, seismic viscosity would increase during quiescence, due to the increase in Δt if all other components in Eq. (3.5) are constant, or during a sequence of higher than average apparent stress events. Consequently, for a fixed ΔV and Δt, viscosity would decrease during a sequence of low apparent stress, or softer, events. The inverse of viscosity is called fluidity. Lower seismic viscosity, or high fluidity, implies easier seismic inelastic deformation and greater stress transfer due to seismicity. The kinematic seismic viscosity is $v_s = \eta_s/\rho$, in m^2/s, where ρ is rock density.

Seismic relaxation time quantifies the rate of change of stress during seismic deformation, $\tau_s = \eta_s/\mu$, and it scales positively with the usefulness of past data in forecasting seismic deformation. The lower the relaxation time the shorter the time span of useful data.

Seismic Deborah number is defined as $De_s = \tau_s/t_f$, where t_f is the time of observation, or the time scale of the process, which in practical applications can be replaced by the expected life span of a structure under consideration. Deborah number, like the relaxation time, may be interpreted as the ratio of elastic to viscous forces, with *De* going to infinity for a perfectly elastic medium.

The concept of Deborah number was presented by Marcus Reiner in his after dinner speech to the Fourth International Congress on Rheology at Brown University in August 1963, and then published in Reiner (1964). In the paper, Reiner quotes the Prophetess Deborah who in the Book of Judges proclaimed "The mountains flowed before the lord". In his speech, Reiner said "Deborah knew two things. First, that the mountains flow, as everything flows. But, secondly, that they flowed before the Lord, and not before man, for the simple reason that man in his short lifetime cannot see them flowing, while the time of observation of God is infinite". If the time of observation is long or the relaxation time of the material is short, then "fluid-like" behaviour is to be expected. Conversely if the relaxation time of the material is large, or the time of observation short, then the Deborah number is high and the material behaves, for all practical purposes, as a solid.

Knowing the relaxation time, one can use the Deborah number to evaluate the potential stability of pillars over time t_f. The higher the seismic Deborah number the more stable it is. Crush pillars, on the other hand, should be designed for shorter relaxation time so that they would soften and drop off the load within the designed t_f.

3.4.3 Seismic Diffusivity

A non-equilibrium process which is moving towards equilibrium at the rate governed by its distance from equilibrium is described by the diffusion equation

$$\frac{\partial \phi}{\partial t} = D\nabla^2 \phi = D\left(\frac{\partial^2 \phi}{\partial x^2} + \frac{\partial^2 \phi}{\partial y^2} + \frac{\partial^2 \phi}{\partial z^2}\right), \tag{3.6}$$

where $\partial \phi / \partial t$ is the change of ϕ with time, D is diffusivity, and $\nabla^2 \phi$ is called the divergence of gradient or Laplacian of ϕ. The solution in the 4D space of the initial value problem $\phi(t = 0, \mathbf{X}) = \delta^3(\mathbf{X})$, where δ here is the Dirac delta function, is

$$\phi(t, \mathbf{X}) = \frac{1}{(4\pi Dt)^{3/2}} \exp\left(-\frac{\mathbf{X}^2}{4Dt}\right). \tag{3.7}$$

If we compare with the standard Gaussian function, $g(x) = \exp(-x^2/\sigma^2)$, then the standard deviation $\sigma = \sqrt{4Dt}$, which points to the linear dependence of the variance with time, $\sigma^2 \sim t$, or $\langle d(t)^2 \rangle \propto t$, where $d(t)$ is the distance travelled over time t. This is the so-called normal diffusion. The diffusivity D has dimension m^2/s and is interpreted in terms of a characteristic distance of the process which varies only with the square root of time.

3.4 Statistical Parameters of Co-seismic Deformation

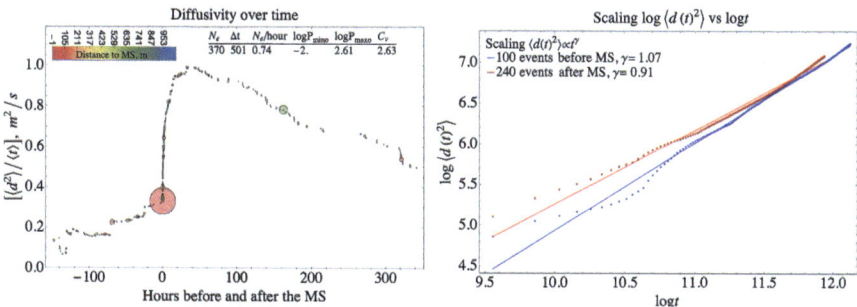

Fig. 3.3 Diffusivity vs. time of events with $\log P \geq -2.0$ before and after the main shock *(left)*. Scaling of $\log \langle d(t)^2 \rangle$ vs. $\log t$ for 100 events before (in blue) and for 240 events after the main shock (in red) is shown on the *right*

Many different experiments though reveal deviations from normal diffusion, in that diffusion is either faster or slower, which is termed an anomalous diffusion. A useful characterisation of the diffusion process is through the scaling of the mean square displacement with time, $\langle d(t)^2 \rangle \sim t^\gamma$, where γ is the scaling index. The case $\gamma = 1$ relates to the normal diffusion, and all other cases are termed anomalous. The cases $\gamma > 1$ form the family of super-diffusive processes, including the particular case $\gamma = 2$ which is called ballistic diffusion. The cases $\gamma < 1$ are the sub-diffusive processes. Plotting $\log \langle d(t)^2 \rangle$ vs. $\log t$ is an experimental way to determine the type of diffusion. Figure 3.3 left shows the diffusivity vs. time during 147.7 hours before and 353.8 after an event with $\log P = 2.61$ shown here as a red circle. In this figure, colour scales with distance to the MS, and the size of the event here represents the radius of the source volume taken as a sphere, $V = P/\Delta\epsilon$, where $\Delta\epsilon$ is the assumed strain change at the source.

There was a steady rate of seismic activity of 0.68 events per hour before and a typical burst and a power law decay of aftershock activity after the main shock. Figure 3.3 right shows $\log \langle d(t)^2 \rangle$ vs. $\log t$ scaling for 100 events before and for the first 240 events after the main shock. It shows a normal diffusion before with $\gamma = 1.07$ and a sub-diffusive process with $\gamma = 0.91$ after the MS.

Mendecki (1997) defined seismic diffusivity as the ratio of the mean distance squared between the reference location, e.g. the main shock or the blast, the point of injection in case of hydrofracturing, and seismic events d_{sr}, or between the consecutive sources of events, $\langle d_{IE}^2 \rangle$, to mean time between these events, $\langle t \rangle$,

$$D_s(\Delta V, \Delta t) = \frac{\langle d^2 \rangle}{\langle t \rangle}. \qquad (3.8)$$

When calculating distances, one can also include half of their source radii.

3.4.4 Seismic Schmidt Number

Similar to the turbulent Schmidt number, which is the ratio of eddy viscosity to eddy diffusivity, Mendecki (1997) defined the seismic Schmidt number as the ratio of kinematic seismic viscosity to seismic diffusivity,

$$Sc_s = \frac{v_s}{D_s} = \frac{4\Delta V \Delta t \langle t \rangle \sum E}{\rho \langle d^2 \rangle \left(\sum P\right)^2}. \tag{3.9}$$

Note that Eq. (3.9) encompasses directly all four independent parameters which describe seismicity, namely: $\langle t \rangle$, $\langle d \rangle$, $\sum P$, and $\sum E$. Looking at Eq. (3.9), the seismic Schmidt number would decrease with an increase in activity rate, with a drop in apparent stress of events and with an increase in the mean distance between the consecutive events that may signify an increase in the spatial correlation range. As mentioned before, the mean distance between the consecutive events in mines may not be the best proxy for growing correlation range. In some cases, we also observe a migration of seismic activity towards the future main shock which pushes the seismic Schmidt number up.

3.4.5 Shape Factor—Sphericity

To take into account the strain localisation that may promote localisation of seismic activity before instability, we can introduce a shape factor of seismicity. Shape factor is an index influenced by the shape of the object but independent of its size. Small or thin objects have a larger surface area compared to the volume, and this gives them a large ratio of surface to volume. As the size of an object increases, without changing shape, this ratio decreases. Wadell (1933) defined sphericity index as the surface area of a sphere of the same volume as the object, A_s, divided by the actual surface area of the object, A, and therefore,

$$A_s^3 = \left(4\pi r^2\right)^3 = 36\pi \left(\frac{4}{3}\pi r^3\right)^2 = 36\pi V^2 \Rightarrow A_s = \left(36\pi V^2\right)^{1/3}$$

$$= \pi^{1/3} (6V)^{2/3}, \Psi = A_s/A = \pi^{1/3} (6V)^{2/3} /A. \tag{3.10}$$

Sphericity measures how closely the shape of an object resembles that of a perfect sphere. Sphericity of the sphere is 1 by definition, and any object which is not a sphere will have sphericity less than 1. The sphericity of a cube is 0.806, tetrahedron 0.671, the common salt is 0.84, crushed coal 0.75, and crushed glass 0.65. To estimate the sphericity, Ψ, of a set of seismic events given by their locations, we calculate the volume and the area of the smallest convex polygon that encloses these events, which is called a convex hull.

3.5 Seismic Stability Analysis

3.5.1 Assumptions and Data Selection

The seismic stability analysis has been developed over the last 30 years. The first formulation was in Mendecki (1993), where Eq. (3.1) was quoted, followed by Mendecki (1997) and numerous applications and modifications, e.g. van Aswegen et al. (1997), van Aswegen (2005), Rebuli and van Aswegen (2013).

The objective of seismic stability analysis is not to predict instability or to manage seismic exposure in the short term, but to guide control measures to mitigate seismic hazard. There is a view that these control measures just delay the inevitable. However, experience shows that scattered and slower rock extraction changes the nature of seismic release by producing more smaller or mid-size events and fewer large ones.

Theoretical considerations, laboratory studies, and seismic observations suggest that an inhomogeneous rock mass subjected to loading shares certain seismic symptoms when approaching instability:

- An overall softening, measured by a decrease in the average value of the apparent stress, σ_A, or energy index, EI.
- Increased rate or accelerating deformation, measured by an increase in the activity rate, $\lambda = 1/\langle t \rangle$. Or, for the same activity rate, by lower apparent stress or larger apparent volume events that would result in increased rates of cumulative potency, $\text{Cum} P$, or cumulative apparent volume, $\text{Cum} V_A$.
- Decrease in the dimensionality, or localisation, of seismic activity, measured by the fractal dimension, correlation dimension, or by the sphericity index of seismic sequences, Ψ.
- Increase in correlation length. In this case, one would expect to observe an increase in the spatial distribution of seismic activity and, in some cases, an increase in the number of mid-size events in different parts of the system. One could also observe an extended spatial distribution of the immediate seismic response to blasting or to mid-size events.

The following ratios depict the expected qualitative changes preceding larger instabilities:

$$\Psi \frac{\text{apparent stress}}{\text{activity rate}} \text{ or } \Psi \frac{\text{apparent stress}}{\text{apparent volume}} = \frac{\searrow}{\nearrow} = \Downarrow$$

$$\Psi \frac{\text{seismic stress}}{\text{seismic strain rate}} \text{ or } \Psi \frac{\text{seismic viscosity}}{\text{seismic diffusivity}} = \frac{\searrow}{\nearrow} = \Downarrow . \quad (3.11)$$

All statistical parameters of co-seismic deformation described above directly or indirectly include ΔV that, in the case of stability analysis, should be the seismogenic volume of rock that generates the next larger event. This is the volume we suppose to select seismic events from to plot a given stability function. However, a seismogenic volume is an expression which is not well defined. If we define it as a volume of rock that includes all interdependent events surrounding the future main shock, then it is a function of space and will be different at different locations. It will also scale positively with the size of the future main shock.

A stability analysis yields plots of a given stability function vs. time for a given site, and having done it for a number of sites, one can contour it on a plane or in space. The values of the stability function are calculated in a moving window. The window can be fixed in time or with a fixed number of events, in which case a variable in time. One way to select events into the moving window is to take what is available in the database. In this case, there may be events far away from the site that may have no influence on stability, or that may obscure the analysis. Alternatively, we may select a polygon based on past experience. Another option is to select events that exceed a given threshold of peak ground velocity, PGV, at the selected site. This influence based selection is more general, because if the threshold is set low it gives the first option and it facilitates automatic plotting.

3.5.2 Stability Example

Data We analysed the last 60 days of seismic history before a $\log P = 2.61$ event at a mine here referred to as MineD, see the seismic hazard case study described in Chapter 5. Figure 3.4 left shows the convex hull span over all 6073 events $\log P \geq -4.5$ available for analysis.

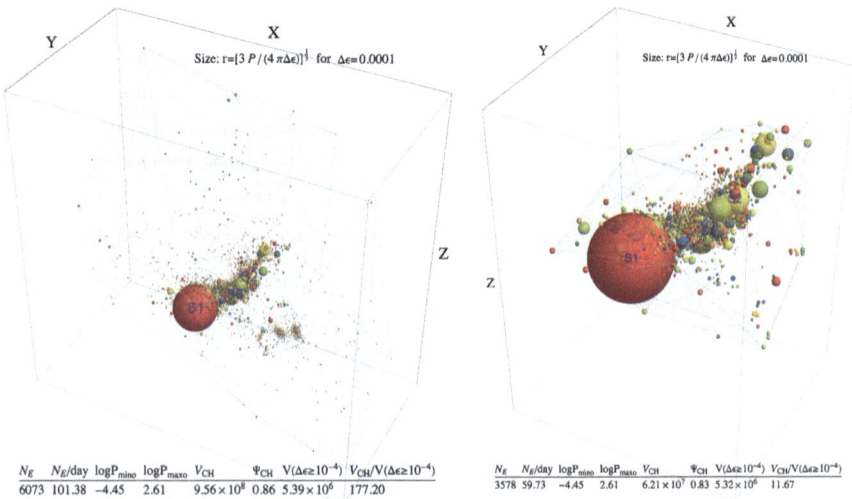

N_E	N_E/day	$\log P_{mino}$	$\log P_{maxo}$	V_{CH}	Ψ_{CH}	$V(\Delta\epsilon \geq 10^{-4})$	$V_{CH}/V(\Delta\epsilon \geq 10^{-4})$
6073	101.38	−4.45	2.61	9.56×10^8	0.86	5.39×10^6	177.20

N_E	N_E/day	$\log P_{mino}$	$\log P_{maxo}$	V_{CH}	Ψ_{CH}	$V(\Delta\epsilon \geq 10^{-4})$	$V_{CH}/V(\Delta\epsilon \geq 10^{-4})$
3578	59.73	−4.45	2.61	6.21×10^7	0.83	5.32×10^6	11.67

Fig. 3.4 Convex hull span over all available 6073 events with sites S1, S2, and S3 *(left)* and over 3578 events that generated $PGV \geq 10^{-7}$ m/s at site S3 *(right)*

3.5 Seismic Stability Analysis

The size of the event scales with the radius of source volume, and the colour indicates the time of the event, from the earliest in blue to the latest in red. Stability was assessed at three sites: at the source of the largest event S1, at the source of the second largest event S2, and at the centre of all 6073 events available for analysis S3. Distances between these sites are: S1 and S2 = 273 m, S1 and S3 = 185 m, and S2 and S3 = 118 m. Figure 3.4 right shows the convex hull span over 3578 events that generated $PGV \geq 10^{-7}$ m/s at the centre of all 6073 events, shown as S3 in blue.

Distances of these events to site S3 range between 12 and 542 m. Tables in these figures give the number of events, N_E, activity rate/day, the log P range, the volume, V_{CH}, and the sphericity index, Ψ_{CH}, of the convex hull, the total volume of seismic sources with strain change = 0.0001, $V(\Delta\epsilon \geq 10^{-4})$, and the ratio of $V_{CH}/V(\Delta\epsilon \geq 10^{-4})$. After the selection of the 3578 events $V_{CH}/V(\Delta\epsilon \geq 10^{-4})$, ratio dropped from 177.2 to 8.85, which indicates that the volume selected for stability is reasonably saturated with co-seismic inelastic deformation. The sphericity index, Ψ_{CH}, here dropped slightly from 0.86 to 0.83, however, while running moving windows through these 3578 events it varied between 0.61 and 0.86.

Figure 3.5 shows the cumulative number of events, CumN, and the cumulative apparent volume, CumV_A in km^3, for all 6073 events. The size of the event represents the radius of the source volume and colour scales with distance to the site S3. There were 17 mid-size events with log $P \geq 0.0$, which gives an activity rate 0.28/day, the coefficient of variation of these 17 events C_v is 0.87, but of all 6073 events $C_v = 1.32$, which indicates a degree of time clustering for the latter.

The CumN vs. time plot is quite steady with almost constant activity rate, while CumV_A shows a bit more structure and a significant increase in rate before the main shock. The largest event with log $P = 2.61$ occurred on 07 July and located 185 m from site S3, the second largest with log $P = 1.24$ on 14 June located 118 m away, and the third largest with log $P = 1.06$ on 08 June 153 m away. The distance between the largest event and the second largest is 273 m, the largest and the third largest 85 m, and the second and the third largest 211 m.

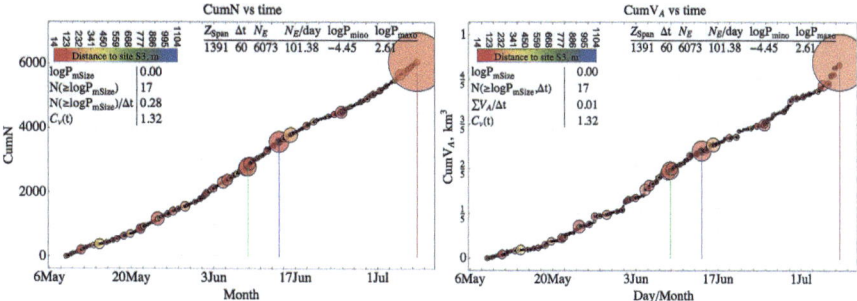

Fig. 3.5 CumN (*left*) and CumV_A (*right*) vs. time plotted for all 6073 events

Results We tested a few different seismic stability functions defined by the combination of statistical parameters of co-seismic deformation described above, and all of them dropped to a low level before the $\log P = 2.61$ main shock. Some of them also dropped before a few mid-size events preceding the main shock. All plots below show the moving window-based qualitative history of a given stability function in black and, as a reference, the CumV_A of the selected 3578 events marked by their sizes and colour indicating distances to site S3. The vertical axes are for CumV_A in km^3. We use a moving window of 119 events, which is 2 days of seismic activity of the selected events. All elements of the stability functions were normalised between 1 and 10.

Note that the CumV_A vs. time plot of the 3578 events does not show the same change in rate before the main shock as that shown by the 6073 events. This increase is caused by three little clusters to the right of the main shock (see Fig. 3.4 left), and these events were too small and/or too far to generate $PGV \geq 10^{-7}$ at Site 3 and were automatically excluded.

Figure 3.6 shows two seismic stability functions derived for site S3. The vertical red, blue, and green lines indicate times of the three largest events respectively. (1) $\Psi_{CH} \cdot \text{Median}[\log \sigma_A]/\lambda$, where σ_A apparent stress, λ the activity rate, and Ψ_{CH} sphericity of the convex hull span over 119 events in each moving window. This function dropped to a low level before all three largest events. (2) $\Psi_{CH} \cdot \text{Median}[\log \sigma_A]/(\lambda \cdot \sum u_{CAD})$, where $\sum u_{CAD}$ is the sum of the cumulative absolute displacements, CAD, in a given time window. CAD is derived from the GMPE for a given mine or area. This function also dropped to a low level before all three largest events.

Figure below shows the main components of these two stability functions.

Figure 3.7 left shows the time history of the Median$[\log \sigma_A]$ that drops convincingly, as desired in terms of the stability criteria described above, before the main shock, less so before the second and third largest events. Figure 3.7 right shows the time history of the activity rate, λ, that is at high level before all three largest events and corrects the undesired increases in the Median$[\log \sigma_A]$.

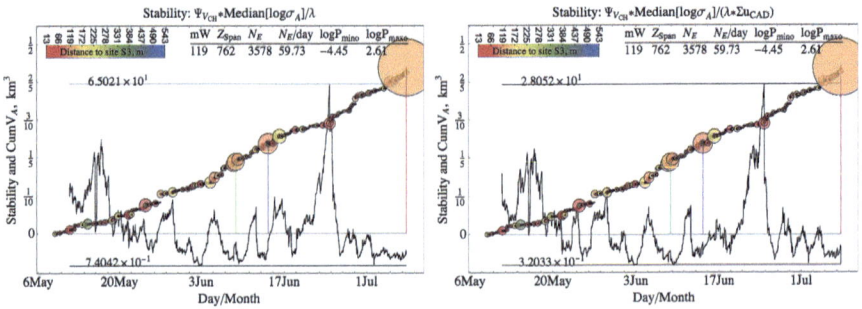

Fig. 3.6 Time history of the first two seismic stability functions, in black, with CumV_A of the selected data set

3.5 Seismic Stability Analysis

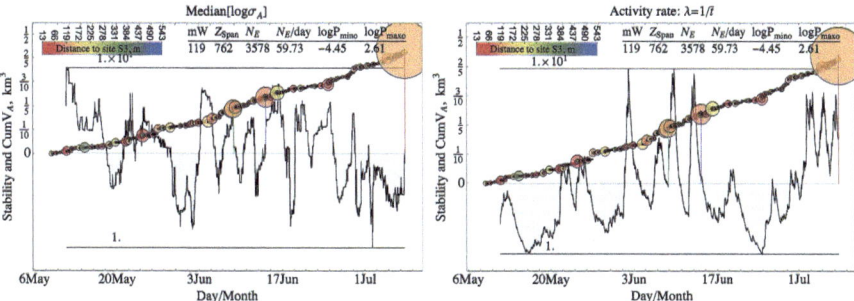

Fig. 3.7 Time history of Median[log σ_A] *(left)*, and the activity rate, λ, *(right)*

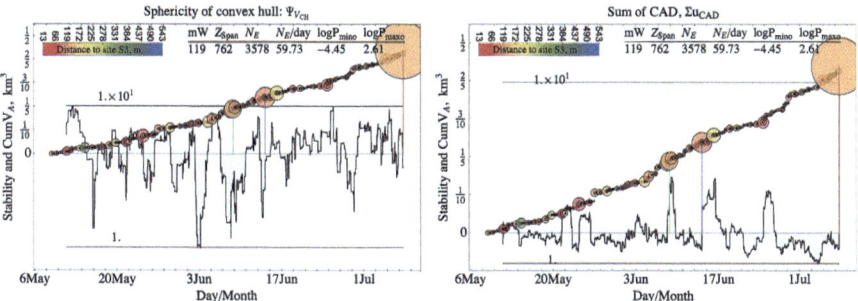

Fig. 3.8 Time history of sphericity, Ψ *(left)*, and the sum of the cumulative absolute displacements, $\sum u_{CAD}$ *(right)*

Figure 3.8 left shows the time history of the sphericity of the convex hull, Ψ_{CH}, that is low before the second and third largest events, not so before the largest one. Figure 3.8 right shows the time history of the $\sum u_{CAD}$ that is undesirably low before the largest event but complies with the assumptions before the second and third ones.

Figure 3.9 left shows a Schmidt number based stability function for site S3, Ψ_{CH}. log Sc_s (V_{CH}), defined by Eq. (3.9) with ΔV taken as the volume of convex hull, V_{CH}. This function indicates low stability before all three largest events. Figure 3.9 right shows seismic diffusivity that is increasing, as assumed in Eq. (3.8), before the largest and the third largest events, no so before the second largest though.

Comparison Between Sites The stability analysis described above was applied to site S3. Figure 3.10 below shows the same stability functions applied to sites S1, S2 and again for S3, for completeness.

Each site has a different set of events that generated $PGV \geq 10^{-7}$, with different distances to a given site. They also have different activity rates: 31.9 events/day at sites S1, 43.72 at site S2, and 59.73 at site S3, and therefore, they have a different length of the moving window.

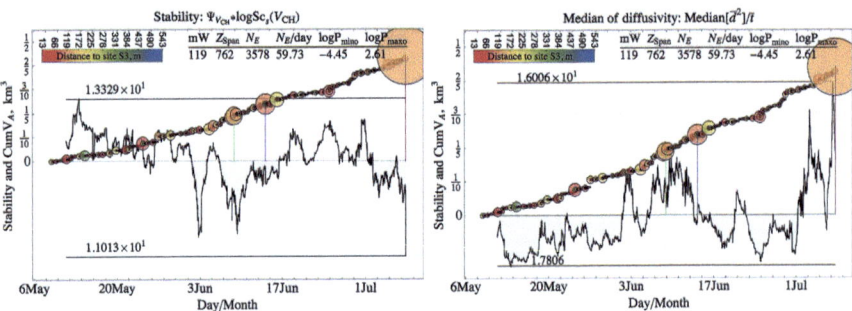

Fig. 3.9 Time history of the Schmidt number *(left)* and diffusivity *(right)*

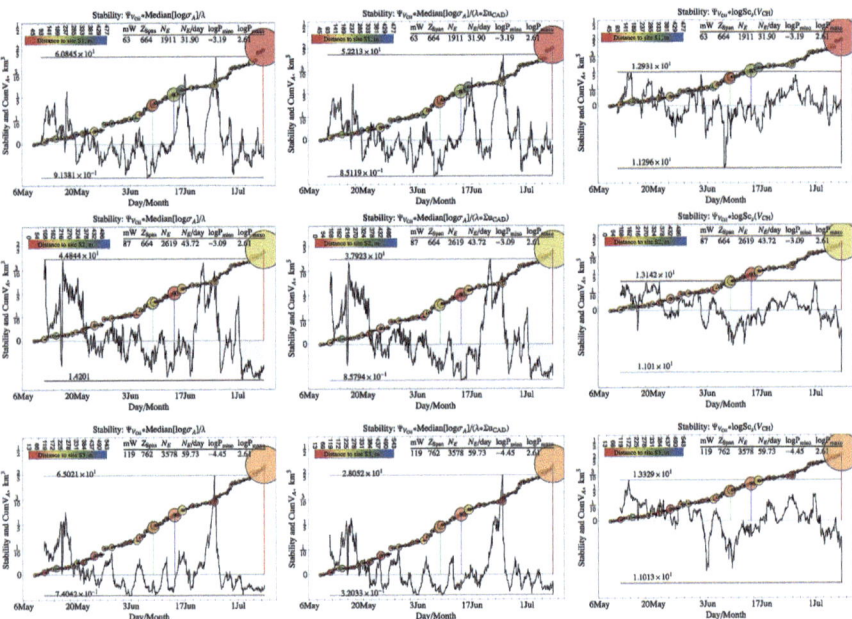

Fig. 3.10 Time histories of the same three stability functions for site S1 *(top row)*, S2 *(middle)*, and S3 *(bottom)*

In this case study, there are no significant differences in qualitative behaviour of stability functions at these three sites, and they are all at a reasonably low level before the three largest events. The main reason is that all three sites are located relatively close to each other and also close to the main volume of seismic activity. In most cases studied to date stability function are reasonably similar for sites at distances less than 200. The notable exception would be if there is a flurry of small events close to one site that does not exceed the assumed PGV threshold at the other site.

Somewhat surprisingly, in this case, sites S2 and S3 behave reasonably similar and gave better indication of instability before the largest event than site S1 that was deliberately located at site S1. One possible explanation is that site S1 attracted the lowest number of events. In general, the described stability analysis, or any other seismic alerts, performs better with higher activity rate.

3.5.3 Stability—General Comments

When analysing seismic stability, one should take into account the following limitations:

1. Seismic stability analysis is based on theoretical considerations, laboratory experiments, and case studies. Therefore, it will always be influenced by local factors associated with a particular way of mining and by geological conditions.
2. The utility of seismic stability analysis is in guiding control measures to mitigate seismic hazard. If the value of a given stability function is systematically dropping or stays low in a given area, then the mine may change the spatial and temporal manner of rock extraction, e.g. scatter the production blasts and/or slow down the rate of mining. Alternatively, they may try to trigger the potential seismic event, while people are not in the area.
3. Events induced by production blasts may not be preceded by instability indicators. On the upside, people are not in the area during blasting. Production blasting with associated increase in seismic activity and strain softening will push the stability function lower. In a reasonably stable system, it should, however, recover relatively quickly. A simple stability function based on median apparent stress and seismic activity could be useful to monitor for safe re-entry.
4. At any particular mine, tests should be carried out to select the most suitable stability function or functions, and the reference strain needs to be calibrated for a given GMPE at different sites.
5. In some mining scenarios, we observe the migration of seismic activity towards the source of the future main shock. In these cases, the distances between consecutive events, $\langle d^2 \rangle$, may decrease before the main shock. If this is the expected outcome, one constructs the appropriate stability function.
6. The seismic Schmidt number has four independent parameters which describe seismicity and, in theory, should be the most suited for stability analysis. In simpler mining scenarios, it is the case, however, in more complex environment, one or two parameters fail to comply with our assumptions and it may give inferior results.

References

Atiyah, M. (1988). *Collected papers*, Oxford.
Bazant, Z. P., & Cedolin, L. (1991). *Stability of structures: Elastic, inelastic, fracture, and damage theories* (1st edn.). Oxford University Press.
Brady, B. (1977a). An investigation of the scale invariant properties of failure. *International Journal of Rock Mechanics & Mining Sciences, 14*, 121–126.
Brady, B. T. (1977b). Anomalous seismicity prior to rock bursts: implications for earthquake prediction. In M. Wyss (Ed.), *Stress in the Earth* (vol. 115, pp. 357–374). Pure Applied Geophysics.
Brune, J. N. (1968). Seismic moment, seismicity and rate of slip along major fault zones. *Journal of Geophysical Research, 73*(2), 777–784.
Cox, S. J. D., & Paterson, L. (1990). Damage development during rupture of heterogeneous brittle materials: a numerical study. In R. J. Knipe, & E. H. Rutter (Eds.), *Deformation mechanisms, rheology and tectonics* (vol. 54, pp. 57–62). Geological Society, Special Publication.
Crutchfield, J. P., & Young, K. (1989). Inferring statistical complexity. *Physical Review Letters, 63*, 105–108.
Cundall, P. A. (1990). Numerical modelling of jointed and faulted rock. In H. P. Rossmanith (Ed.), *Mechanics of jointed and faulted rock* (pp. 11–18). Balkema, Rotterdam.
Darot, M., & Gueguen, Y. (1986). Slow crack growth in minerals and rocks: Theory and experiments. *Pure and Applied Geophysics, 124*, 677–692.
Dieterich, J. (1994). A constitutive law for rate of earthquake production and its application to earthquake clustering. *Journal of Geophysical Research, 99*(B2), 2601–2618.
Dieterich, J. H. (1992). Earthquake nucleation on faults with rate- and state-dependent strength. *Tectonophysics, 211*(1-4), 115–134.
Eddington, A. S. (1928). *The nature of the physical world*. University Press, Cambridge.
Forsee, A. (1967). *Albert Einstein: Theoretical physicist*. The Macmillan Company.
Fredrich, J., Evans, B., & Wong, T. F. (1989). Micromechanics of the brittle transition in Carrara marble. *Journal of Geophysical Research, 94*, 4129–4145.
Hadamard, J. S. (1903). *Lecons sur la Propagation des Ondes et les Equations de l'Hydrodynamique - Chapitre 6* (241–262 pp.). Librairie Scientifique A. Hermann, Paris.
Handley, M. F., de Lange, J. A. J., Essrich, F., & Banning, J. A. (2000). A review of the sequential grid mining method employed at Elandsrand Gold Mine. *The Journal of The Southern African Institute of Mining and Metallurgy, 100*(3), 157–168.
Hill, R. (1958). A general theory of uniqueness and stability in elastic-plastic solids. *Journal of the Mechanics and Physics of Solids, 6*, 236–249.
Hirata, T., Satoh, T., & Ito, K. (1987). Fractal structure of spatial distribution of microfracturing in rock. *Geophysical Journal of the Royal Astronomical Society, 90*(2), 369–374.
Hobbs, B. E., Mulhaus, H.-B., & Ord, A. (1990). Instability, softening and localization of deformation. In R. J. Knipe, & E. H. Rutter (Eds.), *Deformation mechanisms, rheology and tectonics* (vol. 54, pp. 143–165). Geological Society, Special Publication.
Jessell, M. W., & Lister, G. S. (1991). Strain localisation behaviour in experimental shear zones. *Pure and Applied Geophysics, 137*(4), 421–438.
Kaneko, K., & Tsuda, I. (2001). *Complex systems: Chaos and beyond - A constructive approach with applications in life sciences* (273 p.). Springer.
Kato, N., Yamamoto, K., Yamamoto, H., & Hirasawa, T. (1992). Strain rate effect on frictional strength and the slip nucleation process. *Tectonophysics, 211*, 269–282.
Keilis-Borok, V. I. (1994). Symptoms of instability in a system of earthquake-prone faults. *Physica D, 77*, 193–199.
Kostrov, B. V. (1974). Seismic moment and energy of earthquakes and seismic flow of rock. *Izvestiya, Physics of the Solid Earth, 13*, 13–21.
Kostrov, B. V., & Das, S. (1988). *Principles of earthquake source mechanics*. Cambridge University Press.

References

Labuz, J. F., Shah, S. P., & Dowding, C. H. (1985). Experimental analysis of crack propagation in granite. *International Journal of Rock Mechanics and Mining Sciences, 22*(2), 85–98.

Main, I. G., Meredith, P. G., & Sammonds, P. (1992). Temporal variations in seismic event rate and b-values from stress corrosion constitutive laws. *Tectonophysics, 211*(1-4), 233–246.

Mendecki, A. J. (1993). Real time quantitative seismology in mines: Keynote Address. In R. P. Young (Ed.), *Proceedings 3rd International Symposium on Rockbursts and Seismicity in Mines, Kingston, Ontario, Canada* (pp. 287–295). Balkema, Rotterdam.

Mendecki, A. J. (1997). Quantitative seismology and rock mass stability. In A. J. Mendecki (Ed.), *Seismic monitoring in mines* (1st edn., chap. 10, pp. 178–219). Chapman and Hall.

Mendecki, A. J. (2001). Data-driven understanding of seismic rock mass response to mining: Keynote Address. In G. van Aswegen, R. J. Durrheim, & W. D. Ortlepp (Eds.), *Proceedings 5th International Symposium on Rockbursts and Seismicity in Mines, Johannesburg, South Africa* (pp. 1–9). South African Institute of Mining and Metallurgy.

Mendecki, A. J. (2005). Persistence of seismic rock mass response to mining. In Y. Potvin, & M. R. Hudyma (Eds.), *Proceedings 6th International Symposium on Rockburst and Seismicity in Mines, Perth, Australia* (pp. 97–105). Australian Centre for Geomechanics.

Muhlhaus, H. B., Hobbs, B. E., & Ord, A. (1992). Evolution of fractal geometries in deforming material. In J. R. Tillerson, & W. R. Wawersik (Eds.), *Rock Mechanics* (pp. 681–690). Balkema, Rotterdam.

Nicolis, G., & Prigogine, I. (1977). *Self-organization in nonequilibrium systems. From dissipative structures to order through fluctuations* (491 p.). John Wiley and Sons.

Nicolis, G., & Prigogine, I. (1989). *Exploring complexity.* W. H. Freeman and Company.

Ohnaka, M. (1992). Earthquake source nucleation: A physical model for short-term precursors. *Tectonophysics, 211*(1-4), 149–178.

Olson, W. A. (1992). The formation of the yield surface vertex in rock. In H. R. Tillerson, & W. A. Wawersik (Eds.) *Rock mechanics* (pp. 701–705). Balkema, Rotterdam.

Pearson, G. E. (1956). General theory of elastic stability. *Quarterly of Applied Mathematics, 14*(2), 133–144.

Prigogine, I. (1980). *From eeing to becoming. Time and complexity in the physical sciences.* W. H. Freeman and Company.

Prigogine, I. (1996). Time, chaos and the laws of Nature. In P. Weingartner, & G. Schurz (Eds.), *Law and prediction in the light of chaos research, Lecture Notes in Physics* (vol. 473, pp. 3–9). Springer.

Prigogine, I. (1997). *The end of certainty: Time, chaos and the laws of nature.* The Free Press.

Rebuli, D., & van Aswegen, G. (2013). Sort term seismic hazard assessment in s.a. gold mines. In A. Malovichko, & D. A. Malovichko (Eds.), *Proceedings 8th International Symposium on Rockbursts and Seismicity in Mines, St Petersburg-Moscow, Russia.* ISBN:978-5-903258-28-4.

Reiner, M. (1964). The Deborah number. *Physics Today,* 62.

Rudnicki, J., & Rice, J. (1975). Conditions for the localization of deformation in pressure sensitive dilatant materials. *Journal of the Mechanics and Physics of Solids, 23,* 371–394.

Rudnicki, J. W. (1977). The inception of faulting in a rock mass with a weakened zone. *Journal of Geophysical Research, 82,* 844–854.

Rudnicki, J. W. (1988). Physical models of earthquake instability and precursory processes. *Pure and Applied Geophysics, 126*(2-4), 531–554.

Scholz, C. H. (1990). *The mechanics of earthquakes and faulting.* Cambridge University Press.

Stuart, W. D. (1981). Stiffness method for anticipating earthquakes. *Bulletin of the Seismological Society of America, 71*(1), 363–370.

Truesdell, C., & Noll, W. (2004). *The non-linear field theories of mechanics* (3rd edn., 602 p.). Springer-Verlag.

van Aswegen, G. (2005). Routine seismic hazard assessment in some South African mines. In Y. Potvin, & M. Hudyma (Eds.), *Proceedings 6th International Symposium on Rockbursts and Seismicity in Mines, Perth, Australia.*

van Aswegen, G., & Butler, A. G. (1993). Applications of quantitative seismology in South African gold mines. In R. P. Young (Ed.), *Proceedings 3rd International Symposium on Rockbursts and Seismicity in Mines, Kingston, Ontario, Canada* (pp. 261–266). Balkema, Rotterdam. ISBN: 90-5410320-5.

van Aswegen, G., & Mendecki, A. J. (1999). Mine layout, geological features and seismic hazard. *Final report gap 303*, Safety in Mines Research Advisory Committee, South Africa (pp. 1–91).

van Aswegen, G., Mendecki, A. J., & Funk, C. (1997). Application of quantitative seismology in mines. In A. J. Mendecki (Ed.), *Seismic monitoring in mines* (1st edn., chap. 11, pp. 220–245). Chapman and Hall.

Wadell, H. (1933). Sphericity and roundness of rock particles. *The Journal of Geology, 41*(3), 310–331.

Wallerstein, I. (1999). *The end of the world as we know it: social science for the twenty first century.* University of Minnesota Press.

Wawersik, W. R. J. W. R., Olsson, W. A., Holcomb, D. J., & Chau, K. T. (1990). Localization of deformation in brittle rock: Theoretical and laboratory investigations. In S. P. Shah (Ed.), *Proceedings of the International Conference on Micromechanics and Failure.* Elsevier.

Wiener, P. P. (1951). *Leibniz selections.* Charles Scribner's Sons.

Zoller, G., Hainz, S., & Kurths, J. (2001). Observation of growing correlation length as an indicator for critical point behavior prior to large earthquakes. *Journal of Geophysical Research, 106*(B2), 2167–2175.

Open Access This chapter is licensed under the terms of the Creative Commons Attribution 4.0 International License (http://creativecommons.org/licenses/by/4.0/), which permits use, sharing, adaptation, distribution and reproduction in any medium or format, as long as you give appropriate credit to the original author(s) and the source, provide a link to the Creative Commons license and indicate if changes were made.

The images or other third party material in this chapter are included in the chapter's Creative Commons license, unless indicated otherwise in a credit line to the material. If material is not included in the chapter's Creative Commons license and your intended use is not permitted by statutory regulation or exceeds the permitted use, you will need to obtain permission directly from the copyright holder.

Chapter 4
Size Distribution and Seismic Hazard

Abstract This chapter starts with a general description of extreme events characterised by heavy tailed distributions with a few very large values compared to the other values of the data set. Typical examples of power law scaling are the Gutenberg and Richter magnitude frequency relation and the Omori law for aftershock decay. However, both are open-ended distributions and therefore have no characteristic scale, i.e. there is no upper limit on magnitude in the former and there is no end to the duration of aftershock sequences in the latter. Therefore, a need arises for a more physical description, by introducing either a soft transition to finite energy release or a sharp cut-off by a double-truncated distribution. Another important issue described in this chapter is the so -called m_{max} or log P_{max}, which in earthquake seismology is the maximum magnitude, or potency, earthquake that a given seismogenic region can deliver. However, in mines the maximum possible size event scales with the footprint of the mine and with the degradation of rock mass stiffness, both increasing as mining progresses creating conditions conducive for ever larger events to occur. Therefore, the maximum size event associated with mining is not an ultimate number but needs Mendecki (March 11, 2025)—Elements of Seismic Hazard in Mines, Springer Preface 2 to be estimated periodically. It is the next record breaking seismic potency, energy, or magnitude which can be estimated and that needs to be managed.

4.1 Fat Tails, Power Laws, Extreme, and Unexpected Events

Extreme events can be considered as large deviations from the average behaviour in an evolving system. They are governed by the thickness of the tail of a probability distribution that defines the occurrence of events of a given size, i.e. their probability increases with the thickness of the tail of the underlying size distribution. One characteristic of heavy tailed distributions is that there are usually a few very large values compared to the other values of the data set. The extreme events are actually a part of the nonlinear dynamics of many complex systems.

The open-ended power law, $N (\geq P) = \alpha P^{-\beta}$, with $\beta > 0$, with the probability density function $f(P, \beta) = \beta P_{min}^{\beta} P^{-\beta-1}$, see Eq. (4.3), is an example of a fat tailed distribution that for large potency P falls off more slowly than an exponential and much more slowly than a Gaussian, which is thin tail. Sooner or later an event at the end of a thick tail is going to happen, and it is going to be surprisingly big. These extreme events, defying the normal probability, are also called Black Swan events (Taleb, 2007). While unexpected at first, they tend to be rationalised in hindsight creating the impression that they can be predicted or avoided.

Laherrere and Sornette (1999) and Sornette (2009) identified a number of data sets showing power laws with outliers that they claim are the result of positive feedback mechanisms. They call these events Dragon Kings. They document their presence in six different examples: distribution of city sizes, distribution of acoustic emissions associated with material failure, distribution of velocity increments in hydrodynamic turbulence, distribution of financial drawdowns, distribution of the energies of epileptic seizures, and distribution of earthquake energies.

The main aspects of the heavy tail distribution are that the historical averages are unreliable for prediction, differences between successively larger observations increase, and the ratio of successive record values does not decrease. An important lesson to be learned is that risk management strategies and planning based on a normal distribution can lead to serious under preparation if in fact the problem at hand follows a power law distribution—the risk is in the tail of the distribution. The thicker the tail the higher the probability to be surprised, so we should expect the unexpected.

If we order a data set from a power law distribution, then the ratio between two consecutive observations also has a power law distribution. The sum of power law distributions, $f(P, \beta_1)$ and $f(P, \beta_2)$, is a fat tail distribution very close to the power law with the exponent $\min(\beta_1, \beta_2)$, but only for large values of P, see Fig. 4.1 left.

The power law has the property of scale invariance, i.e. the relative change of $N(\geq kP)/N(\geq P) = k^{-\beta}$ is independent of P, and therefore it lacks characteristic scale. The scale transformation changes all lengths by the same factor. The question is are the laws of nature invariant with respect to scales? This was already answered in the negative by Galileo in (1638),[1] see also a very good discussion on the subject in Weingartner (1996) who stated that since atoms cannot be enlarged or reduced therefore laws in which basic physical constants play a role are not scale invariant. Larger systems do not contain larger atoms but just more atoms. The earthquake size distribution scaling can only exist within a certain range of sizes, then either the exponent of the distribution or the nature of the distribution needs to change to secure a finite energy or potency release.

---

[1] From what has already been demonstrated, you can plainly see the impossibility of increasing the size of structures to vast dimensions either in art or in nature; likewise the impossibility of building ships, palaces, or temples of enormous size in such a way that their oars, yards, beams, iron bolts, and, in short, all their other parts will hold together; nor can nature produce trees of extraordinary size because the branches would break down under their own weight; so also it would be impossible to build up the bony structures of men, horses, or other animals so as to hold together and perform their normal functions if these animals were to be increased enormously in height; for this increase in height can be accomplished only by employing a material which is harder and stronger than usual, or by enlarging the size of the bones, thus changing their shape until the form and appearance of the animals suggest a monstrosity.

4.2 Open-Ended Power Law (OE)

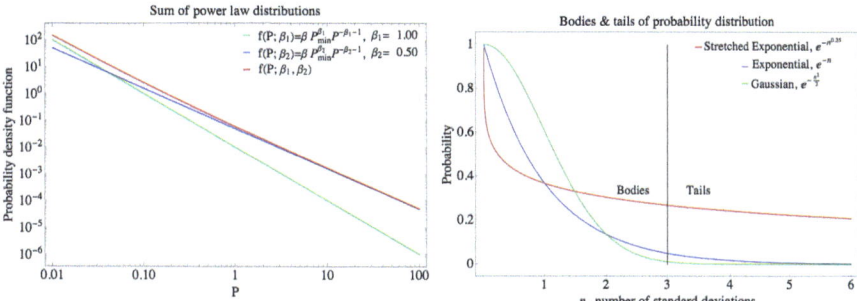

Fig. 4.1 Sum of power law distributions *(left)* and bodies and tails of selected thin- and thick-tailed probability distributions *(right)*

Typical examples of power law scaling are the Gutenberg and Richter magnitude frequency relation and the Omori law for aftershock decay. Both lack a characteristic scale, i.e. there is no upper limit on magnitude in the former and there is no end to the duration of aftershock sequences in the latter.

The distance from the expected value of a given probability distribution is defined by ns_d, where s_d is the standard deviation, see Fig. 4.1 right as an example. The probability of having an unexpectedly large event in a system governed by the Gaussian distribution of sizes is $\Pr(ns_d)_{Gauss} \sim \exp\left[-(ns_d)^2 / (2s_d^2)\right] = \exp(-n^2/2)$. The probability of having an unexpectedly large event in a system governed by the exponential distribution of sizes is $\Pr(ns_d)_{Exp} \sim \exp(-|ns_d|/s_d) = \exp(-n)$, and by the stretched exponential distribution, $\Pr(ns_d)_{SE} \sim \exp\left[-(|ns_d|/s_d)^q\right] = \exp(-n^q)$, where $q < 1$. The ratio $\Pr(ns_d)_{Exp} / \Pr(ns_d)_{Gauss} = \exp(-n + n^2/2)$ and the ratio $\Pr(ns_d)_{SE} / \Pr(ns_d)_{Gauss} = \exp\left[(n^2 - 2n^q)/2\right]$.

Thus, the probability of a $6s_d$ or 6-sigma event in a system described by the exponential distribution is 10^5 times more likely than in a system described by the Gaussian distribution. The tail of the stretched exponential distribution is fatter than the exponential one, and, for $q = 1/2$, it gives a 35 times higher probability of a $6s_d$ event than the exponential. The power law tail is yet fatter than the stretched exponential.

4.2 Open-Ended Power Law (OE)

In a hard rock mass with random heterogeneities and with some regular geological structures undergoing loading, there are patches of rock resisting deformation where stresses are increasing, patches of diffusion where stresses are decreasing, and there are some passive volumes not influenced by loading. Locally, stress build-up and/or strength degradation may lead to fast relaxation via deformation jumps, or seismic events, that radiate and dissipate energy across the system. Over time, the size distribution of these events will, within a certain range, follow the power law, $N (\geq P) = \alpha P^{-\beta}$, where $N (\geq P)$ is the number of events not smaller than P, where P is seismic potency. The same relation applies to seismic moment, M, and the radiated energy E. The parameter α measures the level of seismic activity, and β is the exponent.

Remarkably, the distribution of sizes of earthquakes for Southern California is observed to be a power law with a constant exponent over more than six orders of magnitude (e.g. Christensen et al., 2002). This scaling is the property of regional dynamics rather than individual faults where it is more complex (e.g. Wesnousky et al., 1983; Main & Burton, 1984; Kijko & Stankiewicz, 1987; Wesnousky, 1994; Wiemer & Wyss, 2002). However, some deviations from the power law may lie within the 95% confidence limits of the Poisson process and therefore may not be significant enough to be regarded as "characteristic" or "bi-modal" (Jackson & Kagan, 2006; Kagan et al., 2012).

In mines, the power law scaling has been observed for seismic events as large as $m = 5.0$ (Mendecki et al., 1988) and recently for small fractures with $m = -4.0$ to -0.3 (Kwiatek et al., 2010), although with varying exponents.

The open-ended (OE) potency size distribution power law can be written as

$$N (\geq P) = \alpha P^{-\beta}, \qquad (4.1)$$

where P is potency (or energy, or seismic moment), α estimates the number of seismic events with potency not smaller than one, $\alpha = N(P \geq 1)$, and $\beta > 0$ is the exponent—the lower the β the heavier the tail of the distribution. Taking the logarithm of equation (4.1) gives $\log N (\geq P) = \log \alpha - \beta \log P$, which is the well-known (Gutenberg & Richter, 1944) relation

$$\log N (\geq m) = a - bm, \qquad (4.2)$$

where $a = \log \alpha$, $b = \beta$, and magnitude $m = \log P$. The OE, or the Gutenberg-Richter, relation has no upper limit on the event size. The assumption that solving $\alpha P_{max1}^{-\beta} = 1$ gives the so-called the one largest possible event or the next largest event, $P_{max1} = \alpha^{1/\beta}$ or $m_{max1} = a/b$, is incorrect. Note that the probability of having an event greater than or equal to P_{max1} in the OE distribution is small, but finite, and it increases as β decreases, $\Pr (\geq P_{max1}) = P_{min}^{\beta}/\alpha$.

The cumulative distribution function, $F(P)$, i.e. the probability of having an event with a potency smaller than P, is $\Pr (\leq P) = N(\leq P)/N (\geq P_{min}) = 1 - P_{min}^{\beta} P^{-\beta}$, where P_{min} is the minimum observed potency that fits the power law. The survival function $S(P) = \Pr (\geq P) = 1 - \Pr (\leq P) = P_{min}^{\beta} P^{-\beta}$. Parameter P_{min} is limited by the sensitivity of the monitoring system. One can expect that an increase in system sensitivity by ΔP will increase the number of recorded events by $N (\geq P_{min} - \Delta P)/N (\geq P_{min}) = P_{min}^{\beta} (P_{min} - \Delta P)^{-\beta}$. For an order of magnitude drop in P_{min}, i.e. $P_{min} - \Delta P = P_{min}/10$, one can expect to record $(1/10)^{-\beta}$ more events, which for $\beta = 1$ gives a ten-fold increase. The probability density function for the OE relation is

4.2 Open-Ended Power Law (OE)

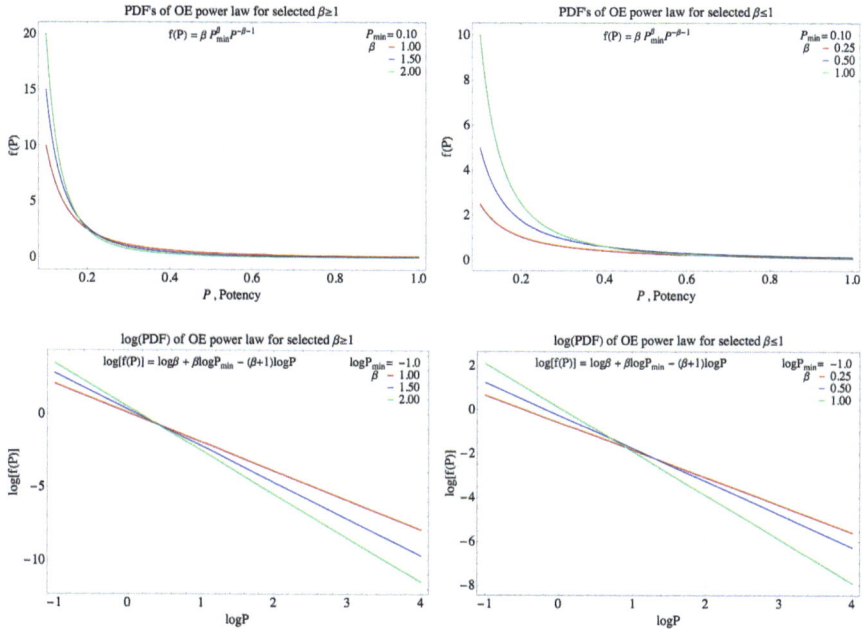

Fig. 4.2 Illustration of the PDF of the OE power law for different values of β

$$f(P) = dF(P)/dP = \beta P_{min}^{\beta} P^{-\beta-1} = \left(\frac{\beta}{P}\right)\left(\frac{P_{min}}{P}\right)^{\beta}. \quad (4.3)$$

Figure 4.2 illustrates the behaviour of probability density function of the open-ended power law for different values of β.

The number of events within the potency range $P_1 < P_2$ is $N(P_1, P_2) = \alpha \left(P_1^{-\beta} - P_2^{-\beta}\right)$, and the probability of having this event is $\Pr(P_1, P_2) = F(P_2) - F(P_1) = P_{min}^{\beta}\left(P_1^{-\beta} - P_2^{-\beta}\right)$. The mean value of the distribution $\langle P \rangle = \int_{P_{min}}^{\infty} P f(P) dP / \int_{P_{min}}^{\infty} f(P) dP = \beta P_{min}/(\beta - 1)$, which is finite for $\beta > 1$. Note that the mean of any finite sample is always finite, but as we draw more samples from the power law distributed population, this sample average tends to increase. One might mistakenly conclude that there is a time trend in such data.

4.2.1 Selection of P_{min}

P_{min}, or $\log P_{min}$, can be interpreted technically as the minimum observed potency that fits the data. A correctly selected P_{min} should also define the completeness of the catalogue, i.e. the lowest potency at which all events within a volume and time of interest, $(\Delta V, \Delta t)$, are detected. While different methods for an objective and/or automatic threshold detection have been proposed, in practice the most reliable is the visual inspection of the potency frequency data. The potency frequency can be presented as a cumulative plot or an incremental plot. In the cumulative plot, the number of events increases or stays the same from high to low potencies. The incremental plot shows the number of events in each potency bin, ΔP, including the empty ones.

It is advisable to inspect both the cumulative and the incremental plot, since each value in a cumulative plot depends on all the preceding values and it may hide fine details. It is important not to underestimate $\log P_{min}$ since it may lead to underestimation of β and thus overstating hazard. The properly determined P_{min} in a well-behaved data set is the minimum potency above which the estimates of β are stable. Data below the P_{min} threshold should not be used in fitting the parameters of the power law distribution. However, in some applications, e.g. short term hazard, dynamic triggering detection, and stability analysis, data below P_{min} is very useful.

4.2.2 Estimation of α and β of the OE Relation

The β-value in the OE potency frequency relation $N (\geq P) = \alpha P^{-\beta}$ can be estimated by maximising the likelihood function (Utsu, 1964; Aki, 1965),

$$L(P_j; \beta) = \prod_{j=1}^{n} f(P_j) = \prod_{j=1}^{n} \beta P_{min}^{\beta} P_j^{-\beta-1}, \qquad (4.4)$$

where $f(P)$ is the probability density function, n is the number of events above P_{min}, and P_{min} is the potency selected to fit the distribution. The values of the probability density function $f(P_j)$ are very small, and when multiplied together the result may be too small to be accurately represented by a computer. This is why it is advisable to maximise the log likelihood function, $\log L(P_j; \beta) = \sum_{j=1}^{n} \log f(P_j)$, where multiplications are replaced by summation.

The $\log L(P_j; \beta)$ is monotonic giving the same maximum as the likelihood function $L(P_j; \beta)$, and it is also easier to differentiate to find its maximum. The point at which the $\log L(P_j; \beta)$ is maximum with respect to β is the solution of the equation $\partial \log L(P_j; \beta)/\partial \beta = n/[\beta \ln(10)] + \sum_{j=1}^{n} \log P_{min} - \sum_{j=1}^{n} \log P_j = 0$, which gives $(n \log e)/\beta + n \log P_{min} - \sum_{j=1}^{n} \log P_j = 0$, and after simple algebra, the estimate of β of the open-ended relation is

$$\hat{\beta}_{OE} = \log(e) / \left(\overline{P} - \log P_{min}\right), \qquad (4.5)$$

where $\log(e) = 0.4343$ and $\overline{P} = (1/n) \sum_{j=1}^{n} \left(\log P_j\right)$. It follows that studying β, or the b-value, is effectively equivalent to studying the mean $\log P$ or mean

4.3 Open-Ended Tapered Power Law (OET)

magnitude. To correct for bias, one can multiply the $\hat{\beta}_{OE}$ by $(n-1)/n$, which makes a difference only for small data sets.

For a sufficiently large data set, the standard deviation can be obtained from the second derivative of the log likelihood, $s_d(\beta) = \left[-\partial^2 \log L(\beta)/\partial \beta^2\right]^{-1/2} = \hat{\beta}_{OE}/\sqrt{n \log(e)}$ (Aki, 1965). Shi and Bolt (1982) gave a useful expression for the standard error of β when it varies slowly in time, which is very much the case in mines, and for large n,

$$s_d(\beta_{OE}) = \frac{\hat{\beta}_{OE}^2}{\log(e)} \sqrt{\frac{\sum_{j=1}^{n}\left(\log P_j - \overline{P}\right)^2}{n(n-1)}}. \tag{4.6}$$

The parameter $\hat{\alpha}_{OE}$ can be estimated from $\log \hat{\alpha}_{OE} = \hat{\beta}_{OE} \overline{P} + (1/n) \sum_{j=1}^{n} \log N (\geq P_j)$.

4.3 Open-Ended Tapered Power Law (OET)

Kagan (1993) suggested the OE relation that tapers to zero by multiplying the survival function of the OE power law by the exponential function, as suggested by Vere-Jones et al. (2001),

$$\Pr(\leq P) = F(P) = 1 - P_{min}^{\beta} P^{-\beta} \exp\left(\frac{P_{min} - P}{P_c}\right), \tag{4.7}$$

where P_c is the soft upper cut-off. The survival function, $S(P) = 1 - F(P)$, for different exponents β is illustrated in Fig. 4.3 left, and the probability density function is

$$f(P) = \left(\frac{\beta}{P} + \frac{1}{P_c}\right)\left(\frac{P_{min}}{P}\right)^{\beta} \exp\left(\frac{P_{min} - P}{P_c}\right). \tag{4.8}$$

The mean and the standard deviation can be obtained from a general formula the higher order moments given by Kagan and Schoenberg (2001),

$$E\left(P^k\right) = P_{min}^k + k P_{min}^{\beta} P_c^{k-\beta} \exp\left(\frac{P_{min}}{P_c}\right) \Gamma\left(k - \beta, \frac{P_{min}}{P_c}\right), \quad \text{for} \quad k = 1, 2, \tag{4.9}$$

where $\Gamma\left(k - \beta, \frac{P_{min}}{P_c}\right) = \int_{P_{min}/P_c}^{\infty} \exp(-t) t^{k-\beta-1} dt$ is the upper incomplete Gamma function. The probability of having an event greater than or equal to P_c as a function of P_c/P_{min} is finite, $\Pr(\geq P_c) = (P_c/P_{min})^{-\beta} \exp\left[(P_c/P_{min})^{-1} - 1\right]$, which is illustrated in Fig. 4.3 right. If we assume $\beta = \hat{\beta}_{OE}$, then the soft cut-off can be estimated from data by

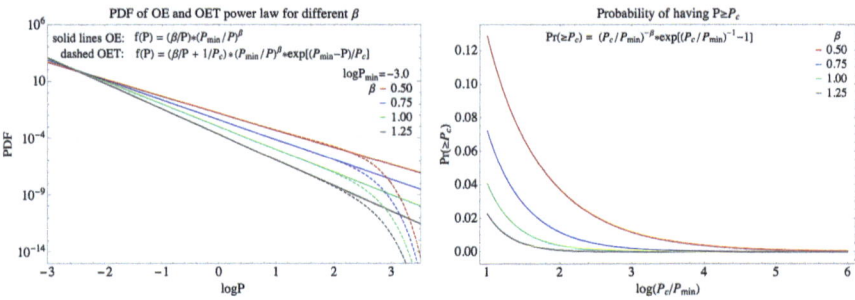

Fig. 4.3 Probability density functions for OE and OET power law *(left)* and the probability of having an event greater than or equal to P_c as a function of P_c/P_{min} *(right)* for selected β

$$P_c = \left(\frac{1}{n}\sum_{i=1}^{n} P_i^2 - P_{min}^2\right) / \left[2\beta P_{min} + 2\overline{P}(1-\beta)\right], \tag{4.10}$$

where $\overline{P}$ is the mean value of P, Kagan and Schoenberg (2001).

4.4 Upper Truncated Power Law (UT)

The upper truncated (UT) power law can be written as

$$N(\geq P) = \alpha\left(P^{-\beta} - P_{max}^{-\beta}\right), \tag{4.11}$$

where $N(\geq P)$ is the number of events not smaller than P. The number of events within the potency range (P_1, P_2), where $P_2 > P_1$ is $N(P_1, P_2) = \alpha\left(P_1^{-\beta} - P_2^{-\beta}\right)$. Parameter α measures the level of seismic activity, β is the exponent, and P_{max} is the limit of the maximum expected event size for a given data set. The probability density and the cumulative distribution functions $F(P)$ are

$$f(P) = \beta P^{-\beta-1} / \left(P_{min}^{-\beta} - P_{max}^{-\beta}\right), \qquad \Pr(\leq P) = 1 - \frac{\left(P^{-\beta} - P_{max}^{-\beta}\right)}{\left(P_{min}^{-\beta} - P_{max}^{-\beta}\right)}, \tag{4.12}$$

where $\Pr(\geq P_{max}) = 0$, and the survival function $\Pr(\geq P) = 1 - \Pr(\leq P)$, see Fig. 4.4 (e.g. Page, 1968; Cosentino et al., 1977; Burroughs & Tebbens, 2001; Kijko, 2004). Note that P_{max} needs to be greater than the maximum observed potency P_{maxo} for $N(\geq P)$ to be positive. The probability of having an event in a given potency range is $\Pr(P_1, P_2) = \left(P_1^{-\beta} - P_2^{-\beta}\right) / \left(P_{min}^{-\beta} - P_{max}^{-\beta}\right)$. For example, for

4.4 Upper Truncated Power Law (UT)

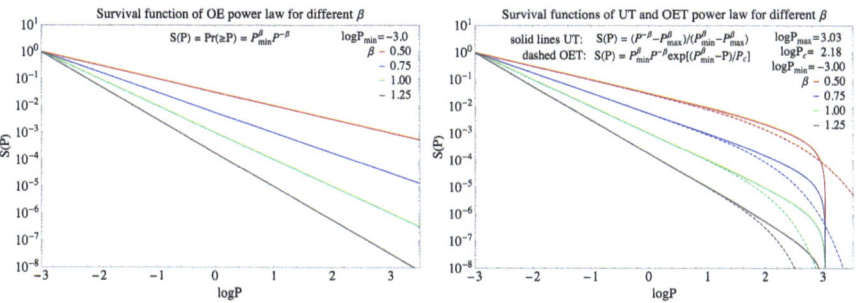

Fig. 4.4 Survival function of the OE power law *(left)* and the UT and OET *(right)* for a selected β

$\beta = 0.75$, $\log P_{min} = 0$, and $\log P_{max} = 3$, we have $\Pr(0.0 \leq \log P \leq 1.0) = 0.827$, $\Pr(1.0 \leq \log P \leq 2.0) = 0.147$, and $\Pr(2.0 \leq \log P \leq 3.0) = 0.026$.

The mean value of the UT power law is $\langle P \rangle = [\beta/(1-\beta)]\left(P_{max}^{1-\beta} - P_{min}^{1-\beta}\right)/\left(P_{min}^{-\beta} - P_{max}^{-\beta}\right)$ for $\beta \neq 1$, and $\langle P \rangle = \ln\left[(P_{max}/P_{min})/\left(P_{min}^{-1} - P_{max}^{-1}\right)\right] = \ln\left[P_{max}^2/(P_{max} - P_{min})\right]$, for $\beta = 1$. The variance, $Var(P) = \langle P^2 \rangle - \langle P \rangle^2$, where $\langle P^2 \rangle = \left[\beta/(2-\beta)\left(P_{max}^{2-\beta} - P_{min}^{2-\beta}\right)/\left(P_{min}^{-\beta} - P_{max}^{-\beta}\right)\right]$.

4.4.1 Estimation of α and β of the UT Relation

For the sake of convenience, the probability density and the survival functions of the upper truncated distribution may be written as

$$f(P) = \beta P_{min}^\beta P^{-\beta-1}/(1-R^\beta) \quad \Pr(\geq P) = P_{min}^\beta \left(P^{-\beta} - P_{max}^{-\beta}\right)/(1-R^\beta), \tag{4.13}$$

where $R = P_{min}/P_{max}$. The log likelihood function is $\log L(P_j;\beta) = \sum_{j=1}^n \log f(P_j)$, and taking its derivative with respect to β gives

$$\frac{n}{\hat{\beta}_{UT}} + \frac{1}{\log(e)}\frac{nR^{\hat{\beta}_{UT}}\log R}{1-R^{\hat{\beta}_{UT}}} - \frac{1}{\log(e)}\sum_{i=1}^n (\log P_i - \log P_{min}) = 0, \tag{4.14}$$

which can be simplified to

$$\hat{\beta}_{UT} = \log(e)\left[\overline{\log P} - \log P_{min} - \left(R^{\hat{\beta}_{UT}}\log R\right)/\left(1-R^{\hat{\beta}_{UT}}\right)\right]^{-1}, \tag{4.15}$$

where $\overline{\log P}$ is the geometric mean of the data, $\overline{\log P} = (1/n)\sum \log P_i$. Assuming the P_{min} and P_{max} are known, the Eq. (4.15) can be solved numerically for $\hat{\beta}_{UT}$ (Page, 1968). The standard deviation is given by

$$s_d(\beta_{UT}) = \frac{1}{\sqrt{n}} \left[\frac{\log(e)}{\hat{\beta}_{UT}^2} - \frac{R^{\hat{\beta}_{UT}} (\log R)^2}{\log(e) \left(1 - R^{\hat{\beta}_{UT}}\right)^2} \right]^{-1/2} \tag{4.16}$$

An approximate solution was given by Kijko and Funk (1994), by truncating the Taylor expansion of equation (4.15) at the second term,

$$\hat{\beta}_{UT} = \hat{\beta}_{OE} - \frac{\hat{\beta}_{OE}^2 R^{\hat{\beta}_{OE}} \log(1/R)}{\log(e) \left(1 - R^{\hat{\beta}_{OE}}\right)}, \tag{4.17}$$

where $\hat{\beta}_{OE}$ is derived from Eq. (4.5). Note that for the same data set the estimate of β for the UT formulation will always be lower than β for the OE one. The estimate $\hat{\alpha}_{UT}$ can be taken as $\log \hat{\alpha} = (1/n) \sum_{j=1}^{n} \log N (\geq P_j) - (1/n) \sum_{j=1}^{n} \log \left(P_j^{-\hat{\beta}_{UT}} - P_{max}^{-\hat{\beta}_{UT}}\right)$.

4.4.2 Comments on Data Selection and Parameter Estimation

It is important to select a data set and to determine the parameters of the power law correctly since a small change in the exponent β influences significantly the expected number of larger events. Typical mistakes and potential problems in fitting the power law to data:

1. Using a data set polluted by blasts.
2. Using a small data set, see Fig. 4.5.
3. Using data over a short time span. The typical power law distribution develops under slow constant loading over longer period of time.

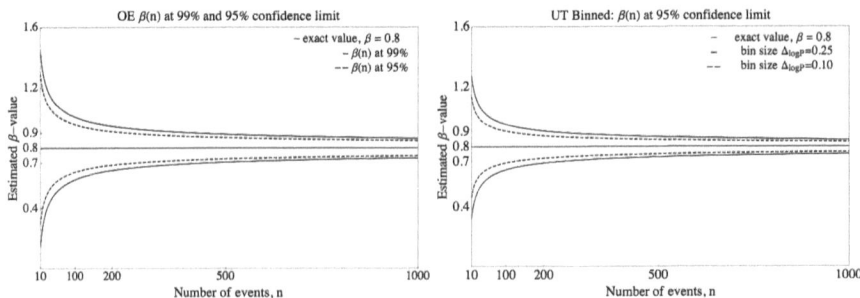

Fig. 4.5 99% and 95% confidence limits for β as a function of the number of events (*left*). 95% confidence limits for β as a function of the number of events for UT power law for two bin sizes, $\Delta_{\log P}$ (*right*)

4. Selecting the time span of data during which the way of mining changed significantly over a considerable period of time, e.g. the introduction of backfill, stabilising pillars, more disordered less concentrated mining method, preconditioning by hydrofracturing or blasting.
5. Selecting a data set which is the result of mixing two independent different seismogenic areas.
6. Fitting data with the linear least squares method (LS) as opposed to the maximum likelihood (ML). The LS method gives a standard deviation of β 2 to 3 times larger than ML.
7. Fitting parameters to the cumulative data which, by the very nature of its construction, is correlated. This violates the basic assumption of independent observations.
8. Underestimating P_{min}, which delivers lower β, i.e. overestimates hazard.
9. Incorrect binning of data. Linear binning produces unacceptable results. Logarithmic binning is better, but results depend on bin size. Too coarse binning produces inaccurate results for two reasons. Firstly, because the average computed from binned data is systematically overestimated. Secondly, because the centre of the first bin is different to P_{min}, and this influences the difference $\langle \log P \rangle - P_{min}$, which is very small and sits in the denominator, and therefore is influential. Small bins are more accurate but frequently empty, and this leads to a biased estimate. If the seismic system delivers source parameters of reasonable accuracy, there is no reason for binning.

4.5 Confidence Limits

A confidence interval is the range of values of a sample statistic that, at a given level of probability called a confidence level, is likely to contain a population parameter. This is the interval that will include the population parameter a certain percentage of the time during repeated sampling. A confidence level is the degree of certainty that one wants to be able to place in the confidence interval. This is the probability that the parameter being estimated by the statistic falls within the confidence interval. The confidence limits are the upper and lower values of a confidence interval.

Confidence intervals and the standard error of the mean serve the same purpose. However, those error bars around the mean that represent the standard error imply that only 66% of them include the parametric mean, which is not always remembered. The error bars with confidence intervals, usually 95%, imply that 95% of them include the parametric means, which is explicit. In addition, for very small sample sizes, which apply to larger events in the size distribution statistics, the 95% confidence interval is larger than twice the standard error. Note that the lower the level of confidence the narrower the confidence region. One would like to have a narrow confidence region with high probability, which is possible for normally distributed data and a great number of observations. For a data set drawn from a power law distribution, e.g. size distribution of seismic events, there are very few observations for larger events and, for the same probability, the confidence region there is wider.

A typical assumption when estimating the confidence limits is that the data is normally distributed. This gives symmetrical confidence limits, which for means near zero or one give negative limits. This is obviously incorrect. Therefore, for data sets which are power law distributed, it is better to calculate the confidence interval based on the binomial or the Poisson distribution.

4.5.1 Confidence Limits for β

If the standard deviation of β is known and if the population is normally distributed, or if the number of events is greater than 30, the 99% confidence limits can be estimated as $\beta \pm 2.58 \cdot sd_\beta$, the 95% confidence limits can be estimated as $\beta \pm 1.96 \cdot sd_\beta$, and the 90% confidence limit as $\beta \pm 1.645 \cdot sd_\beta$. The higher the level of confidence, the wider the interval.

Figure 4.5 (left) shows the 99% and 95% confidence limits for the β-value as a function of the number of events. It shows that one would need well over 1000 events for 95% confidence errors < 0.1. Figure 4.5 (right) shows that, for the same confidence limit, the smaller the bin size the better the estimate.

4.5.2 Confidence Limits for the Power Law Fit

Suppose that the data selected to fit the power law deviates from the model. The deviation may scatter randomly around the model or they may be systematic, e.g. there may be a surplus of events of certain size that, if significant, could be considered as characteristic. To be significant, the deviation has to exceed a certain level of confidence, and therefore, it is important to test to what degree a given data set supports the model.

If we assume that the number of events in a given $\log P$ bin can be estimated from the Poisson distribution, then the probability of having N events in that bin is $\Pr(N) = (\lambda^N/N!) \exp(-\lambda)$, where λ is the mean number of events in that bin. As per the power law, the mean number of events in a potency bin $\Delta_{\log P_j} = P_j - P_{j-1}$ is $N(P_{j-1}, P_j) = \alpha \left(P_{j-1}^{-\beta} - P_j^{-\beta} \right)$, where α and β are derived from the data. According to the Poisson distribution, the probability of having N_j events in that bin is

$$\Pr(N_j) = (1/N_j!) \left[\alpha \left(P_{j-1}^{-\beta} - P_j^{-\beta} \right) \right]^{N_j} \exp \left[-\alpha \left(P_{j-1}^{-\beta} - P_j^{-\beta} \right) \right],$$
$$j = 1, 2, \ldots, n_b, \tag{4.18}$$

where n_b is the number of bins. The upper and lower confidence limits for the centre potency in the jth bin are found by solving Eq. (4.18) for N_j, assuming the required probability $\Pr(N_j)$. The 95% confidence intervals are estimated using $\sum \Pr(N_j) = 0.975$ for upper limit and $\sum \Pr(N_j) = 0.025$ for lower limit, see, for example, Fig. 4.17.

4.6 Utility of Power Law Distributions

4.6.1 Limitations and Benefits

The power law size distribution has some apparent limitations. In general, power laws are asymptotic, and as a consequence, their moments, e.g. mean or variance, exist only for a certain range of exponents. Therefore, there is a need to impose a hard upper limit on the maximum event size and to define the power law within a range, (P_{min}, P_{max}), or to add an exponential term to its tail, to secure a finite energy release.

The power law size distribution disregards the time of seismic events, thus ignoring any potential trend in the data. The bulk of the data are small events which are not necessarily hazardous, and, in some cases, the processes leading to small events may be different from the processes leading to large events.

Inverting α, β, and P_{max} simultaneously from data does not always deliver stable results. Therefore, there is a need to estimate P_{max} independently, e.g. by order statistics which regards the sequence of larger events. Because order statistics and the associated jumps in the history of record breaking events change as mining progresses, P_{max} needs to be reassessed regularly.

The power law also disregards the spatial distribution of seismic events. Subdividing space into sub-volumes without de-clustering, fitting a power law to the data extracted from each sub-volume, and calculating their probabilities are not the best strategies to estimate spatial hazard. Unless one will assume some form of spatial intensity function in the non-stationary Poisson process, there is nothing to extrapolate, and therefore the future hazard will be very much like the past. In addition, there are two potential problems with subdivision of space: (1) Seismic activity within these sub-volumes may not be independent. (2) There is a trade-off between the spatial resolution and the amount of data one can extract from these sub-volumes, and therefore, there may be insufficient data for a reliable power law fit, see Fig. 4.5.

The best way to address the spatial hazard is by numerical modelling of stresses and strains associated with future mining, calibrated with the existing seismic data (Linkov, 2006, Linkov, 2013, Malovichko & Basson, 2014).

Nevertheless, the power law size distribution offers obvious benefits. It provides a useful association between its exponent and the seismic hazard-related factors, e.g. the state of stress, the system stiffness, rock mass and mine layout heterogeneity, potency production, information entropy or the unpredictability of larger events, and the stress transfer due to small and large events. It also gives an insight into the scaling relation between seismic potency and radiated seismic energy.

To take full advantage of the size distribution-based hazard, one needs to apply it carefully. Since the bulk of seismic activity in mines follows production, that can be very intermittent, the traditional recurrence times, $\bar{t} = \Delta t / N (\geq P)$, and associated probabilities may not be as useful as in crustal seismology where loading is steady.

The other traditional size distribution parameters, namely, the activity rate, the β-value, "the one largest event, P_{max1}", also do not measure seismic hazard consistently and reliably. The supposition that solving $\alpha P_{max1}^{-\beta} = 1$ gives *the one largest possible event or the next largest event*, $P_{max1} = \alpha^{1/\beta}$, or $m_{max1} = a/b$ in the Gutenberg-Richter relation, is incorrect. The fact that the data follows the OE power law closer than the upper truncated one indicates the potential for even larger events and larger jumps in record breaking events. Only when the largest observed events are significantly smaller than that predicted by the OE relation can one infer that the size distribution hazard may be contained or controlled.

If rock extraction is not a linear function of time, one should resort to parameters based on volume mined, e.g. an average inter-event volume mined to generate an event above a certain size, $\bar{V}_m (\geq P) = V_m / N (\geq P)$, the probability that a larger seismic event will occur while extracting a given volume of rock, $\Pr (\geq P, \Delta V_m)$, or the volume of ground motion $V_{GM} (\geq v, \Delta V_m)$.

As the extraction ratio increases and the overall stiffness of the rock mass is being degraded, there is a tendency for the β-value to decrease and α to increase. If such a trend can be detected, quantified, and extrapolated, then seismic hazard assessment is less dependent on the assumption of stationarity (Mendecki, 2008).

4.6.2 Missing Potency and β

Seismic monitoring systems record events above their overall sensitivity level P_{min}, therefore there is missing potency below that level, and the ratio of the missing to recovered potency depends on the slope or the β-value of the observed potency frequency distribution.

The potency release by seismic events within the potency range P_1 to P_2 is $P(P_1, P_2) = N(P_1, P_2) \int_{P_1}^{P_2} P f(P) dP / \int_{P_1}^{P_2} f(P) dP$, which for $\beta \neq 1$ gives

$$P(P_1, P_2) = \alpha\beta \left(P_2^{1-\beta} - P_1^{1-\beta} \right) / (1 - \beta).$$

For $\beta = 1$, one can integrate within the finite potency range, $P(P_1, P_2) = \alpha \ln(P_2/P_1)$. For $\beta < 1$, one can integrate from $P_1 = 0$ to a finite potency $P(0, P_2) = \alpha\beta P_2^{1-\beta} / (1 - \beta)$, and for $\beta > 1$, one can integrate from a finite potency to infinity, $P(P_1, \infty) = -\alpha\beta P_1^{1-\beta} / (1 - \beta)$. The equations for the number of events, $N(P_1, P_2)$, for both OE and UT relations are similar, with the exception that in the UT case the P_2 can only go as far as P_{max}.

4.6 Utility of Power Law Distributions

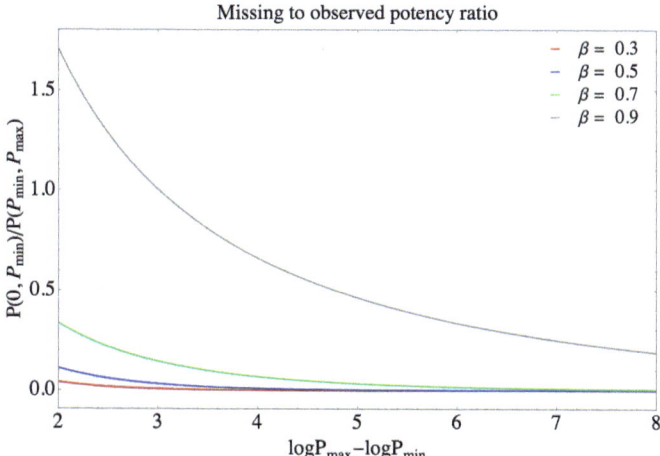

Fig. 4.6 Missing to observed potency ratio as a function of $\log P_{max} - \log P_{min}$ for different β

The following ratio quantifies the UT portion of seismic potency below P_{min} that, if already produced by the rock mass, is missing because it could not have been recorded due to the limited sensitivity of the seismic monitoring system or because it has been produced aseismically,

$$\frac{P(0, P_{min})}{P(P_{min}, P_{max})} = \frac{P_{min}^{1-\beta}}{\left(P_{max}^{1-\beta} - P_{min}^{1-\beta}\right)}, \quad \text{(for } \beta < 1\text{).} \quad (4.19)$$

It can be shown that $P(0, P_{min})/P(P_{min}, P_{max}) = \left[10^{(\log P_{max} - \log P_{min})(1-\beta)} - 1\right]^{-1}$, i.e. it depends only on β and on the difference $\log P_{max} - \log P_{min}$, regardless of P_{min}, see Fig. 4.6. As expected, the higher the β-value the more potency is lost below the threshold level. For $\log P_{max} - \log P_{min} = 3$ and $\beta = 0.9$, there is 50/50 split between the observed and the missed potency.

4.6.3 Power Law and log(Energy) vs. log(Potency) Relation

In practice, the size distribution analysis is carried out in the magnitude and/or in the potency (moment) domain. However, the most appropriate measure of the strength of a seismic source is the radiated seismic energy, E, and, if estimated reliably, it should be the base for size distribution hazard estimation. If both seismic potencies and energies are available for the same data set, assuming the OE power law, we can write $N(\geq P) = \alpha_P P^{-\beta_P} = N(\geq E) = \alpha_E E^{-\beta_E}$, which, after simple algebra, gives

$$\log E = (\beta_P/\beta_E) \log P + (1/\beta_E) \log(\alpha_E/\alpha_P), \quad (4.20)$$

where subscripts P and E stand for potency and energy, respectively (Mendecki, 2013). This equation expresses the scaling relation $\log E = d \log P + c$ via parameters of the potency frequency and the energy frequency distributions, where $d = \beta_P/\beta_E$ and $c = (1/\beta_E) \log(\alpha_E/\alpha_P)$. Parameters d and c are usually derived by fitting a straight line to data using a standard least squares regression or the generalised orthogonal regression (e.g. Mendecki, 1993, Figure 1.2). Since parameters α and β in the potency and the energy frequency distribution are derived by the maximum likelihood (ML) method, equation 4.20 offers an indirect way of ML fitting. Note that the $\log E$ vs. $\log P$ scaling may be affected by the inability of systems to record the wide frequency spectrum radiated from seismic sources. The exponent β_E scales inversely with the number of larger energy events, i.e. lower β_E delivers a larger portion of high-energy events. The exponent β_P scales inversely with the number of larger potency events, i.e. high β_P delivers larger portion of smaller potency events. Since most of seismic potency is delivered at low frequencies and most of seismic energy at higher frequencies, larger events recorded in mines by 4.5 Hz or 14 Hz geophones may underestimate potency more than energy, and therefore parameter d and an increase in apparent stress with potency may be overestimated.

Figure 4.7 left shows the $\log E$ vs. $\log P$ for the MineD data set where the red line represents the fit with coefficients $d = \beta_P/\beta_E$ and $c = (1/\beta_E) \log(\alpha_E/\alpha_P)$, see Equation 4.20, and the green line is by the ordinary least squares. The colour here scales with $\log \sigma_A$. Figure 4.7 right shows apparent stress, σ_A, vs. $\log P$ for the same data set where colour indicates the time of the event.

From $\log E = d \log P + c$, the apparent stress is $\sigma_A = E/P = 10^c P^{d-1}$, which gives

$$\sigma_A = (\alpha_E/\alpha_P)^{1/\beta_E} P^{(\beta_P/\beta_E)-1}, \qquad (4.21)$$

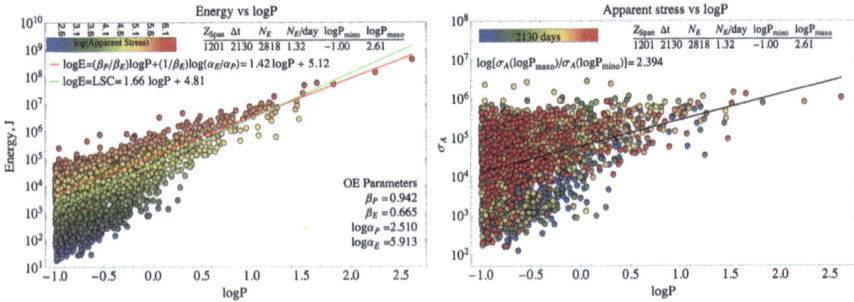

Fig. 4.7 $\log E$, vs. $\log P$ plot of events with $\log P \geq -1.0$. The red line represents the fit with coefficients d and c derived from Eq. (4.20), and the green one is by the ordinary least squares *(left)*. $\log \sigma_A$ vs. $\log P$ for the same data set *(right)*

4.6 Utility of Power Law Distributions

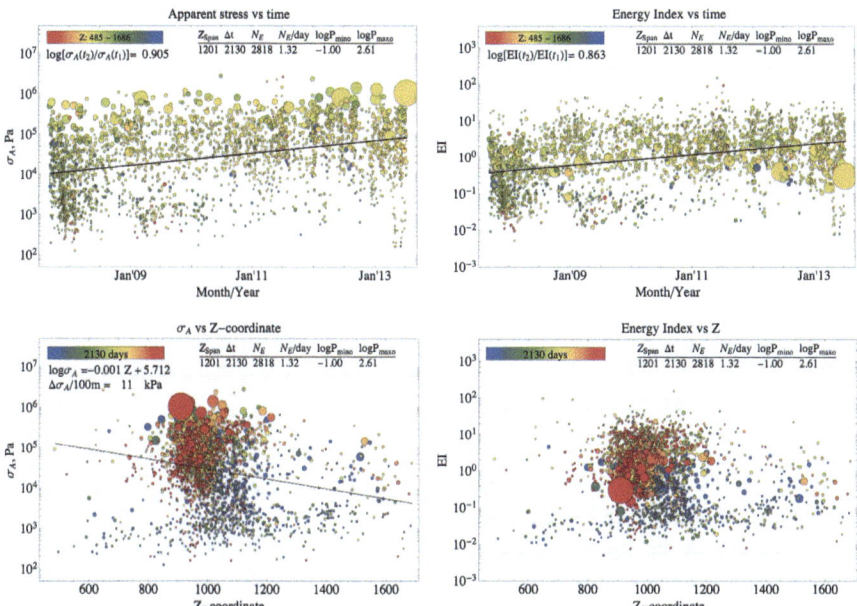

Fig. 4.8 Apparent stress, σ_A, vs. Z-coordinate of the event for the same data set as in Fig. 4.7 (left). Energy index vs. Z-coordinate of the event for the same data set (*right*)

and for $d = 1.0$, or $\beta_P = \beta_E$, the apparent stress is constant and independent of potency $\sigma_A = 10^c$, or $\sigma_A = (\alpha_E/\alpha_P)^{1/\beta_E}$. The higher the driving stress at the source the higher the apparent stress.

The energy index of an event is the ratio of the observed radiated seismic energy of that event E, to the average energy $\bar{E}(P) = 10^{d \log P + c}$ radiated by events of the observed potency P, for a given area of interest, $EI = E/\bar{E}(P) = E/10^{d \log P + c} = 10^{-c} E/P^d$, which for $d = 1.0$ would be proportional to the apparent stress (van Aswegen & Butler, 1993). Energy index can now be written as

$$EI = (\alpha_E/\alpha_P)^{1/\beta_E} E/P^{(\beta_P/\beta_E)}. \tag{4.22}$$

Figure 4.8 top row shows a modest increase in apparent stress, σ_A and energy index, EI, over time. Figure 4.8 bottom row shows σ_A and EI vs. Z-coordinate, where size of the event scales with the radius of source volume and the time of the event from the main shock. Note that here the lower the Z-coordinate the deeper the event. Since 2010, most rock extraction was concentrated at depth 1050 and 950, and this modest increase in apparent stress and energy index over time can be attributed to depth of mining and an increase in the extraction ratio at that depth.

The slope d of the $\log E$ vs. $\log P$ plot of earthquakes is frequently reported to be close to 1.0. However, Choy et al. (2006) reported that earthquakes occurring on

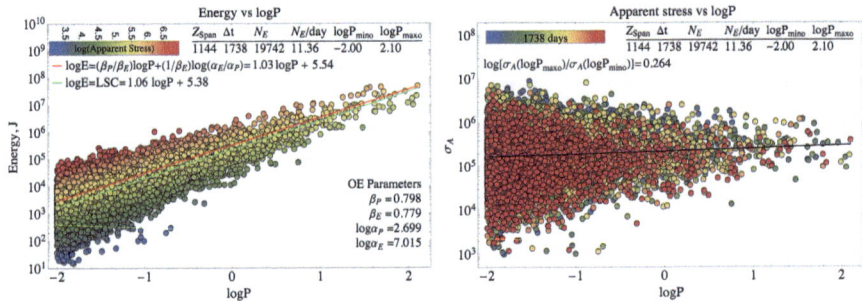

Fig. 4.9 log E, vs. log P plot of events with log $P \geq -1.0$. The red line represents the fit with coefficients d and c are derived from Eq. (4.20) and the green one is by the ordinary least squares *(left)*. log σ_A, vs. log P for the same data set *(right)*

immature faults radiate more a higher frequency energy per unit of moment than earthquakes occurring on mature faults.

The slope d of the log E vs. log P plot of events recorded in mines is frequently reported to be higher than 1.0 (Mendecki, 1993, see also Fig. 4.7). However, there are exceptions. Figure 4.9 shows the log E vs. log P plot of 1738 events recorded during 1738 days between 2015 and 2019 in a deep tabular hard rock gold mine in South Africa. Here the slope d is practically 1.0, consequently, apparent stress is independent of potency, and energy index is proportional to the apparent stress.

4.6.4 Power Law and Stress Transfer

Small seismic events contribute very little to the total energy and potency release and to the cumulative co-seismic deformation, but they can make an important contribution to the spatial and temporal stress transfer within the seismically active rock mass (Hanks, 1992).

The major part of the stress transfer due to inelastic deformation associated with seismic activity takes place within the source volume, $V = P/\Delta\epsilon$. Since the co-seismic stress drop $\Delta\sigma$, and the strain change $\Delta\epsilon = \Delta\sigma/\mu$, associated with larger events are remarkably similar to that of smaller events (Aki, 1972), the overall stress transfer due to small events may be equal to or even dominate that of large events.

Therefore, within the power law distribution, there is the centre potency, $P_{0.5}$, that splits the potency release $P(P_{min}, P_{max})$, and for a constant $\Delta\epsilon$ the cumulative source volume, into two equal parts with similar contributions to stress transfer. Solving $P(P_{min}, P_{0.5}) = P(P_{0.5}, P_{max})$, i.e. $\left(P_{0.5}^{1-\beta} - P_{min}^{1-\beta}\right) = \left(P_{max}^{1-\beta} - P_{0.5}^{1-\beta}\right)$, for $\beta \neq 1$ gives

$$P_{0.5} = \left[0.5\left(P_{max}^{1-\beta} + P_{min}^{1-\beta}\right)\right]^{\frac{1}{1-\beta}}. \tag{4.23}$$

4.6 Utility of Power Law Distributions

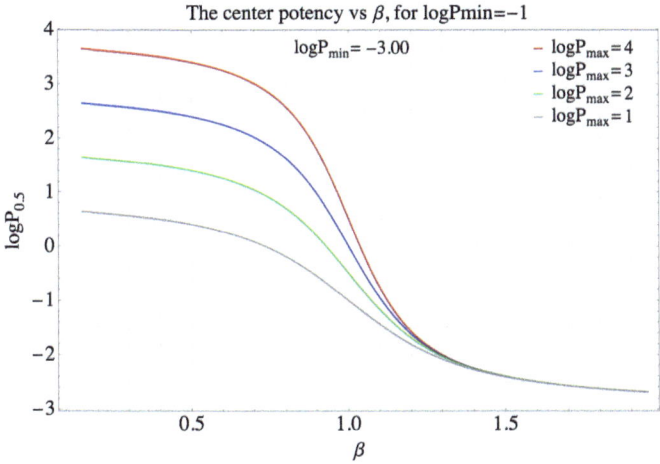

Fig. 4.10 The centre potency, $P_{0.5}$, as a function of β for $\log P_{min} = -3.0$ and for the selected values of $\log P_{max}$

For $\beta = 1$, the centre potency $P_{0.5} = \sqrt{P_{min} \cdot P_{max}}$, which is the right and the left hand limit of the previous equation when $\beta \to 1$. For $\beta < 1$, we can integrate from $P_{min} = 0$, which gives $P_{0.5} = 0.5^{1/(1-\beta)} P_{max}$. For $\beta > 1$, we can integrate from P_{min} to $P_{max} = \infty$ and have $P_{0.5} = 0.5^{1/(1-\beta)} P_{min}$. For example, for $P_{min} = 0$ and $\beta = 0.9$, $\log P_{0.5} = \log P_{max} - 3$, and for $P_{max} = \infty$ (OE relation) and $\beta = 1.1$, $\log P_{0.5} = \log P_{min} + 3$. Figure 4.10 shows that $P_{0.5}$ is lower for lower P_{max} and decreases with increasing β.

While sources of all seismic events smaller than $\log P_{0.5}$ may have the same cumulative volume of inelastic deformation as sources of events greater than $\log P_{0.5}$, the influence of smaller events on stress transfer and on possible triggering of larger events will depend on their spatial distribution. The more clustered they are the more likely that they may trigger a larger event (Helmstetter et al., 2005).

4.6.5 Power Law and Information Entropy

Information is the currency of nature, but not all information is of equal value. The total amount of information in a system is the difference between the system's current entropy and its maximum possible entropy. Following Shannon's second paper (Shannon, 1948), the continuous information entropy of the probability density function of potency can be written as

$$H[f(P)] = \int_{P_{min}}^{P_{max}} f(P) \log [1/f(P)] dP, \qquad (4.24)$$

where $f(P)$ is the probability density function of the upper truncated potency frequency distribution, $f(P) = \beta P^{-\beta-1}/\left(P_{min}^{-\beta} - P_{max}^{-\beta}\right)$.

In Eq. (4.24) $\log[1/f(P)]$ is the information content of event P with probability $f(P)$, and if $f(P)$ is high, then knowledge that event P occurred gives very little information, since it had a high probability of occurrence to start with.

The continuous entropy is not a limit of the discrete entropy when the bin size goes to zero. The entropies of continuous distributions have most, but not all, of the properties of the discrete case.

There is one important difference between the continuous and discrete entropies. In the discrete case, the entropy measures the randomness of the variable in an absolute way. The continuous formulation measures the entropy relative to the coordinate system, and the entropy can be negative. This is not important, though, if one is interested in the differences or in the rate of change between two or more entropies, since they are independent of the frame of reference.

In general, H measures the amount of uncertainty in a given distribution, which is a measure of unpredictability. After integration, the information entropy of the upper truncated potency power law is

$$H = \log\left(\frac{P_{min}^{-\beta} - P_{max}^{-\beta}}{\beta}\right) + 0.43\left(1 + \frac{1}{\beta}\right)$$
$$+ \frac{(\beta+1) P_{min}^{-\beta} \log P_{min} - P_{max}^{-\beta} \log P_{max}}{P_{min}^{-\beta} - P_{max}^{-\beta}}, \qquad (4.25)$$

which monotonically decreases with increasing β. For the OE distribution, where $P_{max} \to \infty$, $H = \log(P_{min}/\beta) + 0.43(1/\beta + 1)$.

Figure 4.11 shows the information entropy H as a function of β for the OE power law, i.e. $\log P_{max} = \infty$, and for the UT one with three different values $\log P_{max}$. It is clear that unpredictability increases with decreasing β for both OE and UT distributions, although more so for the OE one (Mendecki, 2012). Therefore, low β-values imply not only higher hazard, but also a less predictable size distribution of seismic events.

The information entropy should not be confused with the thermodynamical or mechanical interpretation, which states that the amount of entropy increase is proportional to the loss of capacity of the system to deliver work.

4.6.6 Power Law Exponent, Rock Properties, Stress, and Stiffness

For a data set that obeys the power law size distribution, the exponent β is a statistical measure of the ratio of small to large events, and it decreases as the portion of the intermediate and large events increases.

4.6 Utility of Power Law Distributions

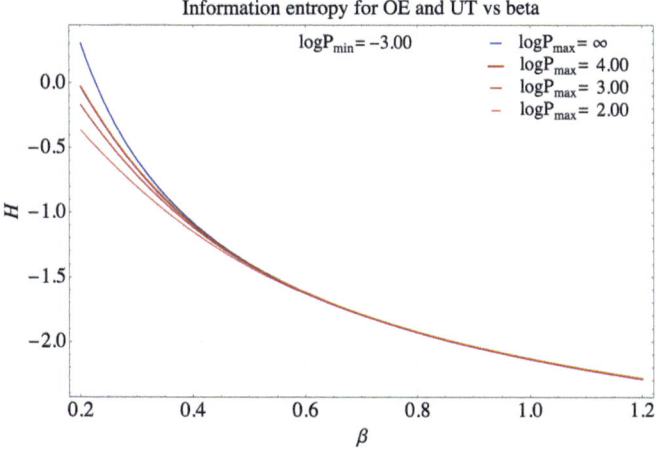

Fig. 4.11 Information entropy, H, as a function of β for the OE power law, i.e. $\log P_{max} = \infty$, in blue and for the UT for selected $\log P_{max}$ in red

In some cases, however, data points on the cumulative frequency plot indicate a double slope convex or concave character. One possible explanation of such deviations is the change in the sensitivity of the monitoring system during acquisition of data. Another possibility is that the data is generated by two spatially separated processes with different size distributions.

Concavity may be generated by a combination of high activity of low-magnitude events induced by mining excavation(s) with a few larger events caused by an existing geological structure. This behaviour is frequently a function of spatial and/or temporal data selection. More frequently observed convexity is caused by a deficit of larger events that may be the result of a limited time span of the selected data. Its persistence, however, indicates that seismic hazard is contained.

In general, the exponent β is positively correlated with the heterogeneity of the rock mass and with its stiffness and negatively with stress.

The rock mass heterogeneity depends on the spatial distribution of sizes and distances between strong or stressed and weak or de-stressed patches of rock, where seismic sources may nucleate and be stopped. An increase in rock heterogeneity results in a higher β, since it is more likely that the initiated rupture be stopped by a soft or hard patch before growing into a larger event (Mogi, 1962; Mori & Abercrombie, 1997). Scholz (1968) stated that the β-value varies inversely with stress. His reasoning is based on a similar argument to heterogeneity, namely, that rupture once initiated grows larger in a high stress regime.

Stiffness measures the rigidity of a system, i.e. its ability to resist deformation in response to an applied load. It scales positively with the ratio of the applied stress to the induced strain. Experimental study on rock samples of equal degrees of heterogeneity in triaxial conditions has shown a decrease in β of acoustic emission

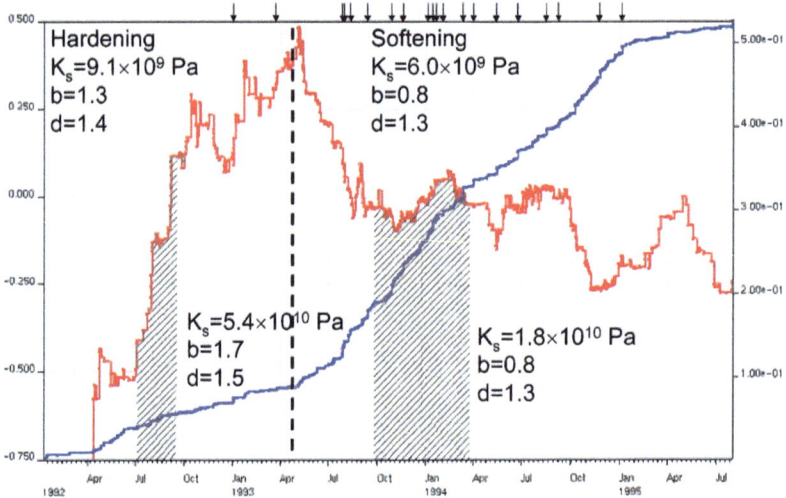

Fig. 4.12 The time evolution of energy index EI (a proxy for stress, in red) and the cumulative apparent volume, V_A (a proxy for deformation, in blue) during shaft pillar extraction progressing from the centre to its outer perimeter

events with both the differential stress and the confining pressure, during all stages of stress-strain regime, including the post-peak strain softening (e.g. Amitrano, 2003).

Observations in mines found a higher β in stiffer systems and lower β in softer systems (Mendecki & van Aswegen, 1998 and van Aswegen & Mendecki, 1999).

Figure 4.12 shows the time evolution of the energy index (a proxy for stress) and the cumulative apparent volume, V_A (a proxy for deformation), during a shaft pillar extraction progressing from the centre to its outer perimeter. Note that almost all events larger than $\log E = 7.5$, marked by arrows, occurred during softening past peak stress. Other parameters are: seismic stiffness K_s, the b-value (or β), and the d-value in $\log E = d \log P + c$.

These observations do not contradict reports on decreasing β with increasing stress during the strain hardening regime, since there is a general loss of stiffness with increasing stress. However, in a strain softening regime, where the strength is decreasing with increasing strain, stress is lower but lower β was observed.

4.7 Expected Maximum Event Size

4.7.1 What Is P_{max} or m_{max}

In earthquake seismology, m_{max} or $\log P_{max}$ is the maximum magnitude, or potency, earthquake that a given seismogenic region can deliver (e.g. McGarr, 1976;

4.7 Expected Maximum Event Size

Ward, 1997; Kijko, 2004; Pisarenko et al., 2008). In regions where the association of earthquakes and geological structures is evident, one can assume that the largest event will occur on one of these structures. The estimate of m_{max} can then be made based on the mapped fault length and geological record of displacement or from the empirical relation between the observed magnitudes and the source size. Alternatively, it can be estimated by numerical modelling of stresses. In regions where connection between geological structures and past earthquakes is less clear, the possible location of a future largest event is uncertain, and the m_{max} needs to be estimated from past observations.

Having an adequately long catalog of data, the problem of estimating the maximum possible potency or magnitude of an earthquake in a given area may be reduced to finding the truncation point of the observed distribution of past events (Robson & Whitlock, 1964; Cooke, 1979; Van Der Watt, 1980; Kijko & Singh, 2011). In such a case, the complex details of the data on small and intermediate events are less important. However, all estimates of m_{max} are highly uncertain, and the upper bounds of the confidence intervals are acceptable only if reliable data over a long time interval, including several earthquakes with magnitudes close to m_{max}, are available (Zoller & Holschneider, 2016). The assessment of the maximum magnitude of a fluid injection-induced earthquake was described by Zoller and Holschneider (2014) and Zoller (2022).

In mines, the geology of the ore body and of the surrounding rock mass is reasonably well explored, and therefore, the maximum possible event size can be estimated by numerical modelling, assuming a complete stress drop on the main geological structures at different mining steps, i.e. different extraction ratios or times, or a complete failure of pillars at different mining steps, i.e. different extraction ratios or times. However, the estimate of reliable m_{max}, defined as the maximum possible event size during the life time of a mine, from the seismic catalogue only is not possible.

At best, it can be guesstimated from the empirical relation between the footprint of a mine and the maximum observed event size. The main reason for the difficulty is that, unlike in a tectonic regime, the rate and the spatial and temporal distribution of loading in mines are highly variable, and this may have significant influence on the size distribution of seismic events, and therefore on seismic hazard. In addition, there are two opposing forces at work.

On one hand, an increase in extraction ratio and consequently the degradation of rock mass stiffness create conditions conducive for larger events to occur. On the other hand, to manage the situation, mines alter the way of mining. There are examples where, after changes to mine layout and/or the introduction of backfill, seismic hazard decreased. Therefore, m_{max} or P_{max} associated with mining is not a fixed parameter and needs to be estimated periodically as mining progresses. It is the next m_{max} or P_{max} which can be estimated and that needs to be managed.

Size of a Mine and the Maximum Event Size In the absence of tectonic forces, the size of the largest possible event induced by mining scales approximately with the characteristic size of the mine, L. The upper bound relation between the maximum magnitude and the linear size of the mine is $m_{max} = 2.0 \log L - 2.0$ (McGarr et al., 2002). For $L = 1500$ m, there is a potential for moment-magnitude $m_{HK} = 4.3$ (or $\log P = 5.07$). For $L = 2500$ m, the maximum $m_{HK} = 4.8$ or $\log P = 5.82$ ($\log P_{max} = 1.5 m_{HK} - 1.38$). There are many mines with characteristic dimension $L \geq 1500$ m that did not generate seismic events of that size.

However, there is also a case where mining triggered an event along an intersecting geological structure which was considerably larger than the characteristic size of the mine. It is speculated that this is more likely in the presence of tectonic horizontal stresses. If mining takes place in an active tectonic regime, then the maximum possible event size within the mine could be the same as the maximum possible size earthquake in the area.

4.7.2 Balance of the Effective Volume Mined and P_{max}

If we assume that the extraction of rock, the backfill placement, and the seismic and aseismic deformation take place within the volume ΔV at discrete times, then one can construct the following step function that reflects a balance of inelastic deformation, $B(t) = V_m(t) - V_b(t) - V_P(t) - V_a(t)$, where: $V_m(t) = \sum_{t_i \leq t} V_m(t_i) \Theta(t - t_i)$ is the volume mined to date, Θ is the Heaviside function, $\Theta(t) = 0$ if $t < 0$, and $\Theta(t) = 1$ if $t \geq 0$, $V_m(t_i)$ is the volume mined at t_i, $V_b(t) = \sum_{t_j \leq t} c_b(t) V_b(t_j) \Theta(t - t_j)$ is the volume of backfill placed to date, where $V_b(t_j)$ is backfill placed at t_j, $V_{meff}(t) = V_m(t) - V_b(t)$ is the effective volume mined to date, $c_b(t)$ is the efficiency of backfill that corrects for shrinkage and compaction, $V_P(t) = \gamma_0 \sum_{t_k \leq t} P(t_k) \Theta(t - t_k)$ is the volume reduction due to seismic inelastic deformation to date, $P(t_k)$ is the seismic potency of the event at time t_k, $P_I(t_k)$ is the isotropic component of the potency $P(t_k)$, $\gamma_0 = P_I(t_k)/P(t_k)$ is the portion of the isotropic component of the total potency associated with seismic activity, which for events close to excavation faces is $0.6 \leq \gamma_0 \leq 0.9$ (McGarr, 1993), $V_a(t) = \gamma_a \sum_{t_k \leq t} P(t_k) \Theta(t - t_k)$ is the volume of aseismic inelastic deformation to date, spatially and temporarily synchronous with seismic activity, where $0.1 \leq \gamma_a = P_a(t_k)/P(t_k) \leq 0.25$, and $P_a(t_k)$ is the aseismic potency at t_k.

The balance function can now be finally written as $B(t) = V_{meff}(t) - \gamma \sum_{t_k \leq t} P(t_k) \Theta(t - t_k)$, where $\gamma = \gamma_0 + \gamma_a$.

Scenario 1 Collapse type closure. Here the current deficit of deformation $B(t)$ is to be closed in a single large seismic event, which may or may not fit to the observed power law size distribution. This can be considered the worst-case scenario.

However, history shows that it is possible to engineer such a pathological mining layout that at some stage it collapses during one large complex seismic event. Without external loading, the maximum potency of such an event would be limited to $P_{max}(t) = B(t)$.

Scenario 2 Power law deficit closure. The deficit of deformation is closed due to seismic activity with a power law size distribution. In this case, P_{max} can be estimated from $B(t) = V_{meff}(t) - \gamma \sum_{t_k \leq t} P(t_k) \Theta(t - t_k)$, by replacing the

4.7 Expected Maximum Event Size

sum of seismic potency by a model of potency production derived from the size distribution, $B(t) = V_{meff}(t) - P_{max}^{1-\beta} \cdot \gamma\alpha\beta/(1-\beta)$. By setting $B(t) = 0$, one can now calculate the P_{max} of the potency frequency distribution that would have closed the deficit of deformation

$$P_{max}(t) = \left[\frac{1-\beta}{\gamma\beta} \cdot \frac{V_{meff}(t)}{\alpha}\right]^{\frac{1}{1-\beta}} \Rightarrow \log$$

$$P_{max} = \frac{1}{1-\beta}\log\left[\frac{1-\beta}{\gamma\beta} \cdot \frac{V_{meff}(t)}{\alpha}\right], \quad (4.26)$$

where α is the number of seismic events with $\log P \geq 0$. Similar equations were derived by Wyss (1973), Smith (1976), McGarr (1976), and Molnar (1979) to estimate the expected maximum magnitude of earthquakes, and by McGarr (1984) for mines.

The interpretation of this equation is straightforward, namely, $P_{max}(t)$ increases with the effective volume mined. For given $V_{meff}(t)$, the more seismic and aseismic potency has been produced to date, i.e. higher α, lower β and higher γ, the less there is to be produced to close the deficit, therefore, the lower $P_{max}(t)$. One way to reduce P_{max}, at least temporarily, is to manage the heterogeneity of the seismogenic volume. Such heterogeneous systems are stiffer and, for a given V_{meff}, maintain higher β for longer and deliver lower V_{meff}/α, therefore lower P_{max}, see Fig. 4.13.

Relation (4.26) is based on a few assumptions that may limit its application. Firstly, it implies that one knows the total potency produced due to mining in a given area. In practice, however, one can at best account only for that portion of seismic potency which has been released since the introduction of the monitoring system. Secondly, it assumes that the only stress acting in the area is the one induced by the particular volume mined, which may not be the case in scattered mining scenarios or in the presence of tectonic stresses. Then the effective volume mined $V_{meff}(t)$ may be smaller than $P_{maxo}^{1-\beta} \cdot \gamma\alpha\beta/(1-\beta)$, where P_{maxo} is the maximum observed potency. Misfitting the power law may also result in the balance $B(t)$ being negative.

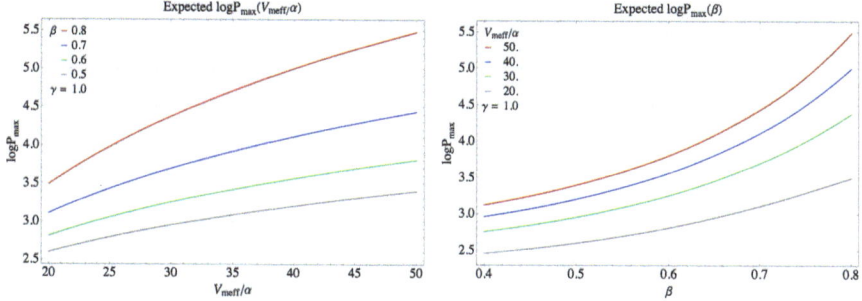

Fig. 4.13 $\log P_{max}(V_{meff}/\alpha)$ (left) and $\log P_{max}(\beta)$ (right)

4.7.3 Order Statistics: P_{max}—The Upper Limit to the Next Largest Event

Order statistics represent the characteristics of a sample after it has been ordered, usually from the smallest observation to the largest observation. For $P_1,\ldots,P_n$, random potencies $P_{(k)}$ is the kth smallest, called the kth order statistic, $P_{(1)} = \min(P_1,\ldots,P_n)$ and $P_{(n)} = \max(P_1,\ldots,P_n)$. Let $P_{(1)} = P_{min},\ldots,P_{(n)} = P_{maxo} < P_{max}$, be the order statistics of a random sample of seismic potency of size n from a population with a continuous distribution function. The question here is: What is the P_{max}—the estimator for the upper bound of this random variable? The probability that any potency is not greater than some P is, by definition, the cumulative distribution function, $\Pr(<P) = F(P)$, but the probability that P_{max} is not greater than P is $\Pr(P_{max} < P) = F_{P_{max}}(P) = \Pr(P_1 < P, P_2 < P, \ldots, P_{max} < P) = \prod_{i=1}^{n} \Pr(P_{(i)} < P) = [F(P)]^n$. The probability density function here is $dF_{P_{max}}/dP = f_{P_{max}}(P) = n[F(P)]^{n-1} f(P)$, where $f(P) = dF(P)/dP$. The mean value of the distribution function $F_{P_{max}}(P)$ is obtained after integrating by parts, $\langle P_{max} \rangle = \int_{P_{min}}^{P_{max}} P\, dF_{max}^n(P) = P_{max} - \int_{P_{min}}^{P_{max}} F_{P_{max}}^n(P)\, dP$, and therefore the estimator of P_{max} can be taken as

$$P_{max} = P_{maxo} + \int_{P_{min}}^{P_{maxo}} F_{P_{maxo}}^n(P)\, dP = P_{maxo} + \sum_{i=1}^{n-1} \left(\frac{i}{n}\right)^n \left[P_{(i+1)} - P_{(i)}\right], \quad (4.27)$$

where $F_{P_{maxo}}(P)$ is the empirical distribution function based on order statistics, $F_{P_{maxo}}(P) = i/n$, for $P_{(i)} \leq P \leq P_{(i+1)}$ and for $i = 1,\ldots,n-1$, $F_{P_{maxo}}(P) = 0$ for $P < P_{min}$, and $F_{P_{maxo}}(P) = 1$ for $P \geq P_{maxo}$, and $dP = P_{(i+1)} - P_{(i)}$ (Cooke, 1979). Rearranging the summation gives

$$P_{max} = 2P_{maxo} - \sum_{i=0}^{n-1} P_{maxo-i}\left[\left(1 - \frac{i}{n}\right)^n - \left(1 - \frac{i+1}{n}\right)^n\right], \quad (4.28)$$

where for $i = 0$ $P_{maxoi-1}$ is the second largest observed potency, for $i = 1$ the third largest observed potency, and so on. For large n, i.e. $\lim_{n\to\infty}(1 - i/n)^n = \exp(-i)$ and $\lim_{n\to\infty}[1 - (i+1)/n]^n = \exp(-i-1)$, it can be simplified to

$$P_{max} = 2P_{maxo} - \left(1 - \frac{1}{e}\right) \sum_{i=0}^{n-1} e^{-i} P_{maxo-i}. \quad (4.29)$$

As expected, the estimator of the upper bound of a random variable is a function of the differences of its largest observations, $P_{maxoi} - P_{maxoi-1}$, and only the first few differences are significant in estimating P_{max}.

4.7 Expected Maximum Event Size

The estimate can also be made on log-order statistics, i.e. $\log P_{(1)},...,\log P_{(n)}$, which is slightly more conservative, since $\langle \log P \rangle \leq \log(\langle P \rangle)$. Then Eq. (4.28) gives

$$\log P_{max} = \log P_{maxo} + \Delta \log P_{max}, \quad (4.30)$$

where $\log P_{maxo}$ is the maximum observed $\log P$ and $\Delta \log P_{max}$ is the maximum expected jump from the history of observed jumps in record $\log P$,

$$\Delta \log P_{max} = 2 \max (\Delta \log P_{maxo})$$
$$- \sum_{i=0}^{n-1} \left[\left(1 - \frac{i}{n}\right)^n - \left(1 - \frac{i+1}{n}\right)^n \right] \Delta \log P_{maxo-i}. \quad (4.31)$$

The order statistics estimate of P_{max} or $\log P_{max}$ is independent of the underlying probability distribution, and therefore, one can make a reasonable estimate even if the data does not conform to any accepted potency frequency power law.

Note that in most cases seismic systems are installed in operating mines, and therefore a part of the history of seismic records is missed. It is very likely then to record larger record jumps at the beginning of the monitoring period that may not reflect the true history of record breaking events. If included in calculations, these false large record jumps will overstate hazard and need to be ignored. Initially, this introduces a degree of discretion, but it fades with time as new records occur.

Figure 4.14 left shows the history of record breaking $\log P$ in MineD. The data set, called here DataSet1, starts on 07 September 2007 and ends on 13 June 2012, just after the largest event with $\log P = 2.24$ ($m_{HK} = 2.41$), and includes 2281 events with $\log P \geq -1.0$ ($m_{HK} \geq 0.25$). Search for the first record started from the mean of all $\log P \geq 0.5$, which was $\log P = 0.88$, and delivered eight forward counting records. The expected upper limit of the next record breaking event estimated by Eq. (4.30) delivered $\log P_{max} = 2.684$. Table 4.1 lists $\log P$

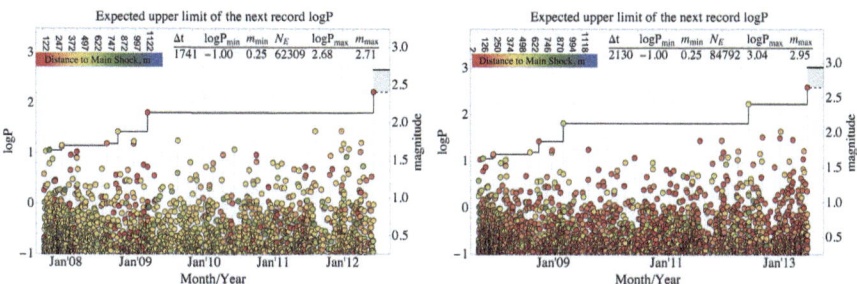

Fig. 4.14 History of the record breaking events and the estimated range of $\log P_{max}$ for MineD based on two overlapping data sets, DataSet1 *(left)* and DataSet2 *(right)*

Table 4.1 List of eight forward counting log P records and eight largest log P events at MineD over the period 07 September 2007 to 13 June 2012

Record log P	1.05	1.07	1.14	1.16	1.20	1.44	1.82	2.24
Largest log P	1.45	1.46	1.47	1.51	1.53	1.65	1.82	2.24

and m_{HK} of all eight forward counting records and the eight largest events. Note that the largest events are not necessarily records.

Figure 4.14 right shows the history of record breaking log P in the same mine for the period 07 September 2007 to 07 July 2013, called here DataSet2, which is 388.8 days longer than DataSet1, and stops just after the next record breaking event on 07 July 2013 with log $P = 2.61$ ($m_{HK} = 2.66$). The new record log $P = 2.61$ is close to the upper limit of the predicted range $2.24 \leq \log P \leq 2.684$.

4.7.4 Record Statistics—The Next Record Breaking Event

A record is an entry in a chronological sequence of data that exceeds all previous entries. Let us ask the following question: Given a set of random observations in time, how often will the record value be surpassed? In a sequence P_j for $j = 1, 2,..., n$ of real independent and identical distributed variables in a time series, a record breaking high occurs at k if $P_k = \max_{j \leq k}(P_j)$. P_1 is always a record. The probability that a record high, or low, occurs at j is $1/j$. The probability of having exactly k records in n samples can be calculated by a recursive formula, $\Pr(k, n) = (1 - 1/n) \Pr(k, n-1) + (1/n) \Pr(k-1, n-1)$ for $k \geq 1$ and $n \geq 2$ with the initial values $\Pr(k = 0, n > 1) = 0$ and $\Pr(k = 1, n = 1) = 1$, see Fig. 4.15 dashed lines. For larger sample size n, the asymptotic result gives $\Pr(k, n) = [\ln(n)]^{k-1} / [n(k-1)!]$. The probability of having at least k records in n samples then is $\Pr(\geq k, n) = 1 - \sum_{j=1}^{k} \Pr(k, n)$, see Fig. 4.15 solid lines.

The expected number of records high (or low), $\langle N_{rb} \rangle$, is $\sum_{j=1}^{n}(1/j)$, which is a harmonic sequence that grows without bound, though rather slowly. The proof that harmonic sequences diverge was provided by Oresme (c.1323 - 1382) in the fourteenth century that was published only in fifteenth century by Johannes de Sancto Martino, Oresme, 1482. This supports the idea that any record can be beaten, but the time for this becomes longer and longer with the increasing number of records. For large n, the number of records can be approximated by $\langle N_{rb}(n) \rangle \approx \ln(n) + 0.577215$, and the variance $Var(N_{rb}) = \sum_{j=1}^{n}(1/j) - \sum(1/j^2) \approx \ln(n) - 1.0677$ (Table 4.2).

The observed frequencies of record highs (or lows) can be used to infer whether or not the data set is random. The number of record breaking events can be determined both with time running forwards and with time running backwards. If future samples behave like earlier samples, then the number of records calculated forwards would be equal to the number of records calculated backwards. Statisti-

4.7 Expected Maximum Event Size

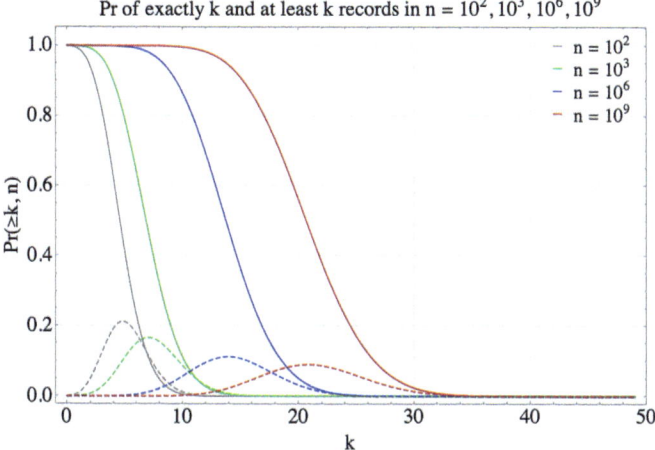

Fig. 4.15 Probabilities of having exactly k records, Pr (k, n), dashed lines, and at least k records, Pr $(\geq k, n)$, solid lines, in n observations

Table 4.2 The expected number of records in n independent observations

n	10^1	10^2	10^3	10^4	10^5	10^6
$\langle N_{rb}\rangle$	2.93	5.19	7.49	9.8	12.1	14.4
$\sqrt{Var}$	1.17	1.88	2.42	2.85	3.23	3.57

cally significantly more forward record breaking events signify an increased hazard and more backward records indicate that hazard is abating. The record sequence is distinctly non-stationary with increasing time, i.e. it becomes exponentially harder to beat the current record.

If P_{maxo} is the maximum potency in a data set of n observations, then, for a random data set, the probability of having a larger potency in the next n_n new observations is Pr$(> P_{maxo}, n_n) = n_n/(n + n_n)$. For large n, the number of new observations needed to beat the current record with probability Pr is given by $n_{Pr} = n \cdot \text{Pr} / (1 - \text{Pr})$. So, there is 60% probability of beating the current record in $1.5n$ new observations.

4.7.5 The Expected Next Record Breaking Potency

Suppose that the selected data set on seismic potency is distributed according to the upper truncated (UT) power law with probability density function $f(P) = \beta P^{-\beta-1} / \left(P_{min}^{-\beta} - P_{max}^{-\beta}\right)$, where P_{min}, P_{max}, and β are parameters. The first randomly selected potency is, by definition, the first record breaking event, and it will most likely be the mean value of that distribution, $P_{nrb(1)} =$

$[\beta/(1-\beta)]\left(P_{max}^{1-\beta}-P_{min}^{1-\beta}\right)/\left(P_{min}^{-\beta}-P_{max}^{-\beta}\right)$ for $\beta \neq 1$, or $P_{nrb(1)} = \ln\left[(P_{max}/P_{min})/\left(P_{min}^{-1}-P_{max}^{-1}\right)\right]$ for $\beta = 1$. The value of the second record breaking event will be the mean value of that portion of that distribution that lies beyond the first record, $P_{nrb(2)} = \int_{P_{rb(1)}}^{P_{max}} P f(P) dP / \int_{P_{nrb(1)}}^{P_{max}} f(P) dP$, and proceeding recursively, the relation between successive typical record potencies can be given by

$$P_{nrb(k)} = \frac{\beta\left(P_{max}^{1-\beta}-P_{nrb(k-1)}^{1-\beta}\right)}{\left[(1-\beta)\left(P_{nrb(k-1)}^{-\beta}-P_{max}^{-\beta}\right)\right]} \quad \text{or} \quad P_{nrb(k)} = \frac{\ln\left[P_{max}/P_{nrb(k-1)}\right]}{P_{nrb(k-1)}^{-1}-P_{max}^{-1}}, \tag{4.32}$$

for $\beta \neq 1$ (left) and $\beta = 1$ (right), and for $k = 2, 3, ...$, where $P_{nrb(k-1)}$ is the potency of the previous (or the last) record breaking event (Mendecki, 2012). Note that $P_{nrb(k)}$ given by Eq. (4.32) are not particularly sensitive to changes in β. Applying a similar procedure, the next record breaking potency for the OET relation is

$$P_{nrb(k)} = P_{nrb(k-1)} + P_{nrb(k-1)}^{\beta} P_c^{1-\beta} \exp\left[P_{nrb(k-1)}/P_c\right] \Gamma\left[1-\beta, \frac{P_{nrb(k-1)}}{P_c}\right]. \tag{4.33}$$

Figure 4.16 shows the histories of records breaking $\log P$ in MineD, as in Fig. 4.14, and the estimated expected next record $\log P$ assuming the UT power law distribution, in red, and OET distribution in green.

The power law exponent for the UT distribution of DataSet1 is $\beta = 0.949$ and for the OET $\beta = 0.951$. For DataSet2 the respected values are $\beta = 0.941$ and $\beta = 0.942$. Assuming the UT distribution, the expected next record breaking $\log P_{nrb} = 2.45$ for DataSet1 and $\log P_{nrb} = 2.81$ for DataSet2. Assuming the OET distribution, the $\log P_c = 1.85$ and the $\log P_{nrb} = 2.36$ for DataSet1 and $\log P_c = 2.41$ and $\log P_{nrb} = 2.77$ for DataSet2.

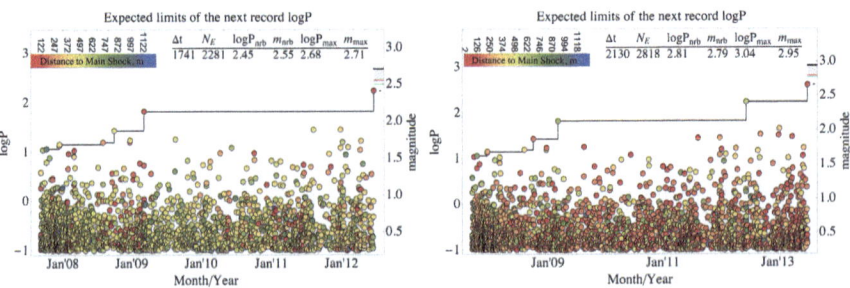

Fig. 4.16 History of the record breaking events and the estimated range of $\log P_{nrb}$ for MineD based on two overlapping data sets assuming UT, in red, and OET distribution in green

4.7 Expected Maximum Event Size

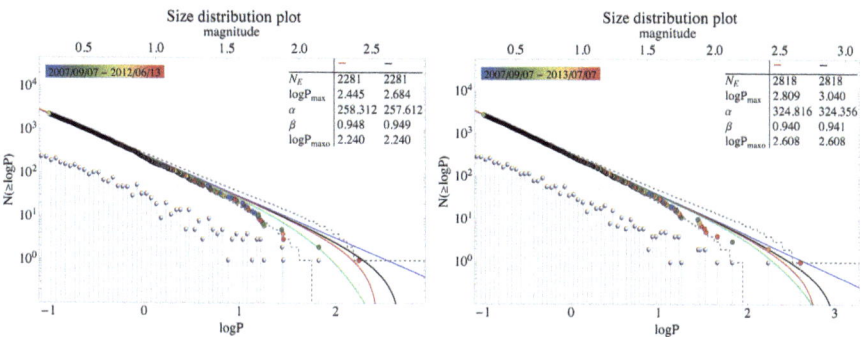

Fig. 4.17 Cumulative number of unbinned data vs. log P for the first *(left)* and second data sets *(right)* in MineD. The three fits marked by black red and green solid lines are described in text

Since it has been published in Mendecki (2012), the method has been applied to data sets from a number of mines in Australia, Africa, South America, Canada, and Europe and, with a few exceptions of extreme outliers, delivered reasonable results. Recently the method has been applied to injection-induced seismicity by Cao et al. (2020) and Verdon and Bommer (2020).

Figure 4.17 left shows the cumulative number of unbinned data vs. $\log P \geq -1.0$ for DataSet1, where colour indicates the time of the event, with three fits. (1) The upper truncated (UT) potency frequency relation assuming $\log P_{max} = 2.684$ and the respective $\alpha = 257.612$ and $\beta = 0.949$, in black. (2) The upper truncated (UT) potency frequency relation assuming $\log P_{max} = 2.45$ and the respective $\alpha = 258.312$ and $\beta = 0.948$ in red. (3) The OET one with the soft upper cut-off $\log P_c = 1.85$ and the respective $\alpha = 256.594$ and $\beta = 0.951$, in green. The blue straight line represents the OE power law and is given as a reference. The light grey vertical spikes below the data illustrate what would be the empirical pdf if the data was binned with bin size 0.05. The dashed grey lines on both sides of the UT power law fit indicate 95% confidence limits, see Eq. (4.18).

Figure 4.17 right shows the cumulative number of unbinned data vs. $\log P \geq -1.0$ for DataSet2 with the three fits. (1) The upper truncated (UT) potency frequency relation assuming $\log P_{max} = 3.04$ with the respective $\alpha = 324.356$ and $\beta = 0.941$ in black. (2) The upper truncated (UT) potency frequency relation assuming $\log P_{max} = 2.81$ with the respective $\alpha = 324.816$ and $\beta = 0.94$ in red. (3) The OET one with the soft upper cut-off $\log P_c = 2.41$ with $\alpha = 323.68$ and $\beta = 0.942$, in green.

4.7.6 Rank Plot

Rank plot of seismic potency, or energy, is a presentation of the potency, or energy, frequency data but with flipped axes, where $\log P$ is on the vertical axis and rank,

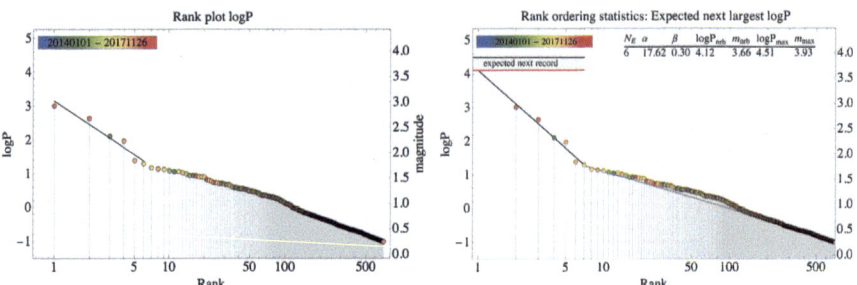

Fig. 4.18 Rank-ordered log P before shifting *(left)* and after with the estimated range of log P_{nrb} *(right)*

or $N (\geq \log P)$ on the horizontal axis. The statement that "the nth largest event has log P" is equivalent to "there are n events greater or equal to log P". Inverting the axes also inverts the exponent from β to $1/\beta$.

Rank statistics is a useful tool to test the continuity of the power law size distribution over the entire range of available data and to estimate the next record breaking event size, log P_{nrb}, in cases where a small number of largest events deviate from the assumed power law (Sornette et al., 1996).

Firstly, all events are ranked so the largest log P observed, log P_{maxo}, is in rank 1, the second largest, log P_{maxo-1}, in rank 2, the third largest log P_{maxo-2} in rank 3, et cetera. Note that the ranked series of events will almost never be in a chronological order. The rank-ordering plot depicts log P vs. the logarithm of the rank. Then, the largest observed event, log P_{maxo}, is shifted to rank 2, the second largest observed event, log P_{maxo-1}, to rank 3, and so on. A fit to the first n largest events in the shifted rank-ordering plot extrapolated to $n = 1$ gives the estimate of the log P_{nrb} event. The log P_{max} is defined as the upper 95% confidence limit of log P_{nrb}. The choice of n depends on the structure of the rank-ordering plot. If the rank-ordering plot has two branches, the cross-over point would be the obvious candidate to select n. In most cases studied, the cross-over point is $10 \leq n \leq 50$, and therefore, log P_{nrb} is constrained by a small number of the largest events.

Figure 4.18 left shows the rank plot of log P recorded at a mine where one can differentiate the two branches of the data and to illustrate the method. The cross-over point is assumed at the 6th largest event. Before shifting, the least square fit to the largest six events gives $\alpha = 24.12$ and $\beta = 0.44$.

Figure 4.18 right shows the rank plot of log P of the same data set after shifting. The least square fit to the six largest events gives $\alpha = 17.62$ and $\beta = 0.30$, which at the intersection with rank 1 gives log $P_{nrb} = 4.12 \pm 19.5$, and therefore one can assume the upper limit of log $P_{nrb} = \log P_{max} = 4.51$.

Note that in this case the upper branch has a steeper slope, i.e. lower beta in the size distribution than the lower branch, which is the opposite to what is expected for tectonic earthquakes. It frequently happens during the end of mining operations when the extraction ratio is high and/or when mining approached larger geological structures with the shear strength comparable to induced stresses.

References

Aki, K. (1965). Maximum likelihood estimate of b in the formula logN=a-bm and its confidence limits. *Bulletin Earthquake Research Institute Tokyo University, 43*, 237–239.

Aki, K. (1972). Earthquake mechanism. *Tectonophysics, 13*(1-4), 423–446.

Amitrano, D. (2003). Brittle-ductile transition and associated seismicity: Experimental and numerical studies and relationship with the b-value. *Journal of Geophysical Research, 108 (B1)*(2044), 1–15. http://doi.org/10.1029/2001JB000680.

Burroughs, S. M., & Tebbens, S. (2001). Upper-truncated power laws in natural systems. *Pure and Applied Geophysics, 158*(4), 741–757.

Cao, N.-T., Eisner, L., & Jechumtalova, Z. (2020). Next record breaking magnitude for injection induced seismicity. *First Break, 38*.

Choy, G. L., McGarr, A., Kirby, S., & Boatwright, J. (2006). An overview of the global variability in radiated energy and apparent stress. In R. E. Abercrombie, A. McGarr, G. D. Toro, & H. Kanamori (Eds.), *Earthquakes: Radiated Energy and the Physics of Faulting* (pp. 43–58). American Geophysical Union.

Christensen, K., Danon, L., Scanlon, T., & Bak, P. (2002). Unified scaling law for earthquakes. *PNAS, 99*(1), 2509–2513.

Cooke, P. (1979). Statistical inference for bounds of random variables. *Biometrica, 66*(2), 367–374.

Cosentino, P., Ficarra, V., & Luzio, D. (1977). Truncated exponential frequency-magnitude relationship in earthquake statistics. *Bulletin of the Seismological Society of America, 67*(6), 1615–1623.

Galileo, G. (1638). *Dialogues concerning two new sciences*. Lodewijk Elzevir.

Gutenberg, B., & Richter, C. F. (1944). Frequency of earthquakes in California. *Bulletin of the Seismological Society of America, 34*, 185–188.

Hanks, T. C. (1992). Small earthquakes, tectonic forces. *Science, 256*, 1430–1431.

Helmstetter, A., Kagan, Y. Y., & Jackson, D. D. (2005). Importance of small earthquakes for stress transfers and earthquake triggering. *Journal of Geophysical Research, 110*(B05S08), 1–13. http://doi.org/10.1029/2004JB003286.

Jackson, D. D., & Kagan, Y. Y. (2006). The 2004 Parkfield earthquake, the 1985 prediction, and characteristic earthquakes: Lessons for the future. *Bulletin of the Seismological Society of America*, S397–96S409. http://doi.org/10.1785/0120050821.

Kagan, Y. Y. (1993). Statistics of characteristic earthquakes. *Bulletin of the Seismological Society of America, 83*(1), 7–24.

Kagan, Y. Y., & Schoenberg, F. P. (2001). Estimation of the upper cutoff parameter for the tapered pareto distribution. *Journal of Applied Probability, 38A*, 158–175.

Kagan, Y. Y., Jackson, D. D., & Geller, R. J. (2012). Characteristic earthquake model, 1884-2011, R.I.P. *Seismological Research Letters, 83*(6), 951–953.

Kijko, A. (2004). Estimation of the maximum earthquake magnitude, Mmax. *Pure and Applied Geophysics, 161*(8), 1655–1681. http://doi.org/10.1007/s00024-004-2531-4.

Kijko, A., & Funk, C. W. (1994). The assessment of seismic hazards in mines. *The Journal of The Southern African Institute of Mining and Metallurgy, 94*(7), 179–185.

Kijko, A., & Singh, M. (2011). Statistical tools for maximum possible earthquake magnitude estimation. *Acta Geophysica, 59*(4), 674–700. http://doi.org/10.2478/s11600-011-0012-6.

Kijko, A., & Stankiewicz, T. (1987). Bimodal character of the distribution of extreme seismic events in Polish mines. *Acta Geophysica Polonica, 35*, 491–506.

Kwiatek, G., Plenkers, K., Nakatani, M., Yabe, Y., & Dresen, G. (2010). Frequency-magnitude characteristics down to magnitude −4.4 for induced seismicity recorded at Mponeng gold mine, South Africa. *Bulletin of the Seismological Society of America, 100*(3), 1165–1173.

Laherrere, J., & Sornette, D. (1999). Stretched exponential distributions in nature and economy: Fat tails with characteristic scales. *European Physical Journal B*, (2), 525–539.

Linkov, A. M. (2006). Numerical modeling of seismic and aseismic events in three dimensional problems of rock mechanics. *Journal of Mining Science, 42*(1), 1–14.

Linkov, A. M. (2013). Numerical modelling of seismicity: Theory and applications, Keynote lecture. In A. Malovichko, & D. Malovichko (Eds.), *8th International Symposium on Rockbursts and Seismicity in Mines, Russia* (pp. 197–218).

Main, I. G., & Burton, P. W. (1984). Physical links between crustal deformation, seismic moment and seismic hazard for regions of varying seismicity. *Geophysical Journal of the Royal Astronomical Society, 79*(2), 469–488.

Malovichko, D., & Basson, G. (2014). Simulation of mining induced seismicity using Salamon-Linkov method. In M. Hudyma, & Y. Potvin (Eds.), *7th International Conference on Deep and High Stress Mining* (pp. 667–680). http://doi.org/10.13140/2.1.1365.0561.

McGarr, A. (1976). Upper limit to earthquake size. *Nature, 262*(5567), 378–379.

McGarr, A. (1984). Some applications of seismic source mechanism studies to assessing underground hazard. In N. C. Gay, & E. H. Wainwright (Eds.), *Proceedings 1st International Symposium on Rockbursts and Seismicity in Mines, Johannesburg, South Africa* (pp. 199–208). South African Institute of Mining and Metallurgy.

McGarr, A. (1993). Keynote Address: Factors influencing the strong ground motion from mining induced tremors. In R. P. Young (Ed.), *Proceedings 3rd International Symposium on Rockbursts and Seismicity in Mines, Kingston, Ontario, Canada* (pp. 3–12). Balkema, Rotterdam.

McGarr, A., Simpson, D., & Seeber, L. (2002). Case histories of induced and triggered seismicity. In W. H. K. Lee, H. Kanamori, P. C. Jennings, & C. Kisslinger (Eds.), *International Handbook of Earthquake and Engineering Seismology* (pp. 647–661). Academic Press.

Mendecki, A. J. (1993). Real time quantitative seismology in mines: Keynote Address. In R. P. Young (Ed.), *Proceedings 3rd International Symposium on Rockbursts and Seismicity in Mines, Kingston, Ontario, Canada* (pp. 287–295). Balkema, Rotterdam.

Mendecki, A. J. (2008). Forecasting seismic hazard in mines. In Y. Potvin, J. Carter, A. Diskin, & R. Jeffrey (Eds.), *Proceedings 1st Southern Hemisphere International Rock Mechanics Symposium, Perth, Australia* (pp. 55–69). Australian Centre for Geomechanics.

Mendecki, A. J. (2012). Size distribution of seismic events in mines. In *Proceedings of the Australian Earthquake Engineering Society 2012 Conference, Queensland*, pp. 1–20.

Mendecki, A. J. (2013). Characteristics of seismic hazard in mines: Keynote Lecture. In A. Malovichko, & D. A. Malovichko (Eds.), *Proceedings 8th International Symposium on Rockbursts and Seismicity in Mines, St Petersburg-Moscow, Russia* (pp. 275–292). ISBN:978-5-903258-28-4.

Mendecki, A. J., & G. van Aswegen (1998). System stiffness and seismic characteristics - A case study. In M. Ando (Ed.), *International Workshop on Frontiers in Monitoring Science and Technology for Earthquake Environments, Japan* (pp. F4–2).

Mendecki, A. J., van Aswegen, G., Brown, J. N. R., & Hewlett, P. (1988). The Welkom seismological network. In C. Fairhurst (Ed.), *3rd International Symposium on Rockbursts and Seismicity in Mines, 08-10 June 1988, Minneapolis, USA* (pp. 237–244), Balkema, Rotterdam, 1990.

Mogi, K. (1962). Magnitude frequency relations for elastic shocks accompanying fractures of various materials and some related problems in earthquakes. *Bulletin Earthquake Research Institute of Tokyo University, 40*, 831–853.

Molnar, P. (1979). Earthquake recurrence intervals and plate tectonics. *Bulletin of the Seismological Society of America, 69*(1), 115–133.

Mori, J., & Abercrombie, R. E. (1997). Depth dependence of earthquake frequency-magnitude distribution in California: Implications for rupture initiation. *Journal of Geophysical Research, 102*(B7), 15,081–15,090.

Oresme, N. (1482). *Tractatus de Configuratione Qualitatum et Motuum*. Johannes de Sancto Martino.

Page, R. (1968). Aftershocks and microaftershocks of the great Alaska earthquake of 1964. *Bulletin of the Seismological Society of America, 58*(3), 1131–1168.

Pisarenko, V. F., Sornette, A., Sornette, D., & Rodkin, M. V. (2008). New approach to the characterization of Mmax and of the tail of the distribution of earthquake magnitudes. *Pure and Applied Geophysics*, *165*(5), 847–888. http://doi.org/10.1007/s00024-008-0341-9.

Robson, D. S., & Whitlock, J. H. (1964). Estimation of a truncation point. *Biometrica*, *51*(1-2), 33–39.

Scholz, C. H. (1968). Microfractures, aftershocks, and seismicity. *Bulletin of the Seismological Society of America*, *58*(3), 1117–1130.

Shannon, C. E. (1948). A mathematical theory of communication. Part 2. *Bell System Technical Journal*, *27*, 623–656.

Shi, Y., & Bolt, B. A. (1982). The standard error of the magnitude-frequency b value. *Bulletin of the Seismological Society of America*, *72*(5), 1677–1687.

Smith, S. W. (1976). Determination of maximum earthquake magnitude. *Geophysical Research Letters*, *3*(6), 351–354.

Sornette, D. (2009). Dragon-kings, black swans and the prediction of crises. *International Journal of Terraspace Science and Engineering*, *2*(1), 1–18.

Sornette, D., Knopoff, L., Kagan, Y. Y., & Vanneste, C. (1996). Rank-ordering statistics of extreme events: Applications to the distribution of large earthquakes. *Journal of Geophysical Research*, *101*(B6), 13,883–13,893.

Taleb, N. N. (2007). *The black swan: The impact of the highly improbable*. Random House.

Utsu, T. (1964). Estimation of b-value in Giutemberg-Ricter formula logN = a - bm, paper read at the meeting of the Seismological Society of Japan.

van Aswegen, G., & Butler, A. G. (1993). Applications of quantitative seismology in South African gold mines. In R. P. Young (Ed.), *Proceedings 3rd International Symposium on Rockbursts and Seismicity in Mines, Kingston, Ontario, Canada* (pp. 261–266). Balkema, Rotterdam. ISBN: 9054103205.

van Aswegen, G., & Mendecki, A. J. (1999). Mine layout, geological features and seismic hazard. *Final report gap 303*, Safety in Mines Research Advisory Committee, South Africa, 1–91.

Van Der Watt, P. (1980). Note on estimation of bounds of random variables. *Biometrica*, *67*(3), 712–714.

Verdon, J. P., & Bommer, J. J. (2020). Green, yellow, red, or out of the blue? An assessment of traffic light schemes to mitigate the impact of hydraulic fracturing-induced seismicity. *Journal of Seismology*. http://doi.org/10.1007/s10950-020-09966-9.

Vere-Jones, D., Robinson, R., & Yang, W. (2001). Remarks on the accelerated moment release model: problems of model formulation, simulation and estimation. *Geophysical Journal International*, *144*, 517–531.

Ward, S. N. (1997). More on Mmax. *Bulletin of the Seismological Society of America*, *87*(5), 1199–1208.

Weingartner, P. (1996). Under what transforrnations are laws invariant? *Laws and predictionin the light of chaos research*. Springer.

Wesnousky, S. G. (1994). The Gutenberg-Richter or characteristic earthquake distribution, which is it? *Bulletin of the Seismological Society of America*, *84*(6), 1940–1959.

Wesnousky, S. G., Scholz, C. H., Shimazaki, K., & Matsuda, T. (1983). Earthquake frequency distribution and the mechanics of faulting. *Journal of Geophysical Research*, *88*(B11), 9331–9340.

Wiemer, S., & Wyss, M. (2002). Mapping spatial variability of the frequency-magnitude distribution of earthquakes. In R. Dmowska, & B. Saltzman (Eds.), *Advances in Geophysics* (vol. 45, pp. 259–302). Elsevier.

Wyss, M. (1973). Towards a physical understanding of the earthquake frequency distribution. *Geophysical Journal of the Royal Astronomical Society*, *31*(4), 341–359.

Zoller, G. (2022). A note on the estimation of the maximum possible earthquake magnitude based on extreme value theory for the Groningen gas field. *Bulletin of the Seismological Society of America*, *112*(4), 1825–1831.

Zoller, G., & Holschneider, M. (2014). Induced seismicity: What is the size of the largest expected earthquake? *Bulletin of the Seismological Society of America*, *104*(6), 3153–3158. http://doi.org/10.1785/0120140195.

Zoller, G., & Holschneider, M. (2016). The earthquake history in a fault zone tells us almost nothing about mmax. *Seismological Research Letters*, *87*, 132–137. http://doi.org/10.1785/022015017.

Open Access This chapter is licensed under the terms of the Creative Commons Attribution 4.0 International License (http://creativecommons.org/licenses/by/4.0/), which permits use, sharing, adaptation, distribution and reproduction in any medium or format, as long as you give appropriate credit to the original author(s) and the source, provide a link to the Creative Commons license and indicate if changes were made.

The images or other third party material in this chapter are included in the chapter's Creative Commons license, unless indicated otherwise in a credit line to the material. If material is not included in the chapter's Creative Commons license and your intended use is not permitted by statutory regulation or exceeds the permitted use, you will need to obtain permission directly from the copyright holder.

Chapter 5
Time Distribution and Seismic Hazard

Abstract This chapter starts with the time characteristics of seismic activity, the coefficient of variation, and proportional variability and progresses to stationary Poisson processes, probabilities of exceedance, and mean recurrence times. In this section, there are two practical examples. The first one compares seismic hazard between three mines with different volumes of rock extracted over the same time period at different depths, and it is done in the $\log E$ domain. It shows the utility of plotting the cumulative potency vs. the cumulative volume of rock extracted as an indication of seismic hazard. The second example compares the seismic hazard characteristics of two mostly overlapping seismic and rock extraction data sets selected from the same mine, and it is done in the $\log P$ domain. Having partial data on the volume of rock extraction, it postulates that apart from increasing extraction ratio, the more concentrated production blasting at greater depth may have contributed to the observed increase in seismic hazard.

The rest of this chapter is dedicated to the seismic rock mass response to step loading, mainly by production blasting and by larger seismic events. Seismic rock mass response to blasting is driven strongly by the stress level in rock surrounding the blast and by the volume of rock blasted. In most cases, rock extraction by blasting induces and triggers seismic events immediately and in close proximity to the blast. In some cases, the rate of activity follows typical aftershock sequences that can be described by the simple Omori law or by the stretched exponential relaxation function. Here, the non-stationary Poisson process needs to be invoked to assess the probabilities of having larger events during the aftershock sequences. Two examples of seismic hazard assessment after large events are presented.

In practice, mines frequently group blasts to optimise the required exclusion time after blasting. It is important then to schedule blasting sequences to limit the overlap of seismic relaxation. If larger blasts are too frequent and/or too close to each other, i.e. if the next blast is still during the excitation phase of the previous one, it may push a larger and larger part of the system into the sub-critical stage. There is also the possibility that such blasting may induce a larger seismic event that would not have happened otherwise, as opposed to advancing the clock for events that are almost ready to be triggered. This issue is addressed in the section Seismic Rock Mass Response and Blasting Sequence, where the defined proximity index may help to test whether the preferred time differences and distances between blasts sequence of a planned sequence are at an acceptable level.

5.1 Time Characteristics of Seismic Activity

Stationarity is a property of an underlying stochastic process, and not of observed data. They would be characterised by a constant mean and a constant standard deviation. Non-stationary processes have statistical properties that are deterministic functions of time. All natural processes are non-stationary, and therefore, the question is whether the underlying non-stationarity is strong enough to justify building a complex deterministic model of the process. In many cases over longer time, a simple stationary stochastic model may represent the process adequately.

A renewal process is a point process characterised by the fact that the successive inter-arrival times are distributed with the same probability density function. A Poisson process is a renewal process in which the inter-arrival times are exponentially distributed, $f(t) = \lambda \exp(-\lambda t)$, where the rate of events, λ, is constant. In general, the superposition of a number of mutually independent renewal processes converges to Poisson process.

A homogeneous Poisson process results in a random series of events occurring at a constant rate, λ, in time or at a constant density in space. It is characterised by independence, stationarity, and orderliness. Independence means that the occurrence of a given event is not influenced by the occurrence of any other event in the group—they are not correlated. Stationarity means that the rate λ, or the underlying probability distribution, is constant over time or space, but it does not mean that all time or space intervals are equally likely.

While in a Poisson process occurrences are very irregular, the probability of an event remains constant regardless of how much time elapsed since a previous event. Orderliness precludes the possibility of multiple events at a single point in time or space or the possibility of an infinite number of events in a finite interval of time or space.

Over the short term, small seismic events in mines tend to occur in clusters in space and in time in response to rock extraction. Such processes are neither stationary nor independent. However, if a number of such processes are superimposed over a longer time and over a larger area, the outcome may not be far from Poissonian. Assumption of stationarity is very useful because it facilitates statistical analysis where probabilities are well defined as limits of frequency of occurrence.

Larger seismic events in mines are less clustered in time and in space than smaller events. Hence, the Poisson process may provide an asymptotic fit to the distribution of larger events in the time or volume mined domains. However, there are cases where larger events are clustered in time, specifically at the later stages of mining when the extraction ratio is higher and mining is deeper. While stationary models may apply to the intermediate and long term hazard assessment, over the short term one needs to resort to non-stationary models.

5.1 Time Characteristics of Seismic Activity

Note that in mines the main problem in analysing the spatial and temporal characteristics of seismicity is gaps in the data mainly due to frequent power failures. Therefore, it is advisable to test the continuity of data and, in some cases, split data into uninterrupted sections.

5.1.1 Coefficient of Variation and Proportional Variability

Coefficient of Variation In 1896, Pearson introduced a number of statistical measures, among them the coefficient of variation, C_v, as the ratio of the standard deviation to the mean. For the data set $x = x_1, \ldots, x_n$,

$$C_v = \frac{s_d}{\bar{x}} = \sqrt{\frac{1}{n}\sum_{i=1}^{n}(x_i - \bar{x})^2} \bigg/ \left(\frac{1}{n}\sum_{i=1}^{n} x_i\right) = \sqrt{\frac{1}{n}\sum_{i=1}^{n}\left(\frac{x_i - \bar{x}}{\bar{x}}\right)^2}, \quad (5.1)$$

where s_d is the sample standard deviation and $\bar{x}$ is the sample mean. If $t_1, t_2, \ldots, t_n$ are the times of n consecutive seismic events and $t_d = t_2 - t_1, \ldots, t_n - t_{n-1}$ are the $n-1$ respective time differences between them, then $C_v(t) = s_d(t_d)/\bar{t}_d$ can be used to test for the time distribution of these events.

The time distribution of seismic activity can be either random, quasi-periodic, or clustered. If $C_v(t) \ll 1$, the process is close to periodic oscillations, and if $0 < C_v(t) < 1$, the process is quasi-periodic. For a Poissonian process, where the standard deviation is equal to the mean, $C_v(t) = 1$. For a clustered process, $C_v(t) > 1$. For a power law distribution of inter-event times, the clustering takes an extreme form, and the mean inter-occurrence time, the standard deviation, as well as the coefficient of variation tend to infinity as the observation time increases. This form of clustering is called fractal.

Note that the fact that in a Poisson process the time intervals between events of a given size are exponentially distributed means that short intervals are more probable than long ones and events tend to cluster in small groups, separated by longer spacings. This apparent clustering does not indicate any greater dependence between the closely spaced events than between the more distant ones. In statistical physics, the stationary Poissonian process is considered as a prototype of disorder.

If the inter-event time distribution for events greater than log P is thin tailed, e.g. the periodic, uniform, semi-Gaussian, and the stretched exponential with exponent $q > 1.0$, then the longer it has been since the last event of that size the shorter the expected time till the next one. However, if this distribution is thick tailed, e.g. the power law or the stretched exponential with exponent $q < 1.0$, then the longer it has been since the last event of that size the longer the expected time till the next one (Davis et al., 1989; Sornette & Knopoff, 1997). The case $C_v(t) = 1$ is memoryless; therefore, the expected time until the next event greater than or equal to log P is independent of time.

In practice, we have a data set of n events with time span $\Delta t = t_n - t_1$, all above a given $\log P_{min}$, and wish to calculate the coefficient of variation of events $C_v(t)$ of events with $\log P \geq \log P_{min}$. In this case, the time difference $\Delta t (\geq \log P) = t_n (\geq \log P) - t_1 (\geq \log P)$ may be significantly smaller than $\Delta t = t_n - t_1$. This underestimates the mean recurrence intervals; therefore, it overestimates seismic hazard and needs to be corrected. One way to correct is to "stretch" the observed inter-event times by $\Delta t / \Delta t (\geq \log P)$, so that they would sum to Δt.

The coefficient of variation is not a perfect measure of clustering, and it has its drawbacks: (1) It is problematic when the data are both positive and negative. (2) When the mean value is near zero, the coefficient of variation is sensitive to small changes in the mean, limiting its usefulness. (3) It lacks an upper bound; therefore, it is difficult to interpret. (4) It does not identify the timescales involved, so it is always useful to quote the mean as a reference. (5) It is sensitive to outliers. (6) The estimate of the coefficient of variation is a negatively biased quantity, and the bias increases as the sample size decreases. Haldane (1955) derived a small sample correction, $\hat{C}_v = C_v [1 + 1/(4n)]$, but this correction is always smaller than the expected error. Kvalseth (2016) proposed the ratio,

$$C_{v2} = \frac{s_d}{\sqrt{\overline{x^2}}} = \sqrt{\frac{1}{n}\sum_{i=1}^{n}(x_i - \overline{x})^2 / \frac{1}{n}\sum_{i=1}^{n} x_i^2} = \sqrt{\frac{s_d^2}{s_d^2 + \overline{x^2}}}, \quad (5.2)$$

called the second order coefficient of variation which has the following properties. (1) It is well defined for all real-valued variables and data. (2) Takes values between (0, 1) and $C_{v2} = 0$ if $s_d = 0$ and $C_{v2} = 1$ only if $\overline{x} = 0$. (3) It is less affected by outliers and less sensitive to changes in the mean. For a Poissonian process, the $C_v(t) = 1$ and $C_{v2}(t) = \sqrt{1/2} = 0.7071$.

Proportional Variability The proportional variability, P_v, is quantified by comparing the numbers with each other, requiring no assumptions about central tendency or underlying statistical distributions, i.e. it is non-parametric. Like the C_v, it provides a summary of variation where the chronology of the data is irrelevant. For a given data set of n non-negative values $x_i \geq 0$, there are $n_k = n(n-1)/2$ combinations of (x_i, x_j) for which one can calculate the relative difference $D(x_i, x_j)$,

$$P_v = \frac{1}{n_k} \sum_{n_k} D(x_i, x_j), \quad (5.3)$$

where $D(x_i, x_j) = 0$ if $x_i = x_j$ and $D(x_i, x_j) = |x_i - x_j| / \max(x_i, x_j) = 1 - \min(x_i, x_j) / \max(x_i, x_j)$. The P_v varies between 0 and 1, it is the average percentage difference between all combinations of observed differences, therefore, unlike the C_v, it is independent of the deviation from the mean, and it is less sensitive to outliers. It has also been shown that the P_v is more accurate than the coefficient of variation at estimating long term variability from short term data (Heath, 2006).

5.1 Time Characteristics of Seismic Activity

If the data is constant, $P_v = 0$, the time series is a simple horizontal line. The inverse of P_v can be interpreted as stability; therefore, $P_v = 0$ would represent complete stability. If there is variability, the ordered series will be increasing and its steepness indicating the extent of variability. If the ordered series is linearly increasing, then $P_v = 0.5$. If $P_v > 0.5$, the values in the ordered series are increasing in a non-linear way.

For a continuous probability distribution $f(x)$ of a non-negative real variable x, the proportional variability is defined as $P_v = 1 - 2 \int_0^\infty \int_{x_i}^\infty (x_i/x_j) f(x_i) f(x_j) dx_j dx_i$. For the uniform distribution, $f(x) = 1/(b-a)$, with the width $w = (a-b)/2$, the standard deviation, $s_d = (b-a)/(2\sqrt{3})$, and the coefficient of variation, $C_v = (b-a)/[\sqrt{3}(b+a)]$. For $w \ll (a+b)/2$, the P_v is very similar to C_v, and for $a = 0$, the $C_v = 1/\sqrt{3} = 0.577$, and the $P_v = 0.5$. For the exponential distribution, $\lambda \exp(-\lambda x)$, the coefficient of variation $C_v = 1$, and the $P_v = 2(1 - \ln 2) \simeq 0.6137$. The P_v can be calculated even when a mean is not defined, e.g. in case of the OE power law, $f(P) = dF(P)/dP = \beta P_{min}^\beta P^{-\beta-1}$, where $P_v = 1/(1+\beta)$, for any β and irrespective of P_{min}. As expected, the lower the exponent β, the higher the variability.

Example 1 Figure 5.1 shows the time histories of $\log P$ in a mine over 50 months for $\log P_{min} = -2.5$ and $\log P_{min} = -0.5$. The coefficient of variation for 39044 events with $\log P_{min} = -2.5$ is 2.31, which indicates relatively strong time clustering, and for 983 events with $\log P \geq -0.5$ is 1.29, which is not far from Poissonian. The second order coefficient of variation, C_{v2}, dropped from 0.92 to 0.79, and the proportional variability, P_v, dropped only a fraction from 0.73 to 0.73. However, there are cases in mines where larger events are clustered.

Example 2 Figure 5.2 shows the time clustering of two short time sequences of 150 seismic events with $-1.0 \leq \log P \leq 1.0$.

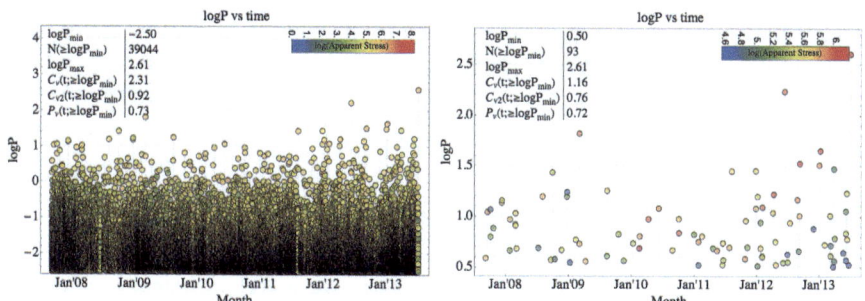

Fig. 5.1 Time histories of $\log P$ in a mine over the time span of 2130 days for two different thresholds of $\log P_{min} = -2.5$ and $\log P_{min} = -0.5$

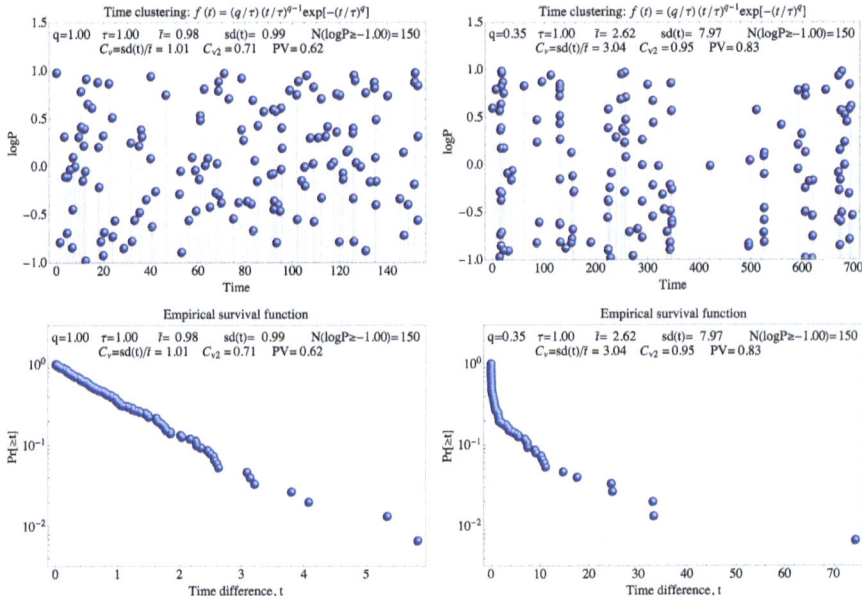

Fig. 5.2 Two time sequences of 150 seismic events simulated by the exponential distribution and the respective empirical survival functions with constant $\tau = 1.0$ for $q = 1.0$ *(left column)* and by the stretched exponential distribution with $q = 0.35$ *(right column)*

Both are simulated by the stretched exponential distribution, $f(t) = (q/\tau)(t/\tau)^{q-1} \exp\left[-(t/\tau)^q\right]$, with $\tau = 1.0$. The first one with $q = 1.0$ that gives the exponential distribution of the inter-event times with $C_v = 1$, $C_{v2} = 0.71$, and $P_v = 0.6$. The second one with $q = 0.35$ which delivers a thick tail power law type distribution in time, therefore strongly clustered with $C_v = 3.02$, $C_{v2} = 0.95$, and $P_v = 0.83$.

The empirical survival function, $\Pr(\geq t)$, for the data set that simulates the exponential distribution is linear for the entire domain of inter-event times, and probabilities of outliers are low. However, the clustered data set with $C_v = 3.02$ delivers a highly non-linear survival function with much higher probabilities of extreme inter-event times.

5.2 Intermediate Term Hazard

The mean recurrence interval between events above a certain potency estimated over the period of time Δt is $\bar{t}(\geq P) = \Delta t / N(\geq P)$, where Δt is the time span of data used and $N(\geq P)$ is the number of events not smaller than P over that time. Seismic activity rate for these events is then $1/\bar{t}(\geq P)$. The mean volume mined between events above a certain size during extraction of V_m volume of rock is $\bar{V}(\geq P) = V_m/N(\geq P)$. The number of events, $N(\geq P)$, per unit of volume of rock extraction is $1/\bar{V}(\geq P)$. Note that the terms "mean inter-event time", "mean

5.2 Intermediate Term Hazard

recurrence interval" may be deceptive since in many cases the dispersion from the mean, as measured by the standard deviation, is comparable with the mean value. If the standard deviation of the observed recurrence intervals is less than 50% of the mean recurrence interval, then the seismic behaviour may be assumed to be periodic rather than episodic, and calculated probability estimates can be considered reasonable. For a very short time interval into the future, $\Delta T < \bar{t} (\geq P)$, the probability of having an event with potency not smaller than P can be estimated as $\Pr(\geq P) \simeq \Delta T/\bar{t} (\geq P)$. The calculated recurrence interval or the mean volume mined that stretch well beyond the span of the data set Δt or V_m should be treated with caution.

5.2.1 Homogeneous Poisson Distribution

The Poisson distribution can be used to forecast the number of random events in a future time, ΔT. There are two random variables that arise from a Poisson process: (1) the number of events in a given time interval, which is a discrete variable, (2) the time until the occurrence of the first event, which is a continuous variable. Having a single event in the time interval Δt, the probability of finding this event in the sub-interval $\Delta T = t_2 - t_1$ is $\Delta T/\Delta t$. Having n events in Δt, the activity rate is $\lambda_t = n/\Delta t$, the recurrence time $\bar{t} = \Delta t/n$, and the expected number of events in ΔT is $\Lambda(\Delta T) = n(\Delta T/\Delta t) = \lambda_t \Delta T$. The probability that exactly $N < n$ events can be found in the interval ΔT is given by the binomial distribution,

$$\Pr(N, \Delta T) = n! [\Lambda(\Delta T)/n]^N \{1 - \Lambda(\Delta T)/n\}^{n-N} / [N!(n-N)!]. \quad (5.4)$$

The mean of the binomial distribution is $n\Delta T/\Delta t$, and the variance is $n\Delta T/\Delta t (1 - \Delta T/\Delta t)$, i.e. the variance is less than the mean. For large n and small $\Delta T/\Delta t$, the binomial distribution can be approximated by Poisson distribution,

$$\Pr(N, \Delta T) = \frac{[\Lambda(\Delta T)]^N}{N!} \exp[-\Lambda(\Delta T)] = \left(\frac{\Delta T}{\bar{t}}\right)^N \frac{1}{N!} \exp(-\Delta T/\bar{t}), \quad (5.5)$$

with the mean value and the variance equal to $\lambda_t \Delta T = n\Delta T/\Delta t$ and the coefficient of variation $C_v = 1/(\lambda_t \Delta T)$. Note that $\Pr(N = 1, \bar{t}) = 0.37$.

The probability that there will be no events in ΔT is $\Pr(N = 0, \Delta T) = \exp(-n\Delta T/\Delta t)$, and the probability that there will be at least one event is $F(\Delta T) = \Pr(N \geq 1, \Delta T) = 1 - \exp(-n\Delta T/\Delta t)$, which is the exponential distribution with the mean $1/\lambda_t$ and the variance $1/\lambda_t^2$, and therefore, the coefficient of variation $C_v = 1$. If $n(\Delta T/\Delta t) < 0.1$, then $\Pr(N \geq 1, \Delta T) \simeq n(\Delta T/\Delta t)$. Note that $\Pr(N \geq 1, \bar{t}) = 0.63$, see Fig. 5.3.

The exponential distribution describes the length of time between events, or the inter-event time distribution, and it does not depend on time. The assumption of stationary and independent increments means that at any point the process probabilistically restarts itself, i.e. it has no memory. The probability density function of an exponential distribution is $f(\Delta T) = \lambda \exp(-\lambda \Delta T)$, and the

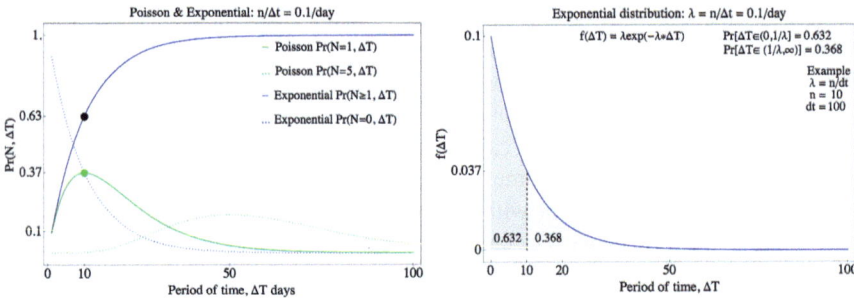

Fig. 5.3 Left. Poisson probabilities of having exactly 1 (green solid line) and exactly 5 (green dotted line) events as a function of waiting time ΔT. Exponential probabilities of having at least one event (blue solid line) and no events (blue dotted line). Right. Probability density function of an exponential distribution (blue) and probabilities that an event will occur in time interval $(0, 1/\lambda)$ or $(1/\lambda, \infty)$

probability that an event will occur in time interval $(0, 1/\lambda)$ is $\Pr(0, 1/\lambda) = \int_0^{1/\lambda} \lambda \exp(-\lambda \Delta T) d(\Delta T) = 0.632$, and the probability that it will occur in the time interval $(1/\lambda, \infty)$ is $\Pr(1/\lambda, \infty) = 0.368$, see Fig. 5.3.

5.2.2 Probabilities of Exceedance and Mean Recurrence Times

If a set of n random variables, $X_1, X_2, \ldots, X_n$, where $X_1 = P_1, X_2 = P_2, \ldots X_n = P_n$ is one possible realisation of each of them, is drawn from the same distribution function $F(P) = \Pr(\leq P)$, then the distribution of the maximum value $P_{max} = \max\{P_1, P_2, \ldots, P_n\}$ is given by $\Pr(\leq P) = \Pr(P_1 \leq P, P_2 \leq P, \ldots, P_n \leq P) = \prod_{j=1}^n \Pr(\leq P_j) = [F(P)]^n$.

If the occurrence of larger events in time Δt is ruled by a homogeneous Poisson process with activity rate $\lambda_t = n/\Delta t$, then the probability of having exactly N events within time ΔT is

$$\Pr(N, \Delta T) = [\Lambda(\Delta T)]^N \exp\{-\Lambda(\Delta T)\}/N!, \qquad (5.6)$$

where $\Lambda(\Delta T) = \lambda_t \Delta T$ is the expected number of events in ΔT. The probability that all P_j within ΔT are smaller than or equal to some large P is

$$\Pr(\leq P, \Delta T) = \sum_{N=0}^{\infty} \left\{ [F(P)]^N [\Lambda(\Delta T)]^N \exp[-\Lambda(\Delta T)]/N! \right\},$$

which gives $\Pr(\leq P, \Delta T) = \exp\{-\Lambda(\Delta T)[1 - F(P)]\}$, or $\Pr(\leq P, \Delta T) = \exp[-\Lambda(\Delta T)\Pr(\geq P)]$, and the probability of having at least one event $\geq P$ within ΔT,

5.2 Intermediate Term Hazard

$$\Pr(\geq P, \Delta T) = 1 - \exp\left[-\Lambda(\Delta T)\Pr(\geq P)\right] = 1 - \exp\left[-(\Delta T/\Delta t)N(\geq P)\right], \tag{5.7}$$

where $\Pr(\geq P) = N(\geq P)/n$. For the UT power law, the Eq. (5.7) gives

$$\Pr(\geq P, \Delta T) = 1 - \exp\left[-\alpha(\Delta T/\Delta t)\left(P^{-\beta} - P_{max}^{-\beta}\right)\right]. \tag{5.8}$$

Epstein and Lomnitz (1966) showed that the OE power law, $N(\geq P) = \alpha P^{-\beta}$, gives $\Pr(\leq P, \Delta T) = \exp\left[-\alpha(\Delta T/\Delta t)\exp(-\beta m)\right]$, where $m = \ln P$, which is the first Gumbel distribution of the extreme values (see also Shakal & Willis, 1972 or Kijko, 1982). Note that replacing ΔT with the recurrence time $\bar{t}(\geq P) = \Delta t/N(\geq P)$ gives $\Pr\left[\geq P, \bar{t}(\geq P)\right] = 0.63$, regardless of P. The mean recurrence time can also be presented as

$$\bar{t}(\geq P) = \frac{\Delta t}{N(\geq P)} = -\frac{\Delta T}{\ln[1 - \Pr(\geq P, \Delta T)]}, \tag{5.9}$$

where $\Pr(\geq P, \Delta T)$ is the probability of exceedance, and ΔT here is called the exposure time. Most seismic building codes are based on the following performance levels that define the ability of a structure to sustain its main functions during and after earthquakes of different strengths. (1) Operational Limit: 50% probability of exceedance over 50 years, which is considered frequent, that according to Eq. (5.9) gives $\bar{t}(\geq P) = 72$ years, (2) Immediate Occupancy: 20% in 50 years (occasional) that gives $\bar{t}(\geq P) = 225$ years, (3) Life Safety: 10% in 50 years (rare) that gives $\bar{t}(\geq P) = 475$ years, and (4) Collapse Prevention: 2% in 50 years (very rare) that gives $\bar{t}(\geq P) = 2475$ years. In general, one expects better structural performances for frequent events of lower intensities, and one can accept higher damage for very rare large events. Building codes also adjust seismic input level to the importance of a given structure, e.g. by increasing the exposure time ΔT in Eq. (5.9) that leads to an increase of the mean recurrence time. For many structures in mines, the exposure times ΔT are shorter. For example, 50% probability of exceedance over 10 years would give $\bar{t}(\geq P) = 14.4$ years, 20% in 10 years would give $\bar{t}(\geq P) = 44.8$ years, and 10% in 10 years would give $\bar{t}(\geq P) = 95$ years.

For a Poisson process, the probability of having N occurrences over a time window equal to the recurrence time, $\Delta T = \bar{t}$, is $\Pr(N) = (1/N!)\exp(-1)$, which gives a probability of 36.8% that such an event will occur once, 18.4% that it will occur twice and 6.1% for three times.

Equation (5.9) can also be presented in terms of potency as a function of recurrence times, which for the upper truncated power law, $N(\geq P) = \alpha\left(P^{-\beta} - P_{max}^{-\beta}\right)$, gives

$$P(\bar{t} \geq P) = \left[\frac{1}{\alpha\,\bar{t}(\geq P)}\Delta t + P_{max}^{-\beta}\right]^{-1/\beta}, \tag{5.10}$$

and for the OET distribution $\bar{t}(\geq P) = \Delta t/\left[N(\geq P_{min})\Pr(\geq P)_{OET}\right]$.

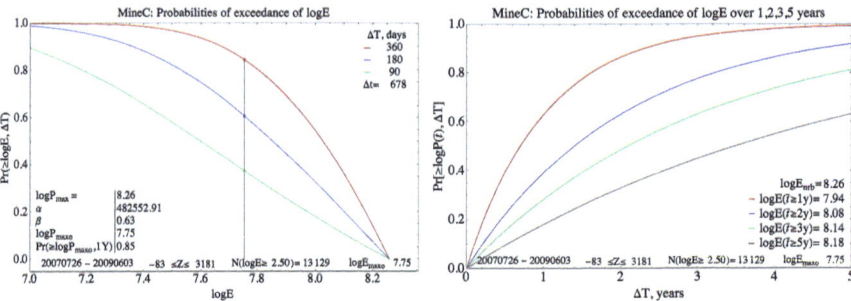

Fig. 5.4 Probabilities of exceedance for MineC derived from Eq. (5.8) for $\log E$ *(left)* and the probabilities of exceedance of $\log E$ with the recurrence times of 1, 2, 3, and 5 year derived from equation 5.10

These equations can be used to calculate the expected potencies over the selected recurrence times and to estimate their probabilities (Eq. 5.7) as a function of ΔT, as mining progresses, to monitor longer term hazard. In this case, P_{max} should be taken as the latest estimate of the maximum expected potency. Figure 5.4 left shows the probabilities of exceedance versus $\log E$ for MineC over 90, 180, and 360 days given by Eq. (5.8), and Fig. 5.4 right shows the probabilities of exceedance of $\log E$ with the recurrence times of 1, 2, 3, and 5 years for the same data set calculated with Eq. (5.10).

Note that the term "the mean recurrence time" or "the return period" as it is frequently called may be misleading unless it also gives the standard deviation around the mean. In practice, the Poisson model is justified if this standard deviation is not far from the mean.

In mines where the volume of rock extraction is not a linear function of time seismic hazard can be expressed in terms of the volume of rock extracted, or volume mined V_m. Then, the probability that there will be at least one event $\geq P$ while extracting an additional volume of rock ΔV_m is

$$\Pr(\geq P, \Delta V_m) = 1 - \exp\left[-\Lambda(\Delta V_m)\Pr(\geq P)\right] = 1 - \exp\left[-\frac{\Delta V_m}{V_m} N(\geq P)\right], \quad (5.11)$$

where $\Lambda(\Delta V_m) = n(\Delta V_m/V_m) = \lambda_m \Delta V_m$ is the expected number of events in ΔV_m. Since $\bar{V}_m(\geq P) = V_m/N(\geq P)$, the Eq. (5.11) gives the ratio $\bar{V}_m(\geq P)/\Delta V_m = -1/\ln[1 - \Pr(\geq P, \Delta V_m)]$. For example, if $\Pr(\geq P, \Delta V_m) = 0.05$, we can mine on average 19.5 times ΔV_m to generate seismic event $\geq P$. For the UT power law, the Eq. (5.11) gives

$$\Pr(\geq P, \Delta V_m) = 1 - \exp\left[-\Delta V_m \frac{\alpha}{V_m}\left(P^{-\beta} - P_{max}^{-\beta}\right)\right], \quad (5.12)$$

which shows that if β is constant and α is proportional to V_m, then seismic hazard can be controlled by the rate of mining.

5.2 Intermediate Term Hazard

Limited Data and Uncertain Activity Rate For limited data and/or for an uncertain activity rate, one can treat the activity rate λ_t as a random variable that can be accounted for by taking the integral $\int_0^\infty \Pr(\geq P, \Delta T) f(\lambda_t) d\lambda_t$, where $f(\lambda_t) = \Delta t (\lambda_t \Delta t)^n \exp(-\lambda_t \Delta t)/n!$ is the probability density function of the activity rate, which gives

$$\Pr(\geq P, \Delta T) = 1 - [1 + (\Delta T/\Delta t) \Pr(\geq P)]^{-n-1}. \quad (5.13)$$

The ratio of expressions (5.13) to (5.7) is approximately $1 + 1/n$, and the probability premium $1/n$ drops quickly with n, and for $n \geq 100$ is less than 1% (McGuire, 1977). Equation (5.13) can also be expressed in the volume mined domain, $\Pr(\geq P, \Delta V_m) = 1 - [1 + (\Delta V_m/V_m) \Pr(\geq P)]^{-n-1}$.

Hazard Trend In mines, the parameters α and β of the potency frequency distribution are not always constant or fluctuating in time as may be assumed in crustal seismology. As mining progresses, the overall stiffness of the rock mass is being degraded, and therefore, the parameters $\alpha\beta$ and P_{max} are expected to be a function of the effective volume mined, V_{meff}. The data shows that the exponent β decreases as the effective volume mined, i.e. extraction ratio of the ore body, increases. P_{max} also tends to increase as mining progresses. In most cases α has been shown to be proportional to V_{meff}, though at different rates in different mining scenarios. However, there are examples where α increases with V_m in a non-linear way, which is indicative of increasing hazard. Figure 5.5 shows the behaviour of α and β versus V_m in two mines.

By extrapolating parameters α and β beyond the observed volume mined or the observed time span, one can estimate future hazard for different mining scenarios from equation,

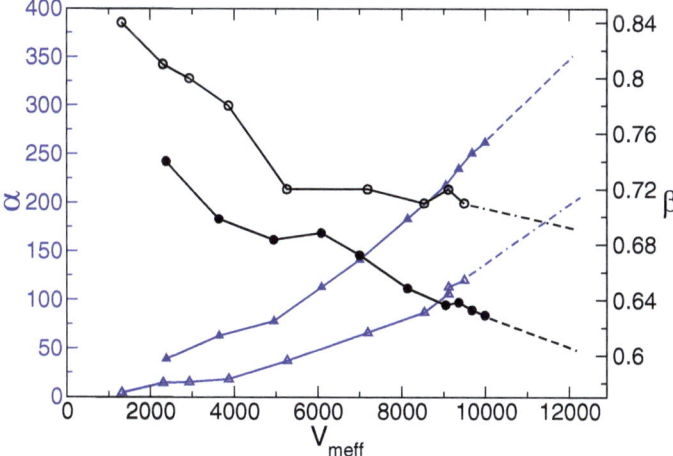

Fig. 5.5 Parameters α (triangles) and β (circles) versus the effective volume mined for two mines

$$\Pr(\geq P, \Delta V_m) = 1 - [1 + \Delta V_m / V_m \Pr[\geq P, \beta(\Delta V_m)]]^{-n[\alpha(\Delta V_m), \beta(\Delta V_m)] - 1},$$
(5.14)

where n is the expected number of events as a function of the future α and β (Mendecki, 2008).

5.3 Empirical Probabilities from the Observed Recurrence Times

Given the latest n observed recurrence intervals of events $\geq \log P$, of which $n_{\Delta T}$ are smaller than or equal to ΔT, one can estimate the empirical probability, $\Pr(\geq \log P, \Delta T)_E$, that a given volume will produce an event $\geq \log P$ within time ΔT after the preceding event of this size,

$$\Pr(\geq \log P, \Delta T)_E = p = \frac{n_{\Delta T} + 1}{n + 2}, \quad s_d(p) = \pm 2\sqrt{\frac{p(1-p)}{n+3}}, \quad (5.15)$$

where $s_d(p)$ is the uncertainty of the estimate. Note that the standard deviation, $sd(p)$, decreases slowly with increasing n, and if the number of observed intervals n is not much greater than 10, the probability generally will not be determined better than ± 0.2 (Savage, 1994). For a given n, the standard deviation, or the uncertainty, is maximum at $p = 0.5$, and it decreases symmetrically to zero as $p \to 0$ and $p \to 1.0$. Equation (5.15) is derived under the assumption that the probability of having an event of certain size within a specified time interval following the preceding event of that size is the same for all event cycles. Empirical probabilities cannot extend to the largest few events and cannot be extrapolated. However, they can be used to test results if there is a sufficient number of the observed recurrence intervals between intermediate and larger events.

Given the latest n observed inter-event volume mined of events $\geq \log P$, of which $n_{\Delta V_m}$ are smaller than or equal to ΔV_m, one can estimate the empirical probabilities in the volume mined domain, $\Pr(\geq \log P, \Delta V_m)_E = (n_{\Delta V_m}) / (n+2)$.

Table 5.1 lists the empirical probabilities $\Pr(\log P \geq 1.2, t_1 + \Delta T)_E$, calculated over the periods of one day, one week and one month, having 15 events with $\log P \geq 1.2$ in the data set with the following 14 recurrence intervals quoted here in hours: 118, 542, 265, 22, 587, 116, 56, 110, 11, 282, 95, 73, 235, 1. The last event with $\log P \geq 1.2$ occurred at $t_1 = $ 14h14' on 25 December 2014. The empirical probabilities are listed in Table 5.1.

Table 5.1 Empirical probabilities

$\Pr(\log P \geq 1.2, \Delta T = $ 1 day$)_E$	$\Pr(\log P \geq 1.2, \Delta T = $ 7 days$)_E$	$\Pr(\log P \geq 1.2, \Delta T = $ 30 days$)_E$
0.25 (± 0.21)	0.62 (± 0.23)	0.94 (± 0.12)

5.4 Example: Seismic Hazard Difference Between Three Mines

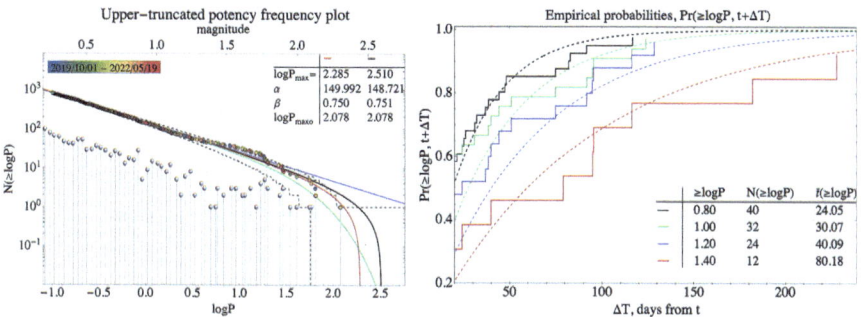

Fig. 5.6 Potency frequency plot *(left)* and the empirical probabilities *(right)* for $\log P \geq 0.8$, marked by the black step line, $\log P \geq 1.0$, green, $\log P \geq 1.2$, blue, and $\log P \geq 1.4$, red. The smooth dashed lines are the respective probabilities estimated by Eq. (5.8) for the same data set

Figure 5.6 left shows an example of the potency frequency data where for the most part it fits the upper truncated power law reasonably well, although it deviates from the 95% confidence limit in the potency range $0.8 \leq \log P \leq 1.4$, mostly at $1.0 \leq \log P \leq 1.2$. Figure 5.6 right shows the empirical probabilities derived from Eq. (5.15). It shows that the UT power law model underestimates seismic hazard in that potency range for the time period of the next 70 days. However, beyond this time range, the UT power law model predicts reasonably well. Note that the empirical probabilities are conditional in nature, i.e. it gives the probability within time ΔT, after the preceding event of this size.

5.4 Example: Seismic Hazard Difference Between Three Mines

Data Sets Data sets from three mines: A, B, and C, were collected over the same two year period $\Delta t = 678$ days, all related to tabular mining with principal vertical stresses but with different geological structures, mining layouts, extraction ratios, depths, and rates of mining.

MineA practiced long-wall mining of highly extracted tabular reef, see Fig. 5.7 left column, and MineB applied the sequential grid method, see Fig. 5.7 centre. For a review of these mining methods, see Vieira et al. (2001). In both cases, the rock extraction took place at practically the same depth of 3300 m. A simple numerical elastic model shows that due to the higher extraction ratio the mean vertical stress calculated over the un-mined areas in MineA is 1.7 times higher than in MineB. MineC is related to the scattered mining imposed by the presence of larger faults at an average depth of 1755 m, see Fig. 5.7 right column.

The depth of mining, the volume mined, the rate of mining, the approximate extraction ratio are listed in Table 5.2. The relative hazard rating from (1) the highest to (3) the lowest, imposed by the author for each parameter, is also quoted

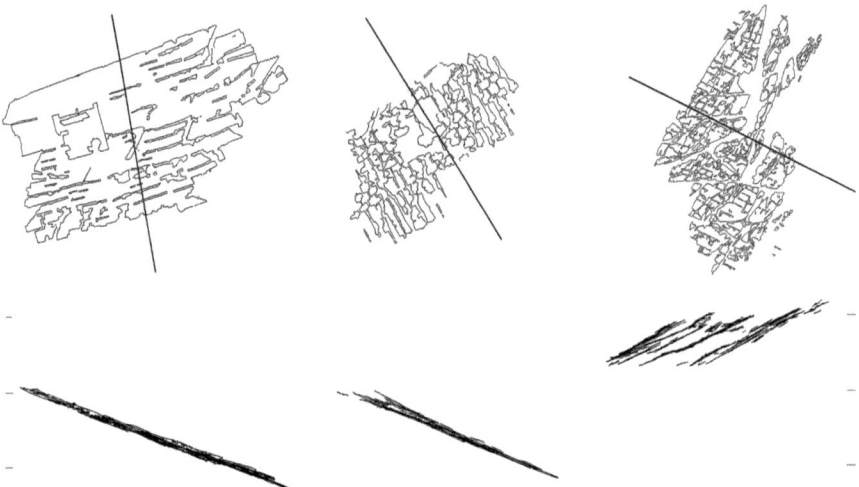

Fig. 5.7 Mine layouts in plan and in section of MineA *(left column)*, MineB *(centre)*, and MineC *(right)*

Table 5.2 Mining factors with hazard rating in parentheses

Parameter	MineA	MineB	MineC
Time of observation, Δt, days	678	678	678
Weighted depth of extraction, m	3294 (1)	3287 (2)	1755 (3)
Volume mined, V_m, m^3	207688 (3)	673209 (1)	386158 (2)
Rate of mining, m^3/day	306 (3)	993 (1)	570 (2)
Approximate extraction ratio, %	80 (1)	70 (2)	60 (3)

in parentheses, where applicable. Here we assumed that seismic hazard scales positively with the depth of mining, the deepest being MineA, the volume of rock extraction, the highest being MineB, the rate of mining, also MineB and with the extraction ratio, the highest being MineA. While these criteria are sensible, they are not sufficient to rate these three mines regarding seismic hazard.

Layout, Sequence of Mining, and Relative Hazard To gain insight into the influence of the mine layout on seismic hazard, their respective fractal dimensions were estimated. Fractals appear similar at any scale of observation. In mathematical terms, fractal objects exhibit fractional dimensionality, that is, they are neither lines, nor surfaces or volumes. Their dimension falls in between the classical dimensions of Euclidean geometry. An object is fractal when its length L is a function of the length λ of the measuring device, $L \sim \lambda^{1-D}$, where D is the fractal dimension (Mandelbrot, 1967, 1975). If $N(\lambda)$ is the number of cubes of size λ needed to cover the object, then the box counting fractal dimension of an object can be estimated by $D = \ln N(\lambda) / \ln(1/\lambda)$ (Barnsley, 1988). Fractal dimension increases with the degree of irregularity, or raggedness of the object. The lowest fractal dimension of

5.4 Example: Seismic Hazard Difference Between Three Mines

Table 5.3 Mining factors, with hazard rating in parentheses

Parameter	MineA	MineB	MineC
D_{ml} of mine layout	1.67 (1)	1.68 (2)	1.71 (3)
D_{sxy} of epicentres	1.60 (2)	1.49 (1)	1.61 (3)
D_{ste} of space-time extraction	1.84 (1)	1.92 (2)	2.20 (3)
D_{sxyt} of epicentres and time	1.84 (3)	1.71 (1)	1.83 (2)

mine layout, $D_{ml} = 1.67$, is associated with long-wall mining in MineA, followed by sequential grid, $D_{ml} = 1.68$ in MineB, and the roughest is the scattered mining imposed by the presence of geological structures $D_{ml} = 1.81$, in MineC. This sequence could easily be inferred just by looking at the smoothness of lines in Fig. 5.7.

Having coordinates and the dates and times of panel, extraction lines between consecutively extracted panels were connected, and the fractal dimension of such a spatial and temporal graph was calculated, D_{ste}. The smoothest sequence of mining, $D_{ste} = 1.84$, is associated with MineA, then MineB with $D_{ste} = 1.92$, and the roughest by MineC with $D_{ste} = 2.20$. By connecting lines between consecutive seismic events, an image of the sequence of seismic activity was created, and then its fractal dimension calculated. This exercise was limited to the (x, y, t) domain since the tabular ore bodies imposed a flat distribution of seismic stations and made the z-coordinates less reliable, specifically at the fringes of the network. Results and the relative hazard ratings are given in Table 5.3. They show that seismic activity does not follow mining exactly. One may speculate that the high stress and high rate of mining associated with data set B aligned most events with the excavation faces lowering the fractal dimension of their spatial distribution, but this has not been tested numerically.

There is also a negative correlation between the fractal dimension of the epicentres and time of events, D_{sxyt}, and β, see Tables 5.3 and 5.5. There are reports stating the positive correlation for earthquakes, but mainly for single fracturing or single fault processes, e.g. Aki (1981), King (1983), Wyss et al. (2004), Chen et al. (2006). Hirata (1989) reported a negative correlation due to different fault systems and Henderson et al. (1999) for induced seismicity where they show a negative correlation for high loading rates and a positive correlation for slowly loaded systems. Amitrano (2003) also reported negative correlation stating that diffused damage is associated with low β, whereas localised damage is associated with high β. The data set B has the highest loading rate of all three and the lowest fractal dimension, followed by data set C, and the lowest loading rate and the highest fractal dimension is associated with data set A.

Observed Seismicity and Relative Hazard Table 5.4 lists the observed seismic parameters and the relative hazard rating from the highest (1) to the lowest (3), imposed by the author for each parameter.

Figure 5.8 shows the cumulative graphs of the volume mined, seismic potency, and energy vs. time for all three mines.

Table 5.4 Observed seismic parameters with hazard rating in parentheses

Parameter	MineA	MineB	MineC
$\log P_{maxo}$; $\log P_{maxo-1}$	3.37; 2.97 (1)	2.70; 2.59 (2)	2.51; 2.43 (3)
N_{obs} ($\log P \geq 2.0$)	23 (1)	23 (1)	13 (2)
$\bar{t} = \Delta t / N_{obs}$ ($\log P \geq 2.0$), days	29.49 (1)	29.49 (1)	52.17 (3)
$\bar{V}_m = V_m / N_{obs}$ ($\log P \geq 2.0$), m^3	9.03 (1)	29.27 (2)	29.70 (3)
$\log E_{maxo}$; $\log E_{maxo-1}$	9.53; 9.49 (1)	8.68; 8.63 (2)	7.75; 7.58 (3)
N_{obs} ($\log E \geq 7.0$)	98 (1)	89 (2)	12 (3)
$\bar{t} = \Delta t / N_{obs}$ ($\log E \geq 7.0$), days	6.92 (1)	7.61 (2)	56.51 (3)
$\bar{V}_m = V_m / N_{obs}$ ($\log E \geq 7.0$), m^3	2119.3 (1)	7564.1 (2)	32179.87 (3)

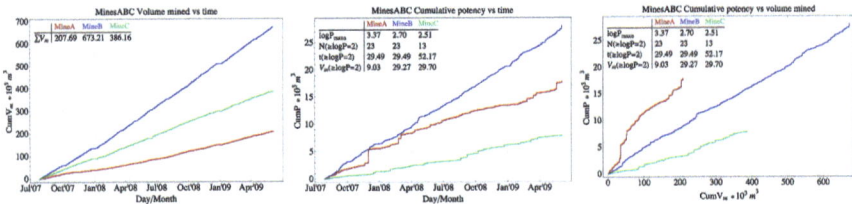

Fig. 5.8 All cumulative for MineA, MineB, and MineC: volume mined versus time, potency release versus time, and seismic energy versus time

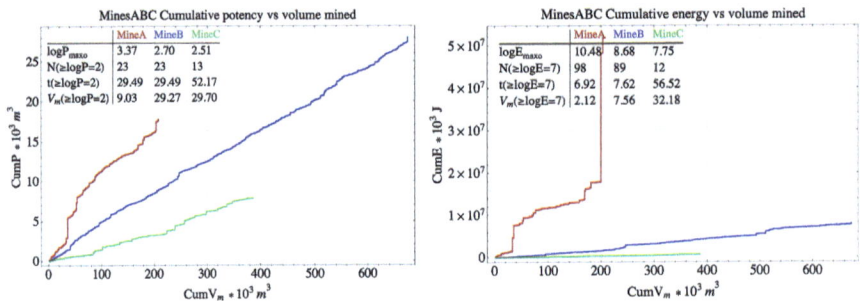

Fig. 5.9 Cumulative potency *(left)* and seismic energy *(right)* vs. volume mined

Figure 5.9 shows the cumulative potency and seismic energy versus volume mined. The potency release and seismic energy versus volume mined plots are evident, and both rate the hazard potential, i.e. assuming the same mining rate, for the three data sets clearly: The highest is MineA (1), the second highest is MineB (2), and the lowest of the three is MineC (3). Note that MineA mined at much lower rate to mitigate seismic hazard.

Size Distribution Parameters and Seismic Hazard Seismic hazard rating is a useful but a rudimentary scale, and it does not differentiate across the size distribution of seismic events. It is frequently the case that one mining area may have higher hazard up to certain event size and lower above it, and if the crossover

5.4 Example: Seismic Hazard Difference Between Three Mines

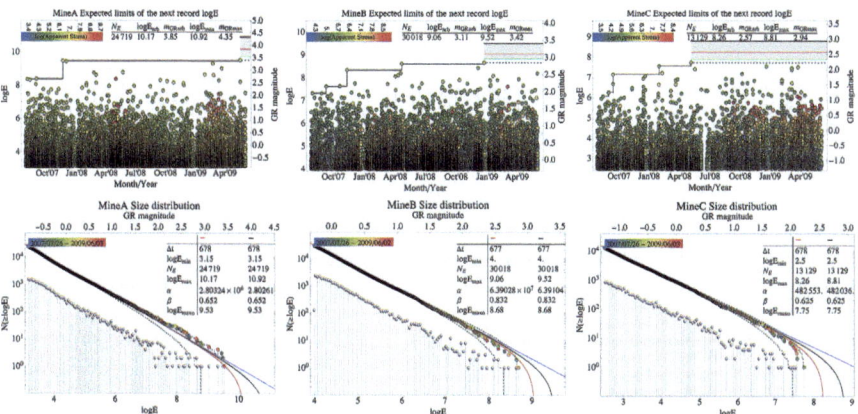

Fig. 5.10 History of records for MineA, MineB, and MineC with the estimated range of $\log P_{nrb}$ *(top row)* and energy frequency plots *(bottom row)*

is at the size that is potentially damaging, it may change seismic risk because the events below the threshold are more frequent.

Figure 5.10 top row shows the history of records for all three mines where $\log E_{max}$, marked by black line, is the upper limit of the next record breaking $\log E$, $\log E_{nrb}$, marked by red line, is the expected next record estimated by the upper truncated distribution (UT). The green line shows the level of the expected next record estimated by the open-ended tapered distribution (OET).

Figure 5.10 bottom row shows the cumulative number of unbinned data versus $\log E$ for all three mines where colour indicates the time of the event, with 4 fits: OE shown in blue, the UT assuming that the upper limit of the next record breaking $\log E = \log E_{max}$ in black, the UT assuming that the upper limit of the next record breaking $\log E = \log E_{nrb}$ in red, and OET in green. The light grey vertical spikes below the data illustrate what would be the empirical pdf if the data was binned with bin size 0.05. The dashed grey lines on both sides of the UT power law fit indicate 95% confidence limits.

The data set for MineB and MineC shows the frequency of the largest observed events deviated from the predicted UT model; therefore, one can infer that the size distribution hazard may be contained or controlled. The frequency of the largest observed events for MineA is close to the OE relation, indicating a potential for even larger events.

In general, seismic hazard should scale positively with α and negatively with β; however, Table 5.5 shows that α indicates the highest hazard for MineB and β for MineC. The expected next record breaking in energy, $\log E_{nrb}$, and its upper limit, $\log E_{max}$, rate seismic hazard correctly.

Figure 5.11 left shows seismic hazard estimated by Eq. (5.8) in terms of $\log E$ for all three mines in the time domain for $\Delta T = 30$ days, assuming their current rate of mining. Seismic hazard at MineC is clearly the lowest. However, there is the

Table 5.5 Size distribution parameters for potency and energy with hazard rating in parentheses

Parameter	MineA	MineB	MineC
$\log \alpha_{EUT}$	6.4476 (2)	7.8055 (1)	5.6835 (3)
β_{EUT}	0.6517 (2)	0.832 (3)	0.6254 (1)
$\log E_{nrb}$; $\log E_{max}$	10.167; 10.917 (1)	9.058; 9.524 (2)	8.257; 8.812 (3)
$\Pr(\log E \geq 7; \ \Delta T = 30 \text{ days})$	0.1966 (2)	0.984 (1)	0.527 (3)
$\Pr(\log E \geq 8; \ \Delta T = 30 \text{ days})$	0.527 (1)	0.418 (2)	0.063 (3)
$\Pr(\log E \geq 7; \ \Delta V_m = 10^4 \text{ m}^3)$	0.975 (1)	0.752 (2)	0.354 (3)
$\Pr(\log E \geq 8; \ \Delta V_m = 10^4 \text{ m}^3)$	0.557 (1)	0.166 (2)	0.038 (3)

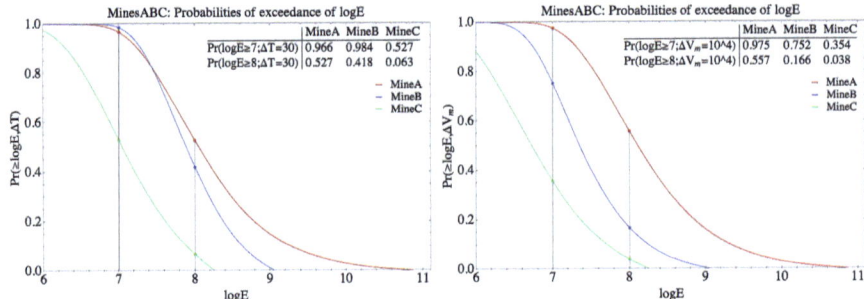

Fig. 5.11 Seismic hazard for the three mines in the time domain *(left)* and in the volume mined domain *(right)*

crossover point at $\log E = 7.4$ between MineA and MineB below which seismic hazard for MineB is higher. Figure 5.11 right shows seismic hazard potential, estimated by Eq. (5.12), assuming all three mines extract the same volume of rock $\Delta V_m = 10000 \text{ m}^3$, otherwise for the same set of size distribution parameters. Clearly, hazard potential at MineA is the highest of all three mines, and the difference between MineA and MineB is significant. Therefore, slowing down rate of mining in MineB would lower seismic hazard considerably.

5.5 Example: Seismic Hazard Difference in Time

In this example, we compare seismic hazard characteristics of two mostly overlapping seismic and rock extraction data sets selected from the same mine:

1. The DataSet1 starts on 07 September 2007 and ends on 13 June 2012.
2. The DataSet2 starts on 07 September 2007 and ends on 07 July 2013, so it is DataSet1 plus an additional 389 days of mining.

Seismic Activity and Production During DataSet1 This data set starts on 07 September 2007 and ends on 13 June 2012 just after a $\log P = 2.24$ ($m = 2.41$) event. It spans 1741 days and includes 2281 events with $\log P \geq -1.0$ ($m \geq 0.25$),

5.5 Example: Seismic Hazard Difference in Time

which gives the rate of 1.31 events/day. These events delivered $\Sigma P = 1630.1$ m³ of seismic potency at the rate of 0.9363 m³/day. The largest event in DataSet1 has $\log P = 2.24$ ($m = 2.4$), and there are 20 events with $\log P \geq 1.0$ ($m \geq 1.59$). The mean recurrence interval, $\bar{t} (\log P \geq -1.0) = 0.7639$ days with the standard deviation of 1.133 days, which gives the coefficient of variation $C_v (\log P \geq -1.0) = 1.48$. The mean recurrence interval, $\bar{t} (\log P \geq 1.0) = 91.634$ days with the standard deviation of 110.75 days, which gives the coefficient of variation $C_v (\log P \geq 1.0) = 1.21$, which indicates that larger events are less clustered in time than small events. The vertical black line in Fig. 5.12 marks the time of the second largest event during that time with $\log P = 1.82$.

During this time, the mine did 1071 blasts and extracted 1092912.9 m³ of rock that gives the average extraction rate of 627.735 m³/day. The ratio $\Sigma V_m / \Sigma P = 670.416$, i.e. on average the rock mass produced 1m³ of seismic potency every 670.416 m³ of volume mined. The minimum single extraction was 19.5 m³, maximum 11116 m³, and the mean 1020.4 m³. The minimum distance between consecutive production blasts was 1 m, the maximum 355, and the mean 108 metres. The minimum time between production blasts was 0.96 days, the maximum 16, and the mean 1.63 days.

Figure 5.12 shows the cumulative number of events, cumulative potency, and cumulative volume mined vs. time for DataSet1. Size of the event scales with the radius of source volume with strain change $\Delta \epsilon \geq 10^{-2}$, i.e. $V = P/10^{-2}$, and the colour indicates the distance of that event from the $\log P = 2.24$ main shock. The size of the production blast scales with the size of the rectangle, and its colour scales with the distance of that blast from the main shock.

There is an increase in the rate of seismic events with $\log P \geq -1.0$ from September 2007 to March 2008. While there is no increase in the rate of seismic activity before the $\log P = 1.82$ on 06 March 2009, there is an increase in the frequency of mid-size events which one could attribute to blasting of large volumes of rock. However, after January 2010, the mine also had a few large extractions, but there were only few mid-size events.

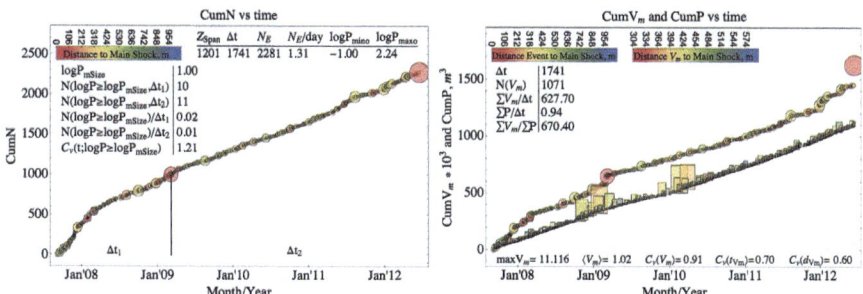

Fig. 5.12 Cumulative number of events *(left)*, cumulative potency, and cumulative volume mined vs. time *(right)* for DataSet1

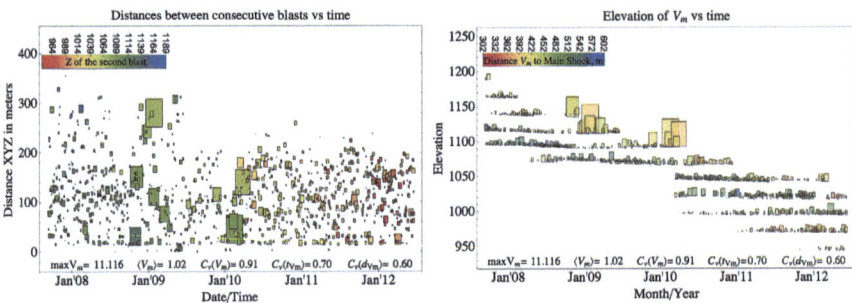

Fig. 5.13 Distances between consecutive blasts *(left)* and elevation of production blasts vs. time *(right)* for DataSet1

From January 2012, there was an increase in frequency of mid-size events, also reflected in the increased rate of CumP, that lasted almost to the $\log P = 2.24$ main shock on 13 June 2012. While the rate of production was steady over that time, the distances between subsequent extractions were smaller (Fig. 5.13 left) and at greater depth (Fig. 5.13 right).

Seismic Activity and Production During DataSet2 This data set also starts on 07 September 2007 and ends on 07 July 2013 just after a $\log P = 2.61\,(m = 2.66)$ event. It spans 2130 days and includes 2818 events with $\log P \geq -1.0\,(m \geq 0.25)$, which gives the rate of 1.32 events/day. These events delivered $\Sigma P = 2526.06$ m³ of seismic potency at the rate of 1.186 m³/day, which is higher than that during DataSet1. The largest event in DataSet2 has $\log P = 2.61\,(m = 2.66)$, and there are 30 events with $\log P \geq 1.0\,(m \geq 1.59)$.

The mean recurrence interval $\bar{t}\,(\log P \geq -1.0) = 0.756$ days with the standard deviation of 1.125 days, which gives the coefficient of variation $C_v\,(\log P \geq -1.0) = 1.49$. The mean recurrence interval $\bar{t}\,(\log P \geq 1.0) = 73.44$ days with the standard deviation of 94.76 days, which gives the coefficient of variation $C_v\,(\log P \geq 1.0) = 1.29$.

During this time there were 1343 production blasts extracting 1346518 m³ of rock that gives the average extraction rate of 632.23 m³/day. The ratio $\Sigma V_m / \Sigma P = 533.06$, i.e. on average the rock mass produced 1m³ of seismic potency every 533.06 m³ of volume mined. The minimum single extraction was 19.5 m³, maximum 11116, and the mean 1002.6 m³. The minimum distance between consecutive production blasts was 1 m, the maximum 355, and the mean 105 metres. The minimum time between production blasts was 0.25 days, the maximum 16, and the mean 1.58 days.

Figure 5.14 shows the cumulative number of events, cumulative potency, and cumulative volume mined versus time for DataSet2. Here the vertical black line marks the end of DataSet1 that was analysed above. After the $\log P = 2.24$ on 13 June 2012, the rate seismic activity was fairly steady but characterised by frequent mid-size events that pushed the rate of cumulative potency up.

5.5 Example: Seismic Hazard Difference in Time

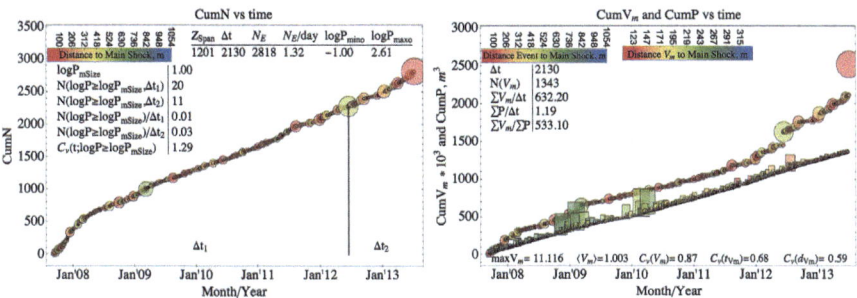

Fig. 5.14 Cumulative number of events *(left)*, cumulative potency, and cumulative volume mined vs. time *(right)* for DataSet2

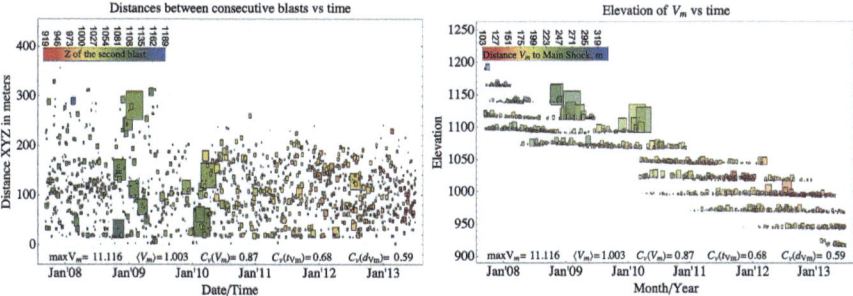

Fig. 5.15 Distances between consecutive blasts *(left)* and elevation of production blasts vs. time *(right)* for DataSet2

Figure 5.15 shows the distances between consecutive events and the elevation of production blasts versus time for DataSet2.

Comments

1. The rate of rock extraction, measured by ΣV_m/day, was also similar, 628.1 m^3/day versus 633.1 m^3/day.
2. The mean span between consecutive production blasts in DataSet1 was 368 m in X, 195 m in Y, and 245 in Z-direction.
3. The mean span of 272 production blasts during the additional 389 days of mining past DataSet1 was 265 m in X, 154 m in Y, and 102 m in Z-direction.
4. The volume mined weighted depth of production was 91 m deeper during the additional 389 days of mining.
5. The rate of seismic activity, $N (\log P \geq -1.0)$/day, was similar, 1.31/day versus 1.32/day, but the rate $N (\log P \geq 1.0)$/day increased from 0.0115/day in DataSet1 to 0.0141/day in DataSet2.

One can postulate that apart from increasing extraction ratio, the more concentrated rock extraction at greater depth may have contributed to the observed increase in seismic hazard.

The Upper Limit to the Next Largest Event The upper limit to the next record breaking potency, $\log P_{max}$, was estimated from $\log P_{max} = \log P_{maxo} + \Delta \log P_{max}$, where $\log P_{maxo}$ is the maximum observed $\log P$ and $\Delta \log P_{max}$ is the maximum expected jump that can be estimated using order statistics based on the history of observed jumps in record $\log P$,

$$\Delta \log P_{max} = 2 \max \left(\Delta \log P_{maxo}\right)$$

$$-\sum_{i=0}^{n-1}\left[\left(1-\frac{i}{n}\right)^n - \left(1-\frac{i+1}{n}\right)^n\right]\Delta \log P_{maxo-i},$$

where $\Delta \log P_{maxo-i}$ are the observed jumps in the history of records and n is the number of observed record jumps. Note that the order statistics estimate of P_{max} or $\log P_{max}$ is independent of the underlying probability distribution, and therefore, one can make a reasonable estimate even if the data do not conform to the potency frequency power law. The upper limit to the next record breaking potency for DataSet1 $\log P_{max} = 2.68$ and for DataSet2 $\log P_{max} = 3.04$, see Fig. 5.17.

Size Distribution Figure 5.16 shows the cumulative number of unbinned data vs. $\log P \geq -1.0$ for DataSet1 (left) and DataSet2 (right). The black solid line represents the UT power law, $N(\geq P) = \alpha \left(P^{-\beta} - P_{max}^{-\beta}\right)$, assuming truncation at $\log P_{max}$ and the red solid line at $\log P_{nrb}$. The straight blue line represents the open-ended power law, $N(\geq P) = \alpha P^{-\beta}$, also known as the Gutenberg-Richter relation, and is plotted as a reference. Data points below $\log P_{min} = -1.0$ are excluded from fitting. The grey dashed lines on both sides are the 95% confidence limits. The light grey vertical spikes below the data illustrate what would be the empirical pdf of the distribution for bin size 0.05.

Note that DataSet2 fits the UT power law a little better than DataSet1, although in both cases it overestimates frequency of events between $\log P = 1.2$ and $\log P = 2.0$.

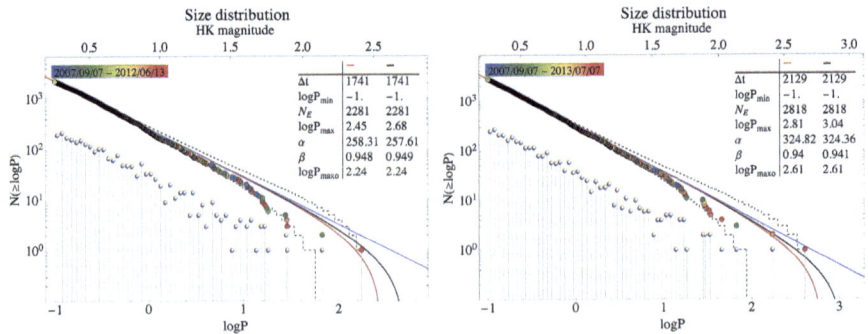

Fig. 5.16 Potency frequency fits to DataSet1 *(left)* and DataSet2 *(right)* recorded at Mine D

5.5 Example: Seismic Hazard Difference in Time

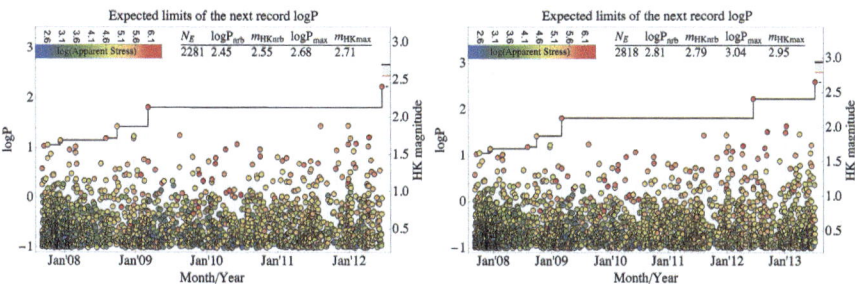

Fig. 5.17 History of the record breaking events and the estimated range of log P_{nrb} for Mine D based on DataSet1 *(left)* and DataSet2 *(right)*

The Expected Next Record Breaking Potency The expected next record potency, P_{nrb}, was calculated from

$$P_{nrb(k)} = \beta \left(P_{max}^{1-\beta} - P_{r(k-1)}^{1-\beta} \right) / \left[(1-\beta) \left(P_{r(k-1)}^{-\beta} - P_{max}^{-\beta} \right) \right],$$

where $P_{r(k-1)}$ is the potency of the previous (or the last) record breaking event and β is the exponent of the potency frequency distribution.

Figure 5.17 (left) shows the history of records and the estimated range of the next record log P for DataSet1 with colour indicating the distance to the main shock. The record breaking event log $P = 2.61$ recorded on 07 July 2013 is within the predicted range of $2.45 \leq \log P_{nrb} \leq 2.68$. There are eight forward counting log P records and one backward counting log P record, which indicates an upward trend in seismic hazard. Figure 5.17 (right) shows the history of records and the estimated range of the next record breaking log P for DataSet2. The expected range of the next record breaking event is $2.81 \leq \log P_{nrb} \leq 3.04$. There are nine forward counting log P records and one backward counting record, confirming an upward trend in seismic hazard.

Intermediate Term Hazard Change in Time The intermediate term hazard, $\Pr\left[\geq \log P; \Delta T\right]$, was calculated assuming the stationary Poisson process, $\Pr(\geq P, \Delta T) = 1 - \exp\left[-(\Delta T/\Delta t) N (\geq P)\right]$, where Δt is the time of observation, $N(\geq P) = \alpha \left(P^{-\beta} - P_{Limit}^{-\beta} \right)$, P_{Limit} is log P_{nrb} for the lower limit (red line) and log P_{max} for the upper limit (black line) of probabilities, and α and β are the respective parameters of the upper truncated potency frequency distribution.

Figure 5.18 shows the intermediate term probabilities of having at least one event greater than or equal to a given log P over 3, 6, and 12 months.

Comments

1. Seismic hazard for the period 07 September 2007 to 07 July 2013 (DataSet2, $\Delta t = 2130$ day) is higher than it was during the period 07 September 2007 to 13 June 2012 (DataSet1, $\Delta t = 1741$ days).

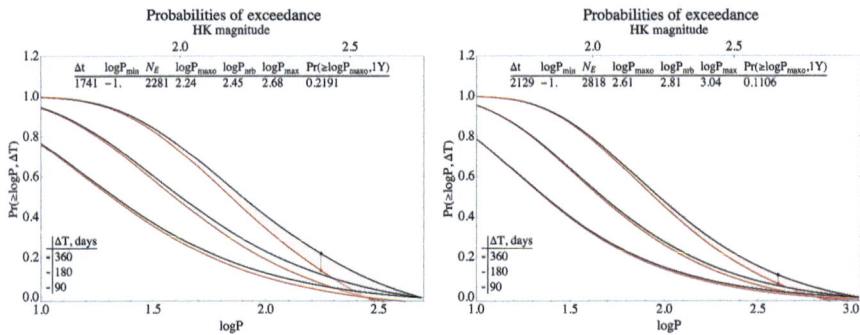

Fig. 5.18 Intermediate term probabilities for DataSet1 *(left)* and DataSet2 *(right)*

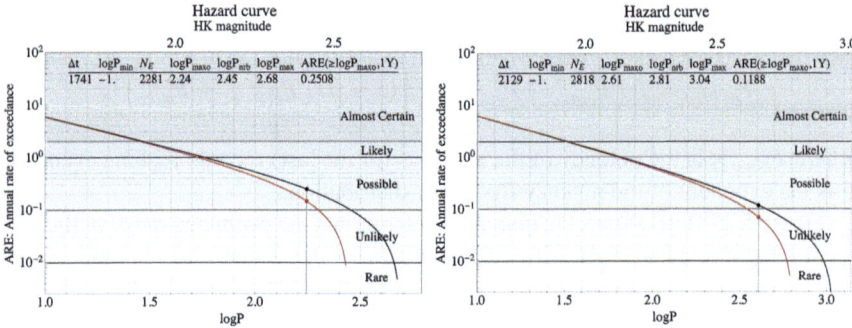

Fig. 5.19 Hazard curves for DataSet1 *(left)* and DataSet2 *(right)*

2. The ratio $\Sigma V_m / \Sigma P$ indicates how much rock needs to be extracted to produce 1 m^3 of seismic potency. During the first 1741 days of mining, this ratio was 670.42, and it dropped to 245.1 during the additional 389 days of mining.
3. During the first 1741 days, the activity rate of events with $\log P \geq 1.0$ was 0.0115/day, and it increased to 0.0283/day during the additional 389 days.
4. During the first 1741 days, the probability of having at least one event with $\log P \geq 2.0$ ($m \geq 2.25$) within 1 year was 0.407, and it increased to 0.475 during the additional 389 days of mining.

Frequently mine management asks for seismic hazard to be presented as seismic hazard curve that shows the annual rate at which a given $\log P$ will be exceeded. Hazard curve is then superimposed on the probability rating scheme, or risk matrix, defined by the mine. Figure 5.19 shows hazard curves, i.e. the expected annual rate of exceedance for both data sets.

5.6 Seismic Response to Step Loading: Short Term Hazard

5.6.1 Introduction

The rock mass dynamics in mines is driven mainly by the transient deformation of nearby excavations in response to extraction of stressed rock and by the near-field deformation of seismic sources, including pillar and abutment failures, and by blasting. To a lesser degree, it is also influenced by waves from remote sources of seismic radiation and by tidal forces. The step loading caused by rock extraction and by seismic events induces large displacements, plastic instability, splitting, buckling, and bursting of rock close to excavations. The energy imparted into the rock mass during step loading is partially converted into strain energy and partly dissipated through friction, local plastic deformation, and strain energy of radiated waves.

After a step loading, one can observe two partly competing stress-related processes occurring in the affected rock mass:
(1) An excitation phase that elevates the average stress level in volumes of rock where loading is faster than the ability to dissipate the excess of strain energy. It is generally short-lived, and it can be observed by monitoring changes in seismically inferred stress and strain by high-resolution networks.
(2) A relaxation phase that cascades the system down from its excited state through intermediate phases toward a non-equilibrium steady state. It is facilitated by different forms of inelastic deformation—seismic and aseismic. It modifies the stress pattern by reducing its elevated levels and moving it further away from excavations. It generates the bulk of the seismic activity after loading, and it lasts much longer than the excitation phase.

Seismic rock mass response to blasting is driven strongly by the stress level in rock surrounding the blast and by the volume of rock blasted. In the case of a preconditioning blast, when there is little or no rock extraction, the stress level plays the major role. In most cases, rock extraction by blasting induces and triggers seismic events immediately and in close proximity to the blast, which helps to set up a proper re-entry protocol, i.e. the exclusion zone and re-entry time.

Aftershock sequences after larger seismic events in mines or after major production blasts are not as well developed and do not last as long as after tectonic earthquakes of similar size. The main reason is faster relaxation facilitated by the presence of openings and extensive fracturing. If the immediate aftershocks are subdued and locate close to the main shock, there is less potential for a large event. As the immediate aftershocks extend further away from the main shock, it is more likely they may trigger larger event, see Fig. 5.20.

However, rock mass convergence associated with large rock extractions or with larger pillar removals may produce elevated seismic response for days or even weeks with sporadic elevated levels of seismic activity away from the place of extraction. Locations and orientations of larger events during aftershock sequences are controlled by geological structures rather than by the elevated stress level.

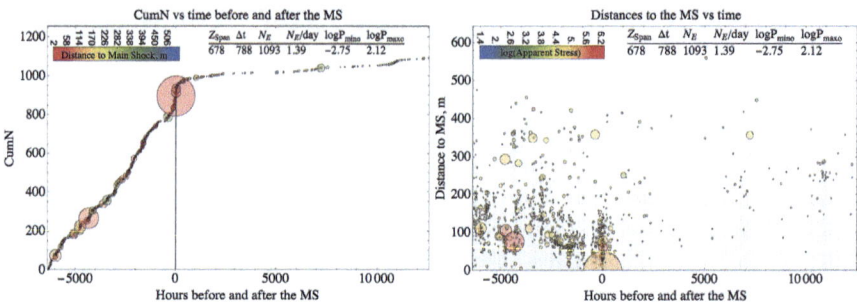

Fig. 5.20 Cumulative number *(left)* and distances *(right)* of seismic events to the main shock (MS) with $\log P = 2.61$

The seismic activity after step loading tends to cluster in space and time. In some cases, the rate of activity follows typical aftershock sequences that can be described by the Omori law or the stretched exponential relaxation function. However, in some cases, the response is more complex and cannot be described by a simple intensity function and therefore needs to be treated with non-parametric statistics.

5.6.2 Seismicity Rate Change

The fundamental premise here is that as the rate of seismic activity increases so does the likelihood that one of these events may be larger and damaging. In many cases, specifically after production or development blasts, the recorded activity is complex and does not follow the typical aftershock sequences that could be described by the Omori law or the stretched exponential relaxation function, and therefore needs to be treated by non-parametric statistics. One way to check if the elevated seismic activity has returned to an acceptable level is to test the null hypothesis of no change. The acceptable level of seismic activity can be defined a priori, or it can be taken as an average activity rate before step loading.

To detect changes in seismic activity rates in two different time intervals, Δt_1 and Δt_2, within the same volume of rock, one can count the respective number of recorded events above a certain potency. If the time intervals are equal and relatively long and the observed number of events is significantly different, then a statement can be made about the relative change. However, the associated uncertainty increases as the time intervals get shorter and as the difference in the event counts becomes smaller. The situation is even more difficult if the time intervals are not equal.

Under normal conditions, the event counts can be considered as outcomes of a Poisson process, and therefore, their occurrence can be very irregular. To measure the seismicity rate change, we need the probability density function of the activity rate, λ. This can be derived by normalising the Poissonian density function (see Eq. 5.5) so that the integral over λ from 0 to ∞ is unity, which gives

5.6 Seismic Response to Step Loading: Short Term Hazard

$$f(\lambda) = \Delta t \, (\lambda \Delta t)^N \exp(-\lambda \Delta t) / N! \tag{5.16}$$

The probability that the seismicity rate in two different time intervals, with densities f_1 corresponding to time interval Δt_1 and f_2 to Δt_2, increased by more than k times is $\Pr[(\lambda_2/\lambda_1) > k] = \int_0^\infty d\lambda_1 f_1(\lambda_1) \int_{k\lambda_1}^\infty d\lambda_2 f_2(\lambda_2)$, which, taking $\lambda_1 = N_1 (\geq \log P)/\Delta t_1$ and $\lambda_2 = N_2 (\geq \log P)/\Delta t_2$, gives

$$\Pr\left(\frac{\lambda_2}{\lambda_1} > k\right) = \frac{1}{N_2! N_1!} \int_0^\infty x^{N_1} \exp(-x) \, \Gamma\left(N_2 + 1, kx \frac{\Delta t_2}{\Delta t_1}\right) dx, \tag{5.17}$$

where $\Gamma(N_2 + 1, kx\Delta t_2/\Delta t_1) = \int_{kx\Delta t_2/\Delta t_1}^\infty \exp(-t) \, t^{N_2+1} dt$ is the upper incomplete Gamma function, and the ratio $\Delta t_2/\Delta t_1$ caters for the unequal time intervals (Marsan, 2003).

Illustrative Example Let us assume that in a given volume of rock, ΔV, during $\Delta t_1 = 10$ days preceding the step loading in a remote section of a mine, the seismic system recorded $N_1 = 10$ events with $\log P \geq -1.0$, i.e. $\lambda_1 = 1.0$. Figure 5.21 (left) shows the probability calculated by Eq. (5.17) that the event rate increased by more than k times, if during $\Delta t_2 = 10$ days after blasting the system recorded $N_2 = 10$, 15 or 20 events with $\log P \geq -1.0$, i.e. $\lambda_2 = 1$, 1.5, or 2.0, respectively. The $\Pr[(\lambda_2/\lambda_1) > k = 1] = 0.5$ indicates an equal chance of activation ($k > 1$) or slow down ($k < 1$) of seismic activity in ΔV. This is the case when $\Delta t_2 = \Delta t_1$ and $\lambda_1 = \lambda_2$, see the green dot in Fig. 5.21 (left).

If, however, it had taken $\Delta t_1 = 20$ days to record $N_1 = 10$ events with $\log P \geq -1.0$ before the step loading, then with all other parameters being equal, the probability of activation would increase to $\Pr[(\lambda_2/\lambda_1) > k = 1] = 0.944$, the probability that seismic activity increased by more than $k = 1.5$ times would be $\Pr[(\lambda_2/\lambda_1) > k = 1.5] = 0.747$, and the probability that seismic activity at least doubled $\Pr[(\lambda_2/\lambda_1) > k = 2] = 0.5$, see the green dots at Fig. 5.21 (right). Then, the probability that 0.944 or 0.747 or more could be obtained by chance if there was

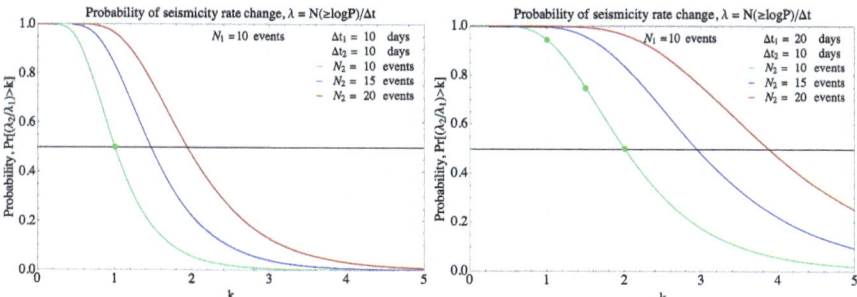

Fig. 5.21 Probabilities of an increase in seismicity rate for equal *(left)* and unequal *(right)* time intervals

no change, which is the null hypothesis, is $1 - 0.944 = 0.056$ or $1 - 0.744 = 0.256$, respectively.

One can also solve Eq. (5.17) for k for a given probability, e.g.

$$(N_1!N_2!)^{-1} \int_0^\infty x^{N_1} \exp(-x)\, \Gamma\,[N_2 + 1, kx\,(\Delta t_2/\Delta t_1)]\, dx = 0.9, \qquad (5.18)$$

to answer the following question: What is the ratio of seismicity rate change that gives 90% certainty. For example, if during $\Delta t_1 = 10$ days preceding the step loading in a remote section of a mine, the seismic system recorded $N_1 = 10$ events with $\log P \geq -1.0$, and $N_2 = 20$ events during $\Delta t_2 = 10$ days after, then there is 90% probability that seismicity rate changed by at least $k = 1.2$ times.

Reference Activity Level Equation (5.17) can be applied to monitor if the elevated rate of seismic activity after production blasts returned to an acceptable reference level. Note that frequently activity rate before larger events is elevated and, if used as a reference, would underestimate the re-entry time. The reference activity rate can be estimated by taking an average over periods of times that satisfy the following criteria: (a) they are outside the influence of blasting, (b) there were no larger events, and (c) there was normal production activity and people working in the area. It is expected that the coefficient of variation of the data selected to estimate the reference activity will not be far from 1.0, so the reference activity would not deviate too far from Poissonian.

Typically, the exclusion zone and time after a large seismic event, or blast, have the following characteristics. (1) The exclusion time increases as the size of the main shock increases. (2) The exclusion zone increases as the size of the main shock increases. (3) The exclusion time decreases as the distance to the main shock increases.

Data One can use the following types of data to measure seismic activity:

1. Magnitude, or $\log P$ or $\log E$, of associated seismic events. This data is delayed since it requires seismological processing, i.e. location and source parameters. Moreover, to provide a reasonable location, the seismic system accepts events that associated with at least five stations, and this removes a great number of small events from the analysis. Data on the activity of associated events also underreports on immediate aftershocks, some of them buried in the coda of the main shock.
2. The peak ground velocity, PGV, and/or the cumulative absolute displacement, CAD, which is the integral of the absolute value of a velocity time series, $CAD = \int_0^t |v(t)|\,dt$, and has units of displacement. Both parameters can be extracted from a stream of continuous data provided by seismic system. This data is available in real time since it does not require processing. An event is declared when during a given time interval, say $\Delta t = 0.25$ second, the GM parameter exceeds a predefined threshold.

5.6 Seismic Response to Step Loading: Short Term Hazard

The data for analysis can be derived from a predefined polygon which is defined as a seismogenic volume that generates seismicity affecting working places. Therefore, all relevant parameters are derived from the data selected from this polygon. This method has been widely applied in mines for many years. The most difficult, and also subjective task, here is the definition of the polygon. Different polygons select different data sets and therefore will produce different seismic characteristics.

Alternatively, one can apply the polygon-less or the influence based approach where one takes into account the influence of all available seismic events, regardless of their location, on a particular working place. The preferred measures of influence can be PGV and/or CAD since their influence is moderated by the distance from the seismic source to the place of potential exposure.

5.6.3 Omori Relaxation Function

In 1893 Fusakichi Omori published the 11 page long note "On aftershocks" (Omori, 1893), which was the abstract of the paper published a year later (Omori, 1894), where he described the activity of aftershocks at any time x after the main shock as $y = k(x + h)^{-1}$, where k and h are constant, which "represents a rectangular hyperbola and implies that the activity varies nearly inversely with time". Hirano (1924) and Utsu (1957) modified Omori's equation by introducing the exponent p, and today the Omori law is presented as

$$dN/dt = k(t + c)^{-p}, \qquad (5.19)$$

where k, c, and p are parameters and t is time. The total number of events $N_t = N(0, \infty)$ is derived from $N_t = \int_0^\infty k(t+c)^{-p} dt$ that gives $N_t = kc^{1-p}/(p-1)$ for $p > 1$ and $N_t = \infty$ for $p \leq 1$ (Fig. 5.22).

The parameter k depends on the magnitude of the main shock and for given c and p scales with the total number of events. The parameter c is the offset time that accounts for the incompleteness of the data set immediately after the main shock— the better the resolution of the seismic network, the smaller the c. Setting $c > 0$ also prevents the infinite N at $t = 0$, and its physical meaning, however, is debatable.

When plotting the observed number of events within a certain interval of time versus time on a log-log scale, the data points should tend to a straight line whose slope is an estimate of p. The parameter p controls the rate of decay of aftershocks, and, for slowly decaying sequences with $p \leq 1$, the Omori law predicts that the total number of events becomes unbounded with increasing time. The parameter p measures the overall relaxation time of the process following the major event. The lower the p-value, the slower the relaxation, which is characteristic of stiffer systems; the opposite would apply to softer systems. As a consequence, the p-value would be expected to correlate negatively with the Gutenberg-Richter b-value, e.g. (Wang, 1994). Note that p and c are positively and almost linearly correlated, which can frustrate the estimation process. In general, the Omori law is an open-ended

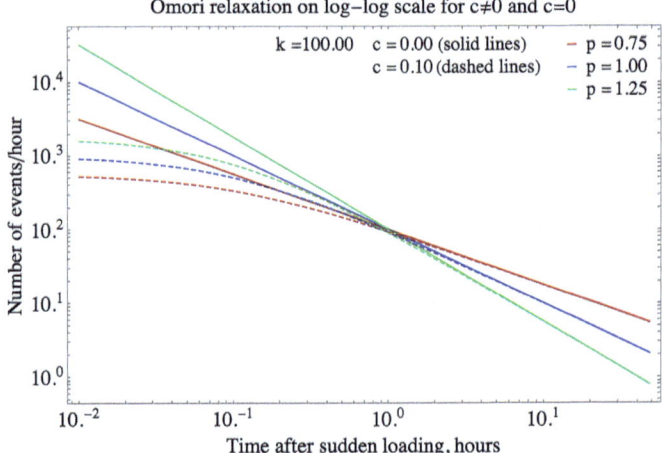

Fig. 5.22 Omori relaxation for $k = 100$, $c = 0.1$ and different values of parameter p on the log-log scale

power law, $N (\geq t) \sim t^{-p}$, has infinite moments and therefore the property of scale invariance, i.e. the relative change of $N (\geq st) / N (\geq t) = s^{-p}$ is independent of t, and therefore it lacks a characteristic scale—there is no characteristic time. To be physical, the power law description of a finite system must be bounded; otherwise, aftershocks would continue forever.

The number of events in a time interval $N(t_1, t_2)$ after the step loading can be calculated by $N(t_1, t_2) = \int_{t_1}^{t_2} k/(t+c)^p \, dt$, which gives

$$N(t_1, t_2) = \begin{cases} \frac{k}{(1-p)} \left[(t_2+c)^{1-p} - (t_1+c)^{1-p} \right], & p \neq 1 \\ k \ln\left(\frac{t_2+c}{t_1+c}\right), & p = 1 \end{cases} \quad (5.20)$$

and the ratio

$$\frac{N(t_1, t_2)}{N(0, t_2)} = \begin{cases} \left[(t_2+c)^{1-p} - (t_1+c)^{1-p} \right] / \left[(t_2+c)^{1-p} - c^{1-p} \right], & p \neq 1 \\ \ln\left[(t_2+c)/(t_1+c) \right] / \ln\left[(t_2+c)/c \right], & p = 1 \end{cases}. \quad (5.21)$$

5.6.4 Omori Distribution, Parameters, and Simple Omori

To get the probability distribution, we need to normalise parameter k so $\int_0^T k(t+c)^{-p} \, dt = 1$, where T is the assumed end of the aftershock sequence, which gives

5.6 Seismic Response to Step Loading: Short Term Hazard

$$\begin{cases} f(t) = (p-1)(t+c)^{-p} / \left[c^{1-p} - (T+c)^{1-p} \right], & p \neq 1 \\ f(t) = (t+c)^{-1} / \ln[T/(1+c)], & p = 1 \end{cases} \quad (5.22)$$

and the cumulative distribution function, $F(t) = \Pr(\leq t) = \int_0^t f(u)\,du$, which gives

$$\begin{cases} \Pr(\leq t) = c^{-p}(T+c)^{-p}[c(T+c) - c^p(T+c)] / \left[c^{1-p} - (T+c)^{1-p} \right], & p \neq 1 \\ \Pr(\leq t) = \ln[(c+t)/c] / \ln[t/(1+c)], & p = 1 \end{cases} \quad (5.23)$$

and $\Pr(\leq t) = (t/T)^{1-p}$ for $c = 0$ and $p \neq 1$. The probability of having an event during the time interval (t_1, t_2) is

$$\begin{cases} \Pr(t_1, t_2) = \left[(t_1+c)^{1-p} - (t_2+c)^{1-p} \right] / \left[c^{1-p} - (T+c)^{1-p} \right], & p \neq 1 \\ \Pr(t_1, t_2) = \ln[(t_2+c)/(t_1+c)] / \ln[T/(1+c)], & p = 1 \end{cases} \quad (5.24)$$

For $p \leq 1$, the end of the aftershock sequence T must be finite; otherwise, the integral diverges, but for $p > 1$ there is a finite number of aftershocks as $T \to \infty$ and the respective equations are as follows:

$$\begin{cases} f(t) = (p-1)(t+c)^{-p} / c^{1-p}, & p > 1 \\ F(t) = -\left[(t-c)^{1-p} - c^{1-p} \right] / c^{1-p}, & p > 1 \\ \Pr(t_1, t_2) = \left[(t_1+c)^{1-p} - (t_2+c)^{1-p} \right] / c^{1-p}, & p > 1 \end{cases} \quad (5.25)$$

The maximum likelihood (ML) method of estimating parameters of the Omori relation is similar to the method of calculating the β-value of the power law size distribution. However, the results are not always stable, because p and c are almost linearly correlated. One can simplify calculations by assuming that parameter $c = 0$, i.e. simple Omori (SO). Then to get the probability density function, we need to normalise parameter k so $\int_{t_1}^{t_2} k t^{-p} = 1$, which gives $k = (1-p) / \left(t_2^{1-p} - t_1^{1-p} \right)$. The ML function then is $L(t, p) = \prod_{j=1}^{N} (1-p) t_j^{-p} / \left(t_2^{1-p} - t_1^{1-p} \right)$, where N is the number of events in the time period $t_2 - t_1$. The parameter p can be calculated from $\partial \ln L(p) / \partial p = 0$, which gives

$$\frac{N}{(1-p)} + \frac{N \left(t_2^{1-p} \ln t_2 - t_1^{1-p} \ln t_1 \right)}{t_1^{1-p} - t_2^{1-p}} + \sum_{j=1}^{N} \ln t_{dj} = 0, \quad (5.26)$$

where t_{dj} is the time of seismic events. Equation (5.26) needs to be solved numerically for p. The parameter k can then be calculated from

$$N = k \int_{t_1}^{t_2} t^{-p} dt = \frac{k}{(1-p)} \left[t_2^{1-p} - t_1^{1-p} \right] \Rightarrow k = \frac{N(1-p)}{t_2^{1-p} - t_1^{1-p}}. \quad (5.27)$$

The standard deviation of p can be obtained from the second derivative of the log likelihood function, $\left[-\partial^2 \ln L(p)/\partial p^2\right]^{-1/2}$, which gives

$$s_d(p) = \left\{N/(1-p)^2 - N(t_2/t_1)^{1-p}\left[\ln(t_2/t_1)\right]^2 / \left[(t_2/t_1)^{1-p} - 1\right]^2\right\}^{-1/2}.$$
(5.28)

The single most important factor influencing estimates of $s_d(p)$ is the number of observations, and for small N, the estimates are unreliable. For $p = 1$, taking $\partial \ln L(p = 1, k)/\partial k = 0$ gives $k = N/(\ln t_2 \ln t_1)$.

5.6.5 Stretched Exponential Relaxation Function

An alternative to the Omori law is the stretched exponential function introduced by von Rudolf Kohlrausch (1854) to interpret charge relaxation in a Leiden jar, which was the first capacitor. The stretched exponential is also known as the Kohlrausch-Williams-Watts (KWW) function, after Williams and Watts (1970) used it to characterise the dielectric relaxation rates in polymers. The stretched exponential function has a fatter tail than the exponential, but less so than a pure power law distribution, offering a compromise between the two descriptions. It can therefore be used to account for both a limited scaling regime and a cross over to non-scaling. While the Omori law is purely empirical, the parameters of the stretched exponential have a clear physical meaning, i.e. the relaxation time of the exponential decay process and the total number of events.

The function is defined as $\exp\left[-(t/\tau)^q\right]$ for $t \geq 0$, where $\tau > 0$ is the relaxation time and $0 < q < 1$ is called the stretching or the shape exponent. The origin of the stretched exponential is assumed to be the result of a competition between two or more exponential processes. The reciprocal of q is called a heterogeneity parameter, $h = 1/q$, which is the number of generations in a multiplicative process, and it gives an insight into the heterogeneity in complex systems. The larger the values of relaxation time τ, the slower the relaxation processes. The smaller the value of the shape exponent $q < 1$, the quicker the decay for short times and slower for long times.

To apply the stretched exponential to aftershock sequences, it is convenient to start at time t_1 after t_0 when the step loading was applied (Kisslinger, 1993). Then the number of events yet to occur in a sequence at time t_1 is

$$N(t_1, T) = N(t_0, T) \exp\left[-(t_1/\tau)^q\right], \quad (5.29)$$

where $N(t_0, T)$ is the total number of events that will eventually occur and $T - t_0$ is the duration of the process (Fig. 5.23). At time $t_1 = \tau$, the number of events still to occur is $N(\tau, T) = N(t_0, T)/e$, regardless of q, so 63% of the work is done during the relaxation time. The number of events that had occurred to time t_1 is $N(t_0, t_1) = N(t_0, T) - N(t_1, T)$, which gives a prediction of the total number of events at time t_1 as

5.6 Seismic Response to Step Loading: Short Term Hazard

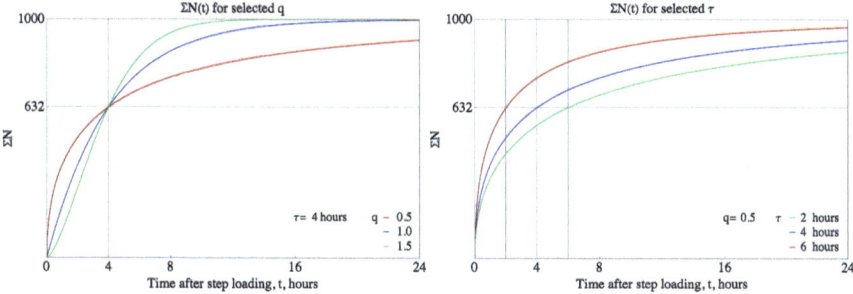

Fig. 5.23 Cumulative number of events vs. time after step loading for $\tau = 4$ hours and for selected q (*left*), and for $q = 0.5$ and for selected τ (*right*)

$$N(t_0, T) = N(t_0, t_1) / \left\{1 - \exp\left[-(t_1/\tau)^q\right]\right\}. \tag{5.30}$$

The expected number of events to occur in the interval $(t_1, t_1 + \Delta T)$ is

$$N(t_1, t_1 + \Delta T) = N(t_0, T) \left\{\exp\left[-(t_1/\tau)^q\right] - \exp\left[-((t_1 + \Delta T)/\tau)^q\right]\right\}. \tag{5.31}$$

5.6.6 Stretched Exponential Distribution and Its Parameters

The cumulative distribution function is

$$F(t) = \Pr(\leq t) = N(t_0, t) / N(t_0, T) = 1 - \exp\left[-(t/\tau)^q\right], \tag{5.32}$$

and the survival function, $S(t) = \exp\left[-(t/\tau)^q\right]$. The probability density function is

$$f(t) = dF(t)/dt = (q/\tau)(t/\tau)^{q-1} \exp\left[-(t/\tau)^q\right]. \tag{5.33}$$

The probability of having an event in the time interval $(t, t + \Delta T)$ is $\Pr(t, t + \Delta T) = F(t + \Delta T) - F(t) = \exp\left[-(t/\tau)^q\right] - \exp\left[-((t + \Delta T)/\tau)^q\right]$, and the expected number of events in this time interval is $N(t, t + \Delta T) = N(t_0, T) \Pr(t, t + \Delta T)$. For $q = 1$, the stretched exponential becomes a simple exponential distribution and for $q < 1$ decays slower than exponential for large times (thick tail). For $q > 1$, it decays faster than exponential and is known as the Weibull distribution. For $q = 2$, it is also known as the Rayleigh distribution.

All moments of the stretched exponential distribution are finite, the mean, or the expected value is $\tau \Gamma(1 + 1/q) = (\tau/q) \Gamma(1/q)$, and variance $\tau^2 \Gamma(1 + 2/q) - \tau^2 \Gamma(1 + 1/q)^2$. The gamma function, $\Gamma(t) = \int_0^\infty x^{t-1} e^{-x} dx$, reduces to $\Gamma(n) =$

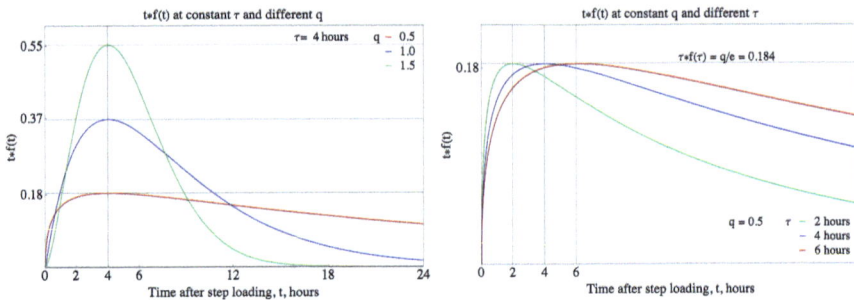

Fig. 5.24 $t \cdot f(t)$ for $\tau = 4$ hours and for selected q *(left)*, and for $q = 0.5$ and for selected τ *(right)*

$(n-1)!$ for positive integers and $\Gamma(t+1) = t\Gamma(t)$. The median is $F(t) = \exp\left[-(t_{0.5}/\tau)^q\right] = 0.5$, which gives $t_{0.5} = \tau (\ln 2)^{1/q}$.

The function $t \cdot f(t) = q (t/\tau)^q \exp\left[-(t/\tau)^q\right]$ has a maximum at the relaxation time τ, regardless of q, and the $\tau \cdot f(\tau) = q/e$ depends only on q, see Fig. 5.24. These two properties of $t \cdot f(t)$ may be utilised while fitting the data.

The maximum likelihood function is

$$L(t_j; \tau, q) = \prod_{j=1}^{n} f(t_j) = \prod_{j=1}^{n} (q/\tau) (t_j/\tau)^{q-1} \exp\left[-(t_j/\tau)^q\right],$$

and $\ln L(t_j; \tau, q) = \sum_{j=1}^{n} \ln f(t_j) = n \ln(q) + q \sum_{j=1}^{n} \ln(t_j) - \sum_{j=1}^{n} \ln(t_j) - nq \ln(\tau) - \sum_{j=1}^{n} (t_j/\tau)^q$. Taking $\partial \ln L(t_j; \tau, q) / \partial \tau = 0$ and $\partial \ln L(t_j; \tau, q) / \partial q = 0$ gives

$$\tau = \left[(1/n) \sum (t_j)^q\right]^{1/q}, \tag{5.34}$$

$$\frac{n}{q} - n \ln(\tau) + \sum \ln(t_j) - \tau^{-q} \left\{\sum \left[(t_j)^q \ln(t_j)\right] - \ln(\tau) \sum (t_j)^q\right\} = 0, \tag{5.35}$$

respectively, where all sums are $\sum_{j=1}^{n}$. Inserting τ given by Eq. (5.34) into Eq. (5.35) gives

$$(1/q) + (1/n) \sum \ln(t_j) - \sum \left[(t_j)^q \ln(t_j)\right] / \sum (t_j)^q = 0, \tag{5.36}$$

which depends only on q but does not have an analytical solution and needs to be solved numerically, and then one can solve for τ using Eq. (5.34). The standard deviation of q and τ can be obtained from the second derivative of the log likelihood function, $\left[-\partial^2 \ln L(q, \tau) / \partial q^2\right]^{-1/2}$ and $\left[-\partial^2 \ln L(q, \tau) / \partial \tau^2\right]^{-1/2}$, which gives

5.6 Seismic Response to Step Loading: Short Term Hazard

$$s_d(q) = \left\{ \frac{n}{q^2} + \frac{1}{\tau^q} \sum t_j^q \left[\ln\left(t_j/\tau\right)\right]^2 \right\}^{-1/2} \quad \text{and}$$

$$s_d(\tau) = \tau \left[q \left(n + \frac{q+1}{\tau^q} \sum t_j^q \right) \right]^{-1/2}. \quad (5.37)$$

5.6.7 Hazard Function, Conditional Probability, and Stretched Exponential

The hazard function is defined as $h(t_1) = f(t_1)/[1 - F(t_1)] = f(t_1)/S(t_1)$, where $f(t_1)$ is the probability density function, $F(t_1) = \Pr(\leq t_1)$ is the cumulative distribution function, i.e. the probability that the event will happen before time t_1, and $S(t_1) = \Pr(\geq t_1) = 1 - F(t_1)$ is the survival function, i.e. the probability that the event will not happen until time t_1.

If the distribution of the recurrence times is a stretched exponential with $f(t_1) = (q/\tau)(t_1/\tau)^{q-1} \exp\left[-(t_1/\tau)^q\right]$ and the cumulative distribution function $F(t_1) = 1 - \exp\left[-(t_1/\tau)^q\right]$, then the hazard function becomes

$$h(t_1) = (q/\tau)(t_1/\tau)^{q-1}, \quad (5.38)$$

where t_1 is the time since the last event. For systems characterised by $q = 1$, the probability of having another event of similar size is independent of time, i.e. the stretched exponential distribution is reduced to the exponential one, and since the system has no memory, events occur randomly, see Fig. 5.25 (left).

For $q > 1$, the probability of having another event of that size is increasing as the time since the last event increases. Interestingly, for $q < 1$, the probability of having another event of that size decreases with an increasing time since the last event.

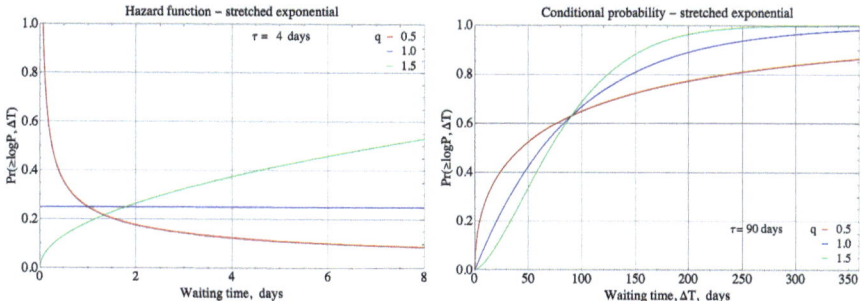

Fig. 5.25 Hazard function *(left)* and conditional probability of the stretched exponential distribution *(right)*

For the intermediate and the long term hazard, one needs to consider the conditional probability that such an event will occur during ΔT, if the time since the last event of that size is t_1, $\Pr(\geq P; t_1 \leq t \leq t_1 + \Delta T \mid t_1 < t) = \int_{t_1}^{t_1+\Delta T} f(t)dt / \int_{t_1}^{\infty} f(t)dt$, which, assuming the stretched exponential distribution, gives

$$\Pr(\geq P; t_1 \leq t \leq t_1 + \Delta T \mid t_1 < t) = F(\Delta T \mid t_1)$$
$$= 1 - \exp\left[\left(\frac{t_1}{\tau}\right)^q - \left(\frac{t_1 + \Delta T}{\tau}\right)^q\right], \quad (5.39)$$

see Fig. 5.25 (right). The waiting time corresponding to a probability of 0.5 gives the median waiting time,

$$\Pr(\geq P; t_1 \leq t \leq t_1+\Delta T \mid t_1 < t) = 0.5 \Rightarrow \Delta T_{0.5} = \tau\left[\left(\frac{t_1}{\tau}\right)^q - \ln 2\right]^{1/q} - t_1. \quad (5.40)$$

The respective probability density function is the derivative $d[F(\Delta T \mid t_1)]d(\Delta T)$, which gives $f(\Delta T \mid t_1) = f(t_1 + \Delta T)/S(t_1)$, where $S(t_1) = 1 - F(t_1)$, and for the stretched exponential,

$$f(\Delta T \mid t_1) = \frac{q}{\tau}\left(\frac{t_1 + \Delta T}{\tau}\right)^{q-1} \exp\left[\left(\frac{t_1}{\tau}\right)^q - \left(\frac{t_1 + \Delta T}{\tau}\right)^q\right]. \quad (5.41)$$

The expected time to the next event then is $\langle \Delta T \rangle = [1/S(t_1)] \int_0^{\infty} \Delta T f(t_1 + \Delta T) d(\Delta T)$, which for the stretched exponential gives

$$\langle \Delta T \rangle = \left(\frac{\tau}{q}\right) \Gamma\left[\frac{1}{q}, \left(\frac{t_1}{\tau}\right)^q\right], \quad (5.42)$$

where $\Gamma\left[(1/q), (t_1/\tau)^q\right]$ is the upper incomplete gamma function.

5.6.8 Information Entropy and Stretched Exponential

The Shannon information entropy measures the amount of uncertainty in a given distribution, which is a measure of unpredictability. Similar to the power law distribution, the continuous information entropy of the stretched exponential distribution is defined as $H[f(t)] = \int_0^T f(t) \log[1/f(t)] dt$, where $f(t)$ is the probability density function, $f(t) = (q/\tau)(t/\tau)^{q-1} \exp\left[-(t/\tau)^q\right]$, where $\tau > 0$ is the relaxation time and $0 < q < 1$ is the shape exponent and t is time. The $\log[1/f(t)]$ is the information content that an event will occur at time t with probability $f(t)$, and, if $f(t)$ is high, then knowledge that event occurred gives

5.6 Seismic Response to Step Loading: Short Term Hazard

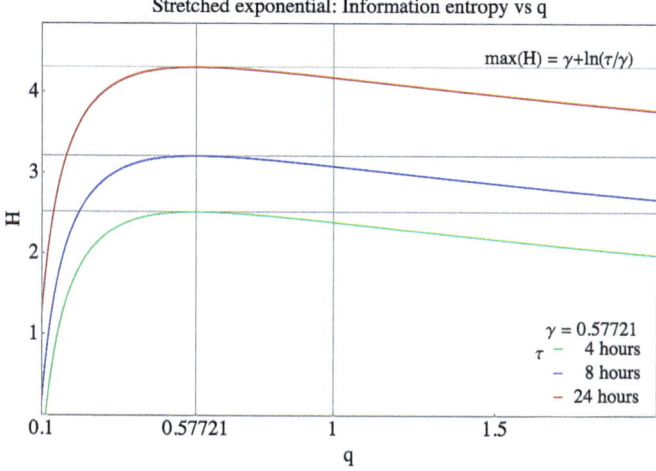

Fig. 5.26 Information entropy vs. q for different τ

very little information, since it had a high probability of occurrence to start with. After integration, the information entropy for the stretched exponential distribution can be expressed as

$$H = \gamma \left(1 - \frac{1}{q}\right) + \ln\left(\frac{\tau}{q}\right) + 1, \qquad (5.43)$$

where $\gamma = 0.57721$ is the Euler-Mascheroni constant. For $q = 1$, i.e. for the exponential distribution $H = \ln(\tau) + 1$, which is independent of q, and for $\tau = 1$, $H = 1$.

The uncertainty is low for low q-values, and it reaches its maximum of $\gamma + \ln(\tau/\gamma)$ at $q = \gamma$, regardless of the relaxation time, and then slowly decays. For a given q, the uncertainty, or the unpredictability, increases with an increase in relaxation time, i.e. systems that are slow to relax are less predictable, see Fig. 5.26. Note that the uncertainty here refers to the forecasting ability of the $f(t)$ and not to the uncertainty or the error in q.

5.6.9 Non-stationary Poisson Process

In a non-stationary, also called a non-homogeneous, Poisson process, the underlying probability distribution and the intensity of the occurrence of events, $I(t)$, vary with time. If $\Lambda(t_1, t_1 + \Delta T) = \int_{t_1}^{t_1+\Delta T} I(t)\,dt$ is the expected number of events in the time interval $(t_1, t_1 + \Delta T)$, then, according to the Poisson distribution, the probability of having exactly N events in that time interval is $\Pr[N; (t_1, t_1 + \Delta T)] = [\Lambda(t_1, t_1 + \Delta T)]^N \exp[-\Lambda(t_1, t_1 + \Delta T)]/N!$. The expected total number of events is $\int_{t_0}^T I(t)\,dt$, where t_0 is the beginning and T is the end of the process. The probability of having zero events during that time is $\Pr[N = 0; (t_1, t_1 + \Delta T)] = \exp[-\Lambda(t_1, t_1 + \Delta T)]$, and the probability of having at least one event is

$$\Pr[N \geq 1, (t_1, t_1 + \Delta T)] = 1 - \exp[-\Lambda(t_1, t_1 + \Delta T)]. \tag{5.44}$$

The non-stationary Poisson process can be applied to model the expected time and size distributions of seismic activity after step loading, e.g. after production blasts, pillar failures, or after larger events. If $I(t, P)$ is the intensity function of the underlying process defined in such a way that $\Lambda[(t_1, t_1 + \Delta T); (P_1, P_2)] = \int_{t_1}^{t_1+\Delta T} \int_{P_1}^{P_2} I(t, P)\,dP\,dt$ is the expected number of events in the time interval $(t_1, t_1 + \Delta T)$ and within the potency range (P_1, P_2), then the probability of having exactly N events in this time interval and in this potency range is

$$\Pr[N; (t_1, t_1 + \Delta T); (P_1, P_2)]$$
$$= \frac{\Lambda[(t_1, t_1 + \Delta T); (P_1, P_2)]^N}{N!} \exp\{-\Lambda[(t_1, t_1 + \Delta T); (P_1, P_2)]\}.$$

If within that time interval the time and size distributions are Independent, then $I(t, P) = I(t)I(P)$ and $\int_{t_1}^{t_1+\Delta T} I(t)\,dt = \Lambda(t_1, t_1 + \Delta T)$ is the expected number of events in this time interval that can be modelled by the Omori law or by the stretched exponential function. The integral $\int_{P_1}^{P_2} I(P)\,dP = \Pr(P_1, P_2)$ gives the probability of having an event in this potency range, and $I(P)$ is the probability density function of the underlying size distribution, e.g. the upper truncated power law. Taking $N = 0$ gives $\Pr[N = 0, (t_1, t_1 + \Delta T); (P_1, P_2)] = \exp[-\Lambda(t_1, t_1 + \Delta T)\Pr(P_1, P_2)]$, which is the probability that the first event in this potency range will occur after time $t_1 + \Delta T$. Therefore, the interval probability that at least one event will occur in potency range (P_1, P_2) in this time interval is $\Pr[N \geq 1; (t_1, t_1 + \Delta T); (P_1, P_2)] = 1 - \exp[-N(t_1, t_1 + \Delta T)\Pr(P_1, P_2)]$, which for $P_1 = P$ and $P_2 = P_{max}$ can be written as

$$\Pr[\geq P; (t_1, t_1 + \Delta T)] = 1 - \exp[-\Pr(\geq P) \cdot N(t_1, t_1 + \Delta T)]. \tag{5.45}$$

The probability $\Pr(\geq P)$ can be given by the UT power law $\Pr(\geq P) = \left(P^{-\beta} - P_{max}^{-\beta}\right)/\left(P_{min}^{-\beta} - P_{max}^{-\beta}\right)$, where β and P_{max} are parameters to be derived from data over the period Δt before the step loading.

Assuming the simple Omori relaxation function, i,e. $c = 0$, the number of events after the step loading is $N(t_1, t_1 + \Delta T) = k\left(t_2^{1-p} - t_1^{1-p}\right)/(1-p)$ for $p \neq 1$ and $N(t_1, t_1 + \Delta T) = \ln(t_2/t_1)$ for $p = 1$. For the stretched exponential relaxation function, the number of events after the step loading is given by Eq. (5.31). The $N(t_0, T)$ in Eq. (5.31) can be estimated in an on-line scheme as the aftershock data is being acquired, $N(t_0, T; t_1) = N(t_0, t_1) / \left(1 - \exp\left[-((t_1 - t_0)/\tau)^q\right]\right)$, where $N(t_0, t_1)$ is the observed number of events at time of forecast, $t_1 > t_0$, and τ and q are estimated at the time t_1 by the maximum likelihood method described above. This process is then repeated as the aftershock sequence progresses to firm up these estimates.

The total number of events to be produced by the sequence, $N(t_0, T)$, can also be estimated by the productivity law of aftershocks, $N(t_0, T) = \alpha_m P_m^\beta$, where P_m is the known potency of the main shock, α_m is the scaling factor to be calibrated, and β is the power law exponent.

Parameters k and p for the Omori relation and τ, q for the stretched exponential can be inverted on-line for the current ongoing sequence by the maximum likelihood method. They can also be taken a priori as average values from previous sequences in the area or, taking a Bayesian-like approach, a weighted average of the a priori and the on-line values (Reasenberg & Jones, 1989).

If after the step loading rock extraction is suspended, the relaxation process would end naturally at time $t[N(t_0, T)]$. If, however, production continues or is restarted long before the relaxation process is completed, the stresses may accumulate and hazard may increase. In principle, mining could restart when seismic hazard dips below a predefined acceptable level and then stays there for a given period of time. This exclusion time would scale positively with seismic hazard before loading and the intensity of sudden loading which scales positively with the stress level in the area.

5.6.10 Aftershock Sequence After $\log P = 2.61$ Event

Here we will discuss the aftershock activity after the main shock (MS) with $\log P = 2.61$ that occurred on 07 July 2013, see description of the data up to the MS in Sect. 5.5. Figure 5.27 left shows the cumulative number of events with $\log P \geq -2.0$ from 8 May to 23 July, where size of the event scales with the source volume and the colour indicates the distance of the event from the MS. The three vertical lines in red, blue, and green mark the times of the three largest events. The rate of events before the MS was steady with no increase in the rate or acceleration of activity before the MS. The first aftershock with $\log P \geq -2.0$ was recorded 3.95, and the second 6.4 seconds after the MS. Within the first 16 hours after the main shock, there were 165 events that give the activity rate just over 10 events/hour, and within the first 171 hours, the rate of activity dropped to 1.4 events/hour.

Figure 5.27 right shows the distances from the MS during the same time with colour scaling with the logarithm of apparent stress, $\log \sigma_A = \log(E/P)$. The bulk of the immediate aftershocks spread 670 metres from the MS. The largest of the immediate aftershock with $\log P = 0.15$ occurred 1.5 minutes after the MS and located 235 metres away. The largest event in the aftershock sequence with $\log P = 1.13$ occurred 162 hours after the MS and located 670 metres away.

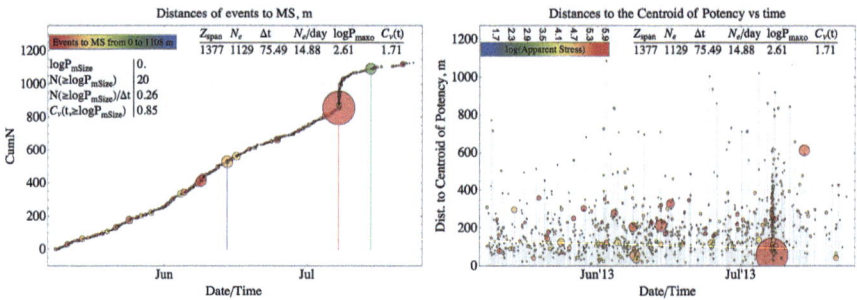

Fig. 5.27 Cumulative number of events *(left)* and distances of events to the MS *(right)* vs. time

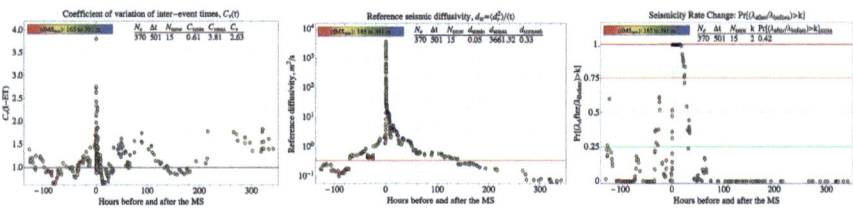

Fig. 5.28 Coefficient of variation of inter-event times *(left)*, seismic diffusivity *(centre)*, and seismicity rate change *(right)* before and after the MS

Figure 5.28 left shows a 15 event moving window of the coefficient of variation of inter-event times, $C_v(t)$, seismic diffusivity $d_s = \langle d^2 \rangle / \langle t \rangle$, and the seismicity rate change, where the colour scales with the mean distance of events in each moving window to the MS, that vary from 165 m in red to 391 m in blue. For details, see Sect. 5.6.2 in this chapter. The high $C_v(t)$ indicates clustering of events in time. Seismic diffusivity, $\langle d^2 \rangle / \langle t \rangle$, increases due to an increase in the rate of seismic activity, i.e. decrease in the mean time between events, $\langle t \rangle$, and/or increase in the spatial extent of events, i.e. increase in the mean distance from the MS or the mean inter-event distance $\langle d^2 \rangle$. Usually, d_s increases when a given system is overloaded and needs to relax. The seismicity rate change gives the probability that there is an increase or a decrease in the rate of events. Probability less than 0.5 means that there was a decrease in the event rate and above 0.5 that there was an increase. Parameter k specifies the magnitude of the change.

The $C_v(t)$ before the MS ranged between 0.79 and 1.4, and it spiked to 3.81 at the time of the MS indicating high clustering, but dropped to below 1.27 within 8 hours. The reference diffusivity started to increase 70 hours before MS and jumped to 3661.32 just after that and it took 63 hours to drop to pre-MS level.

Figure 5.28 right shows the probability that the rate of seismicity increased by at least 2 times with respect to the background level before the MS. The seismicity rate change jumped twice before the MS but reached a maximum 0.61 which is low. It spiked to 1.0 at the time of MS and stayed there for 17 hours dropping below the significant 0.75 level after 25 hours.

5.6 Seismic Response to Step Loading: Short Term Hazard

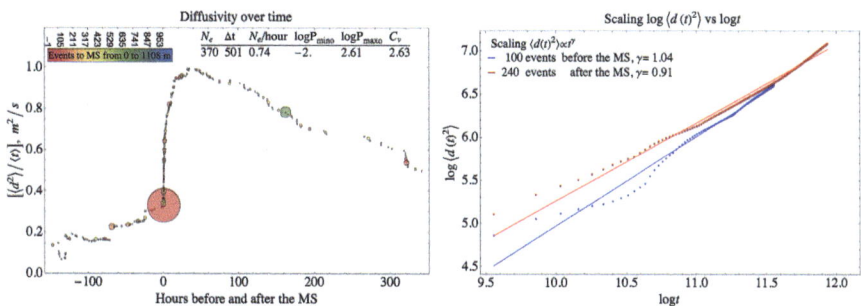

Fig. 5.29 Diffusivity vs. time of events with $\log P \geq -2.0$ before and after the main shock *(left)*. Scaling of $\log \langle d(t)^2 \rangle$ vs. $\log t$ for 100 events before (in blue) and for 240 events after the main shock (in red) is shown on the *right*

Figure 5.29 left shows the diffusivity vs. time during 147.7 hours before and 353.8 after an event with $\log P = 2.61$ shown here as a red circle. In this figure, colour scales with distance to the MS, and the size of the event here represents the radius of the source volume taken as a sphere, $V = P/\Delta\epsilon$, where $\Delta\epsilon$ is the assumed strain change at the source.

There was a steady rate of seismic activity of 0.68 events per hour before and a typical burst and a power law decay of aftershock activity after the main shock. Figure 5.29 right shows $\log \langle d(t)^2 \rangle$ vs. $\log t$ scaling for 100 events before and for the first 240 events after the main shock. It shows a normal diffusion before with $\gamma = 1.04$ and a sub-diffusive process with $\gamma = 0.91$ after the MS.

Figure 5.30 first row shows the cumulative number of events 150 hours before and 200 after the MS with the stretched exponential (SE) fit to the first 16 and 176 hours of aftershock activity in red. Figure 5.30 second row shows the same but with the simple Omori fit (with $c = 0$) in blue. In both cases, the forecast at the time t_f for the next 48 hours is shown in green. All statistics quoted on the right of the plots relate to the aftershocks at the time of forecast.

The SE fitted data well and provided a reasonable forecast at 16 and 171 hours after the MS. The SO fitted data at 16 hours well but failed in forecast and at 171 hours did not fit data well but provided a reasonable forecast. The coefficient of variation of time differences between events, $C_v(t)$, increased from 1.75 in the first fit to 2.47 in the second fit due to the additional 76 intermittent events, but it shows that in both cases there is a temporal clustering and interdependence of events.

Figure 5.30 bottom row shows the interval probabilities, $\Pr[\geq \log P; (t_f, t_f + \Delta T)]$, of having at least one event not smaller than a given $\log P$ within $\Delta T = 48$ hours after t_f, calculated assuming SE, shown in red and SO in blue. The Poissonian probabilities during any 48 hours before the MS are shown in black. In general the SE provided better fit to data.

At 64 hours after the MS, the SE probabilities are considerably higher than before the MS, indicating that the relaxation process is still in progress and seismic hazard is high, and therefore, people should not enter the area affected by aftershock

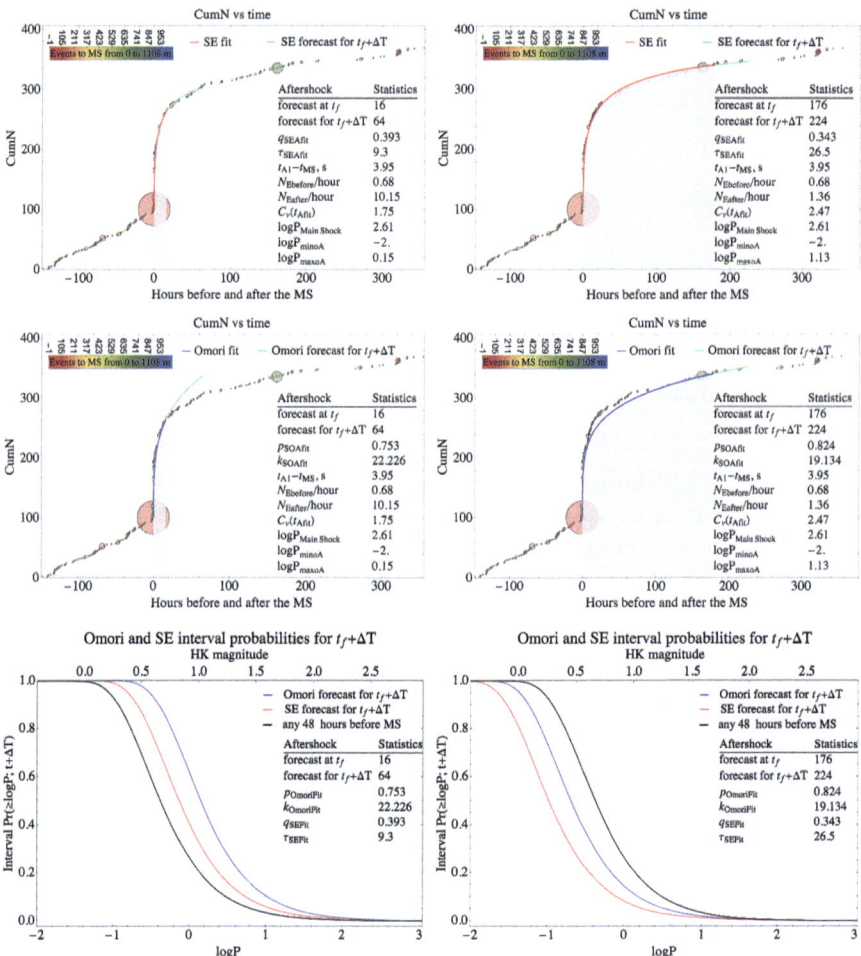

Fig. 5.30 Cumulative number of aftershocks and the SE fit in red *(first row)* and simple Omori in blue *(second row)* to the first 16 and 176 hours of aftershocks vs. relative time in hours. The forecasts for the next 48 hours are plotted in green. The probability of having a seismic event not smaller than $\log P$ within 48 hours after the respective t_f by SE plotted in red and by simple Omori in blue, and the probabilities of having such an event during any 48 hours before the MS are plotted in black *(bottom row)*

activity. For example $\Pr\left[\log P \geq 0.0; 64h\right]$ is 0.4, while before the MS it was 0.265 and $\Pr\left[\log P \geq 1.0; 64h\right]$ is 0.055, and before the MS it was 0.034. At 224 hours after the MS, the relaxation process is mainly completed and seismic hazard dropped. For example, $\Pr\left[\log P \geq 0.0; 240h\right]$ is 0.079, while before the MS it was 0.265, and $\Pr\left[\log P \geq 1.0; 240h\right]$ is 0.009, and before the MS it was 0.034.

Note that after the main shock mining was suspended and this is reflected in the interval probabilities, therefore here we compare seismic hazard before the

5.6 Seismic Response to Step Loading: Short Term Hazard 197

MS driven by production with the interval probabilities after the MS with no production. Obviously, when production will resume, seismic hazard will increase, and therefore, the re-entry into the aftershock area needs to be done with caution.

5.6.11 Aftershock Sequence After $\log P = 5.47$ Event

This example describes aftershock sequence following a large $\log P = 5.47$ ($m_{HK} = 4.6$) and $\log E = 10.6$, mining triggered reverse slip event driven by tectonic sub-horizontal stress. The event was initiated off the bottom edge of a small mine and propagated over one kilometre away from the mine along sub-horizontal geological structure. Waveforms from the mine seismic system, from the close by mines, and regional stations were used to evaluate source parameters and mechanism of the main shock (MS) and aftershocks.

Figure 5.31 left shows three-component velocity waveforms recorded by 14 Hz sensors at distance 280 metres, Fig. 5.31 middle shows waveforms recorded by 4.5 Hz sensors at distance 4384 metres, and Fig. 5.31 right shows 200 seconds of waveforms recorded by broadband sensor at a distance of 92 km from the mine.

The moment tensor solution by the USGS, as well as location and mechanisms of aftershocks, indicates that the event represents a rupture along a plane with strike 240° and dip 24°, in the mine's coordinate system, which extends to the South-South-West of the mine. Analysis of waveforms indicates the two main episodes of rupture, the main shock and 0.7 seconds later the sub-event, that could well be treated as the first aftershock, see Fig. 5.31 left. This sub-event was most likely associated with the rupture of the fault segment about 950 m to the South-South-West from the primary main shock.

The spherical equivalent of the MS is shown in Fig. 5.32 by a large light grey circle that scales with the radius of the source volume with strain change $\Delta \epsilon = 10^{-4}$, $r = [3P/(4\pi \Delta \epsilon)]^{1/3} = 856$ metres, where P is potency. However, the real shape of this event was rather closer to a flat ellipsoid span over the ellipse shown in Fig. 5.32 in blue. The event triggered 3670 aftershocks with $\log P \geq -3.0$ within 156 days which are shown in Fig. 5.33. Most of the aftershocks located away from

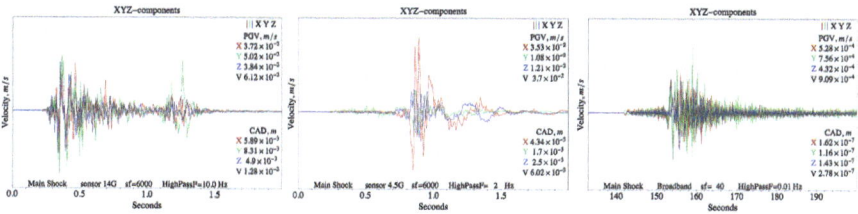

Fig. 5.31 Three-component velocity waveforms recorded by 14 Hz sensors at distance 280 metres (left) recorded by 4.5 Hz sensor at 4384 metres (middle) and by broadband sensor at distance 92 km from the mine (right)

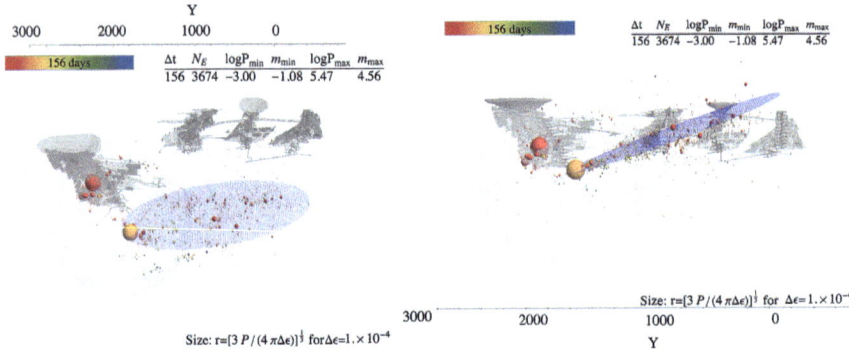

Fig. 5.32 Top *(left)* and section view *(right)* of the mine, including the nearest three mining operations. The main shock is shown in light grey and aftershocks coloured by the time of their occurrence

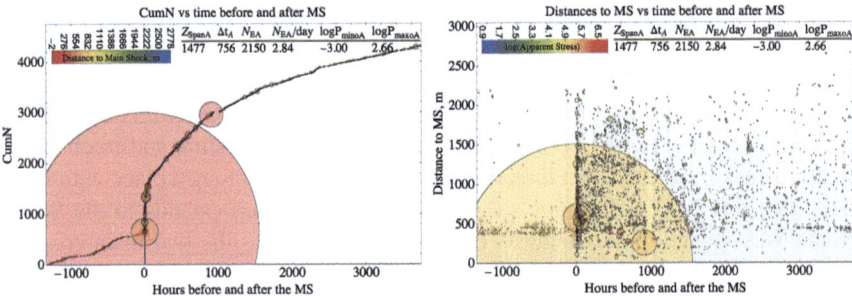

Fig. 5.33 Cumulative number *(left)* and distances to the main shock *(right)* of seismic events with $\log P \geq -3.0$ before and after the main shock

the mine are associated with the sub-horizontal planar structure approximated by ellipse. The largest aftershock with $\log P = 2.66$, shown in Fig. 5.32 in red, occurred within the mine 54 seconds after and 581 metres from the MS. The second largest aftershocks with $\log P = 2.54$ shown in Fig. 5.32 in orange occurred outside the mine 38 days after and 258 metres from the MS. The third largest aftershocks, with $\log P = 1.37$ occurred within the mine 15.6 hours after and 578 metres from the MS.

Figure 5.33 left shows the cumulative number of events with $\log P \geq -3.0$ during 1338.7 hours before and 3730.5 after the MS shown here as the large red circle. Colour here scales with distance to the MS, and the size of the event here represents the radius of the source volume taken as a sphere, $V = P/\Delta\epsilon$, where $\Delta\epsilon$ is the assumed strain change at the source, in this case 10^{-2}. There was a steady rate of seismic activity, $\lambda_B = N(\log P \geq -3.0)/\Delta t_B = 0.484$ per hour before the MS. Figure 5.33 right shows distances of seismic events to the MS over the same period of time where colour scales with the logarithm of apparent stress, $\log \sigma_A = \log(E/P)$. The spatial distribution of events before the MS is mostly concentrated at

5.6 Seismic Response to Step Loading: Short Term Hazard

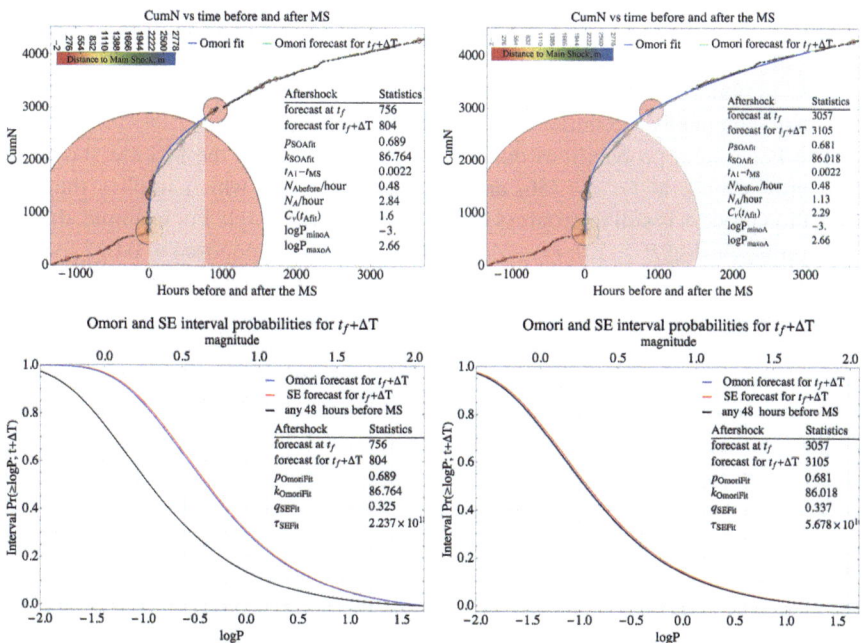

Fig. 5.34 Cumulative number of aftershocks and the SE fit in red *(first row)* to the first 756 and 3057 hours of aftershocks vs. relative time in hours. The forecast for the next 48 hours is plotted in green. The probability of having a seismic event not smaller than $\log P$ within 48 hours after the respective t_f by SE plotted in red and by simple Omori in blue, and the probabilities of having such an event during any 48 hours before the MS is plotted in black *(second row)*

the mine, at distances of 300 to 500 metres from the future MS, with only few events at distances over 500 metres. From the very beginning of the aftershock sequence, the spatial distribution of aftershocks extended to over 2000 metres from the MS. Approximately 2900 hours after the MS, one can see seismic activity at 200 metres from the MS associated with the rehab of mining operations.

Figure 5.34 top row shows the cumulative number of events with the stretched exponential (SE) fit at the time of forecast $t_f = 756$ and $t_f = 3057$ hours after the MS, shown in red, and its extrapolation into the next 48 hours in green. The simple Omori (SO), i.e. with $c = 0$, was also tested for these data sets and gave very similar results. All statistics quoted on the right of the plots relate to the aftershocks at the time of forecast. The first aftershock with $\log P \geq -3.0$ was recorded 8 seconds after the MS.

At $t_f = 756$ hours, the rate of aftershocks was $\lambda_A = 2.84$, the background activity rate $\lambda_B = 0.48$. The coefficient of variation of time differences between events $C_v(t) = 1.6$, which indicates a degree of temporal clustering and interdependence of events. At that time, the mean distance of aftershocks to the MS was 857 metres, extending well beyond the boundary of the mine, which suggests the influence of local tectonics influenced by years of mining in the area.

Figure 5.34 bottom row shows the interval probabilities, $\Pr[\geq \log P; (t_f, t_f + \Delta T)]$, of having at least one event not smaller than a given $\log P$ within $\Delta T = 48$ hours after t_f, calculated assuming SE, shown in red, and SO in blue, which are practically the same.

The Poissonian probabilities during any 48 hours before the MS are shown in dark grey which, at $t_f = 756$, are considerably lower which implies that the relaxation process is still in progress and seismic hazard is high. For example, the SE probability, $\Pr(\log P \geq 1.0; t_f + \Delta T = 48h) = 0.054$ as opposed to 0.022 for any 48 hours before the MS, which is 2.45 times higher. Note that the SE relaxation time, τ, at this stage is large, but it will get smaller as the relaxation process continues.

The forecast at $t_f = 3057$ hours gives a lower hourly rate of aftershocks of 1.13, and the activity rate over the last 1000 aftershocks is 0.51, which is very close to the premain shock background. The coefficient of variation of time differences between events increased to $C_v(t) = 2.29$ due to the intermittency of the recent aftershocks, which indicates a strong degree temporal clustering and interdependence of events. The spatial distribution of aftershocks extended even further beyond the boundary of the mine, which confirms the previous premise of tectonic or other mining influence. Figure 5.34 bottom right shows the interval probabilities, $\Pr[\geq \log P; (t_f, t_f + \Delta T)]$, of having a seismic event not smaller than a given $\log P$ within $\Delta T = 48$ hours after $t_f = 3057$ hours, calculated assuming SE, shown in red, and SO in blue, which are practically the same and equal to the probabilities at any 48 hours before the MS.

At 756 hours, after the MS, the SE probabilities are considerably higher than before the MS, indicating that the relaxation process is still in progress and seismic hazard is high. For example, $\Pr[\log P \geq 0.0; 804h]$ is 0.3, while before the MS it was 0.14 and $\Pr[\log P \geq 1.0; 64h]$ is 0.053, and before the MS it was 0.022.

At 3105 hours after the MS, the relaxation process progressed to the stage that seismic hazard is at the same level it was before the MS.

Note that after the main shock mining was suspended, and this is reflected in the interval probabilities coming back to the pre-shock level rather quickly, and therefore, here we compare seismic hazard before the MS driven by external loading and production with the interval probabilities after the MS with no production. Obviously, when production will resume, seismic hazard will increase, and therefore the re-entry into the aftershock area needs to be done with caution.

5.7 Seismic Rock Mass Response to Blasting Sequence

5.7.1 Introduction

McGarr (1976b) showed that if after extracting a volume of rock V_m, the altered stress and strain field can readjust to an equilibrium state through seismic movements only, the sum of seismic potency released within a given period of time would

5.7 Seismic Rock Mass Response to Blasting Sequence

be proportional to the excavation closure, and in the long term $\sum P = \sum V_m$. Then, assuming the seismic deformation will follow the open-ended power law with parameters α, β, one can estimate the maximum possible event size P_{max} in the seismic sequence, $P_{max} = [(1-\beta) V_m / (\alpha\beta)]^{1/(1-\beta)}$. Similar equations were derived by Wyss (1973), Smith (1976), McGarr (1976a), and Molnar (1979) to estimate the expected maximum magnitude of earthquakes, and by McGarr (1984) for mines. See also section Balance of the Effective Volume Mined and P_{max} in Chap. 4.

Mendecki (2001) studied seismic rock mass response to extraction blasting in a long wall with four 30 m panels at a depth of 3000 m at Tau-Tona mine. The area was covered by a high-resolution microseismic network of five three-component accelerometers that provided enough data to gain insight into the excitation and relaxation processes associated with a sudden transient closure after production blasts. During five days of observations, 15 panels were blasted—minimum 3, maximum 6 panels per day, and no blasts on Sunday. In each stope, there were two rows of 0.9 m long blasting holes separated by 0.5 m vertically and horizontally. They were blasted in pairs top and bottom, with about 200 ms delay. Thus it took approximately 12 to 15 seconds to blast a panel, with face advance by about 0.8 m. A few thousand events were recorded over the five days of observations, but only 1086 with well-developed P- and S-wave signatures were processed and used in analysis. The larger seismic events occurred during Day1 and Day3 after the production blast of at least three contiguous panels. It was concluded that blasting a number of contiguous panels degrades stiffness of the surrounding rock more and for a longer period of time than if the same number of panels were blasted non-contiguously. The loss of stiffness makes the system more vulnerable to larger events.

In another study, Mendecki (2005) analysed the seismic rock mass response to production by blasting at Mponeng mine. The rate of mining $V_m/\Delta t$ ranged from 3 to 33 m^3/day and did not correlate with seismic hazard. Although the highest rate of mining was where non-contiguous panels were blasted, the seismic hazard remained low and larger events occurred during the periods of contiguous blasting.

Martinsson and Torrman (2020) developed a Bayesian hierarchical model that captured the dynamic relationships between seismic activity and production rate, depth, and size of the orebodies. Testing on data from the LKAB Malmberget mine, they inferred that the seismic half-life, defined as the amount of time required for the activity to fall halfway to its steady state value, ranges from weeks up to 2 months for the cases considered. They also found that the effect of the weekly production on the induced seismicity depends exponentially on the average production size in the orebody. The seismic exposure term reveals a dependency on depth, and, for cases considered, an increase in depth by 100 m doubles the seismic activity.

de Beer (2022) tested correlations between production rates, taking into account ring and drawbell blasting and mucking, and seismic potency response in a block cave mine. He found that the rate of mucking is a long term driver of seismic potency release both above and below the undercut. Birch et al. (2024) conducted a similar analysis, fitting a simplified version of a generalised linear model to tons bogged and blasted in weekly intervals over two years at Oyu Tolgoi block cave mine. They found that during this period, which included the start of drawbell development and production bogging, weekly tonnes bogged had much stronger influence on seismicity than blasting.

Most of these papers focus on the seismic rock mass response to production and production rate. Here, we explore the influence of the proximity of group blasting on the intensity of seismic response as measured by the number of larger, potentially damaging seismic events during and after blasting.

5.7.2 Scaled Volume and Proximity Index of Blasts

The intensity of the seismic rock mass response to a single extraction, undercut, or preconditioning blast scales positively with the stress level at the blasting site and the surrounding area and with the volume of rock extracted. In the case of multiple blasts, it also depends on the number and the proximity of these blasts, i.e. the time differences and the distances between them. For a given stress condition and volumes of rock blasted, more clustered blasts result in a more intense seismic response.

Let us denote the volume of the first blast in a sequence of two blasts by V_{b1} and ask what should be the time delay, Δt_{21}, and the space separation, Δd_{21}, to the second blast to avoid superposition of seismic activity. First, let us define the scaled volume of the first blast as

$$V_{sb1} = \frac{V_{b1}}{\Delta t_{21}/t_{\alpha 1} + \Delta d_{21}/d_{\alpha 1}}, \tag{5.46}$$

where the numerator is the volume of rock blasted during the first blast, V_{b1}, and the dimensionless denominator quantifies the proximity in time and in space between the first and the subsequent blast. The proximity in time is quantified by the ratio $\Delta t_{21}/t_{\alpha 1}$, where $t_{\alpha 1}$ is the characteristic seismic relaxation time, and the proximity in space is measured by the ratio $\Delta d/d_{\alpha 1}$, where $d_{\alpha 1}$ is the characteristic size, or the radius, of the seismic relaxation zone. Both $t_{\alpha 1}$ and $d_{\alpha 1}$ are functions of V_{b1}. The best proxies for $t_{\alpha 1}$ and $d_{\alpha 1}$ are the calibrated exclusion or re-entry time, t_r, and the characteristic size of the exclusion zone, d_e. Alternatively, they can be scaled appropriately. For a given Δt_{21} and Δd_{21}, the larger $t_{\alpha 1}$ and/or $d_{\alpha 1}$, the larger V_{sb1}. Since the seismic rock mass response to blasting scales positively with the stress level, so would $t_{\alpha 1}$ and $d_{\alpha 1}$, and therefore, Eq. (5.46) indirectly caters for the stress level.

If the time difference between the first and the second blast Δt_{21} is equal to $t_{\alpha 1}$ and the distance between them, Δd_{21}, to $d_{\alpha 1}$, then

$$V_{sb1}[\Delta t_{21} = t_{\alpha 1}, \Delta d_{21} = d_{\alpha 1}] = V_b/2. \tag{5.47}$$

5.7 Seismic Rock Mass Response to Blasting Sequence

Now, we can define the proximity index of the first blast to the second blast as the ratio,

$$I_{pb1} = \frac{V_{sb1}}{V_{sb1}\left[\Delta t_{21} = t_{\alpha 1}, \Delta d_{21} = d_{\alpha 1}\right]} = \frac{2t_{\alpha 1}d_{\alpha 1}}{t_{\alpha 1}\Delta d_{21} + d_{\alpha 1}\Delta t_{21}}. \quad (5.48)$$

If $\Delta t_{21} = t_{\alpha 1}$ and $\Delta d_{21} = d_{\alpha 1}$, then $I_{pb1} = 1$. For a given $t_{\alpha 1}$ and $d_{\alpha 1}$, the index I_{pb1} measures the deviation from the desired proximity of the second blast from the first one in time and/or in space. The case $I_{pb1} \leq 1$ means the desirable outcome or better, and $I_{pb1} > 1$ means that, for a given V_{b1}, the second blast may be too close in space and/or in time. Since $t_{\alpha 1}$ and $d_{\alpha 1}$ scale positively with V_{b1}, the larger this volume the longer the relaxation time and the larger the relaxation zone. In principle both, t_α and d_α would scale with the cube root of the volume of rock blasted.

Equation (5.48) allows testing whether the preferred time difference and distance between two planned blasts are below the acceptable level of I_{pb} that produces a manageable seismic response.

In practice, mines frequently group more than two blasts to optimise the required exclusion time. The most practical way to test the proximity is to apply the above formulas to each two consecutive blasts. However, while all pairs of consecutive blasts may be reasonably separated in time and space, there may be some non-consecutive blasts missed by this scheme. For example, the case of three blasts carried out within a small Δt where the first and the second blasts are separated enough in space but the third one is very close in space to the first one. Here, if measured in relation to the second blast, the third one would give reasonable space separation resulting in low V_{sb} and I_{pb}, while it would not be the case if the first blast would be measured in relation to the third one.

To detect such cases, one can consider all combinations of two blasts in a given blasting sequence. For n_b blasts, there are $C_2^{n_b} = n(n-1)/2$ combinations of two blasts, and for each, we can calculate I_{pb}. Then, we can sort I_{pb} from the largest to the smallest and compare with the sorted values of consecutive blasts to see if these two lists are the same.

5.7.3 Example: Proximity of Blasting and Seismic Response

The example below is based on the best blasting data available to the author for publication.

Proximity of Blasting Figure 5.35 left shows the cumulative number of 3296 events with $\log P_{min} \geq -3.0$ ($m \geq -1.1$) over 91 days between 6 December and 7 March in an open stopping hard rock mine. In the figure, the size of the event scales with the source volume, and the colour indicates the distance of that event from the $\log P = 2.23$ ($m = 2.4$) main shock (MS) that occurred on 19 February at 13:09:48 at level of 1805. The second largest event with $\log P = 0.84$ ($m = 1.5$) occurred

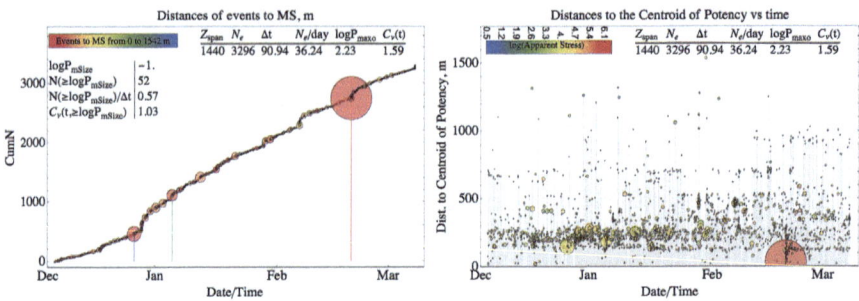

Fig. 5.35 Cumulative number of events (*left*) and their distances to the MS (*right*) vs. time

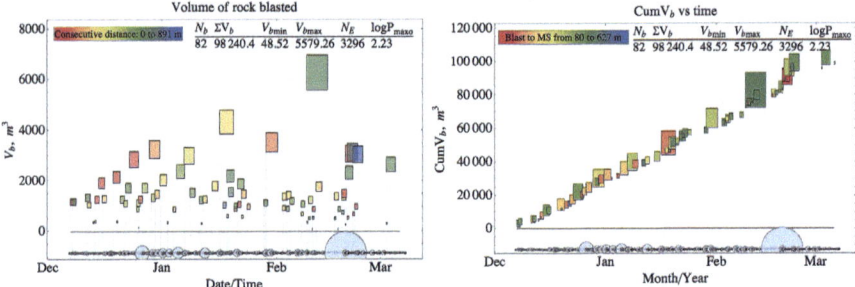

Fig. 5.36 Cumulative number of events and volume of rock blasted (*left*). The cumulative volume of rock blasted with seismic events in light grey plotted at the bottom (*right*)

on 26 December at 14:16:12 at the 1706 level. The three vertical lines in red, blue, and green mark the times of the three largest events. There were 52 mid-size events with $\log P \geq -1.0$ ($m \geq 0.25$), which gives rate of 0.57/day. The mean recurrence interval, $\bar{t}(\log P \geq -1.0) = 34.82$ days with a standard deviation of 35.86 days, which gives the coefficient of variation $C_v = 1.03$. There was no change of rate or acceleration of seismic activity before any of the three largest events. Figure 5.35 right shows the distances of events from the MS during the same time, with colour scaling with the logarithm of apparent stress, $\log \sigma_A = \log(E/P)$. The bulk of these events locate between 200 and 400 metres from the MS, with the exception of the string of events 120 metres away that persisted for days after the MS.

Figure 5.36 left shows the volume of rock blasted where the size of the rectangle scales with the volume of the blast with colour indicating the distance between consecutive blasts varying from 0 metres in red to 891 in blue. The left bottom of the rectangle marks the time of the blast. A timeline of seismic events and they sizes is shown at the bottom in grey. There were 82 blasts extracting 98240 m³ of rock, the smallest 48.5 m³ and the largest 5597.3 m³ at 1728 level, i.e. 77 metres below the MS. All 3321 combinations of two blasts were tested to confirm that they comply with the consecutive order.

5.7 Seismic Rock Mass Response to Blasting Sequence

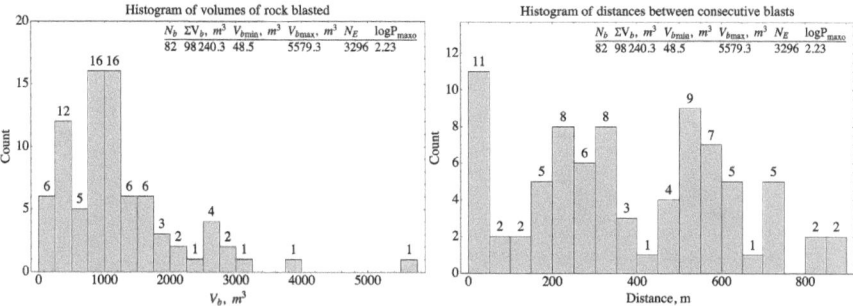

Fig. 5.37 Histograms of volumes of rock blasted *(left)* and distances between consecutive blasts *(right)*

Figure 5.36 right shows the cumulative plot of volume of rock blasted where colour scales with the distance between consecutive blasts. There was an increase in the rate of volume of rock blasted before the second largest event and even more so before the MS. There was also an increase in the number of closely spaced consecutive blasts before these two events.

Figure 5.37 shows the histogram of the volumes of rock blasted and the distances between consecutive blasts. There were 12 blasts with $V_b > 2000$ m³ and three blasts with $V_b > 3000$ m³, the maximum single blast was 5579 m³, and the minimum 49 m³. There were two pairs of blasts practically at the same spot but separated by 12 and 24 hours, respectively. There were three blasts at a distance of less than 10 m from their predecessor, 11 <50 m, and 13 <100 m, 15 <150 m and 20 <200 m.

Figure 5.38 left shows the distances between consecutive blasts where colour scales with the distance to the MS. It also shows the cumulative volume of rock blasted in blue. Note the accelerating rock extraction before the MS.

Since we do not have the calibrated values for the exclusion zone, we scaled it as, $d_{ej} = 10 \cdot S_{bj}$, where $S_{bj} = V_{bj}^{1/3}$ is the characteristic size of rock blasted, which gave $d_{emin} = 36$ metres for $V_{bmin} = 48.5$ m³ and $d_{emax} = 177$ metres for $V_{bmin} = 5579$ m³. The two red horizontal lines in Fig. 5.38 left mark the minimum and the maximum radius of the scaled exclusion zone for all blasts.

We also do not have calibrated re-entry times, and therefore, we scaled them as follows:

$$t_{rj} = t_{rmin} + (t_{rmax} - t_{rmax} \exp\left[-\left((S_{bj} - S_{bmin})/S_{bmax}\right)^q\right]), \qquad (5.49)$$

where t_{rmin} and t_{rmax} are the assumed, for a given range of V_b, minimum and maximum re-entry times, respectively, S_{bj} are the characteristic sizes of the rock blasted, and q is the exponent of the stretched exponential relaxation function. Assuming $t_{rmin} = 4$ and $t_{rmax} = 36$ hours, $S_{bmin} = 3.6$ m, $S_{bmax} = 17.7$ m, and

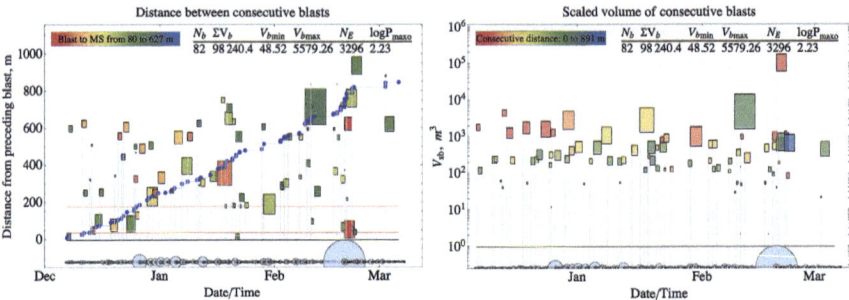

Fig. 5.38 Distances between consecutive blasts with cumulative volume of rock blasted in blue *(left)* and the scaled volume of consecutive blasts *(right)*. Seismic events in light grey vs. time are plotted at the bottom of both figures

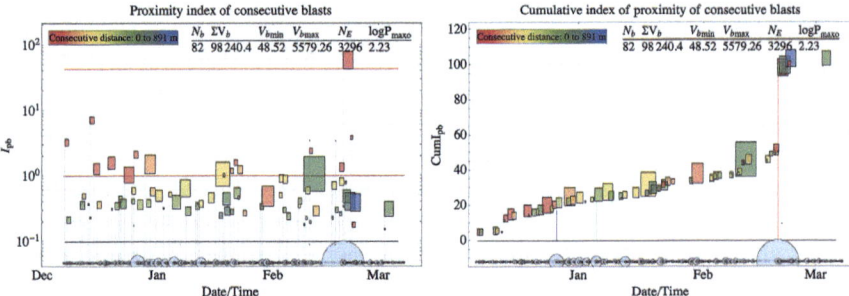

Fig. 5.39 Proximity indices, I_{pb}, of blasts *(left)* and $CumI_{pb}$ vs. time *(right)*. Seismic events in light grey vs. time are plotted at the bottom of both figures

$q = 0.75$, Eq. (5.49) gives the minimum re-entry time of 4 hours for $S_{bmin} = 3.6$ m and 24.5 hours for $S_{bmax} = 17.7$ m.

Figure 5.38 right shows the scaled volume of consecutive blasts where the colour indicates distance between consecutive blasts.

Figure 5.39 left shows the proximity index of all blasts, I_{pb}, where the colour indicates distance between consecutive blasts. The minimum I_{pb} is 0.087 and is associated with $V_{b1} = 49$ m³, with time delay to the next blast $\Delta t_{21} = 0$, but the distance to the next blast $\Delta d_{21} = 836$ m. There are two such blasts in the catalogue, on 10 and 17 of February. The maximum I_{pb} is 43.1 and is associated with $V_{b1} = 2773$ m³ with time delay to the next blast $\Delta t_{21} = 0$ but with $\Delta d_{21} = 7$ m. There were 15 blasts with $I_{pb} \geq 1$, of which 5 were during 19 days before the second largest event with $\log P = 0.84$, and 5 during 9 days before the MS with $\log P = 2.23$. The blast with the maximum $I_{pb} = 43.1$ was carried out 6 minutes before the MS.

Figure 5.39 right shows the cumulative proximity index of blasts, $CumI_{pb}$, vs. time. There was an increase in the index before the MS that started with the blast on 10 February. Table 5.6 lists the proximity parameters of 16 blasts from 10 February before the MS. The data in columns in bold were used to calculate the proximity index of the given blast.

5.7 Seismic Rock Mass Response to Blasting Sequence

Table 5.6 Proximity parameters of 16 blasts before the MS

Month/Day/Time: t_j and t_{j+1}	V_{bj}	d_{MS}	V_{bj+1}	$\Delta t_{j+1,j}$	t_{rj}	$\Delta d_{j+1,j}$	d_{ej}	V_{sbj}	I_{pbj}
02/10/13:03 and 02/10/13:03	84	517	49	0	7.2	26	44	143.7	3.4
02/10/13:03 and 02/10/13:03	49	495	778	0	4.	836	37	2.1	0.09
02/10/13:03 and 02/10/13:03	778	458	471	0	16.3	84	92	850.5	2.19
02/10/13:03 and 02/11/13:03	471	409	1572	24	14.3	709	78	43.6	0.19
02/11/13:03 and 02/16/13:03	1572	467	1216	120	19.2	232	116	190.8	0.24
02/16/13:03 and 02/16/13:03	1216	313	452	0	18.1	346	107	375.2	0.62
02/16/13:03 and 02/17/13:03	452	86	84	24	14.1	588	77	48.2	0.21
02/17/13:03 and 02/17/13:03	84	517	49	0	7.2	26	44	142.8	3.38
02/17/13:03 and 02/17/13:03	49	495	778	0	4.0	836	37	2.1	0.09
02/17 13:03 and 02/18/13:03	778	458	1216	24	16.3	537	92	106.3	0.27
02/18/13:03 and 02/18/13:03	1216	313	656	0	18.1	303	107	428.6	0.71
02/18/13:03 and 02/18/13:03	656	345	1320	0	15.6	714	87	79.8	0.24
02/18/13:03 and 02/19/13:03	1320	409	226	24	18.5	47	110	765.9	1.16
02/19/13:03 and 02/19/13:03	226	440	2733	0	11.3	368	61	37.5	0.33
02/19/13:03 and 02/19/13:03	2733	86	2066	0	21.5	7	140	58893.5	43.1
02/19/13:03 and 02/20/01:03	2066	84	470	12	20.4	588	127	397	0.38

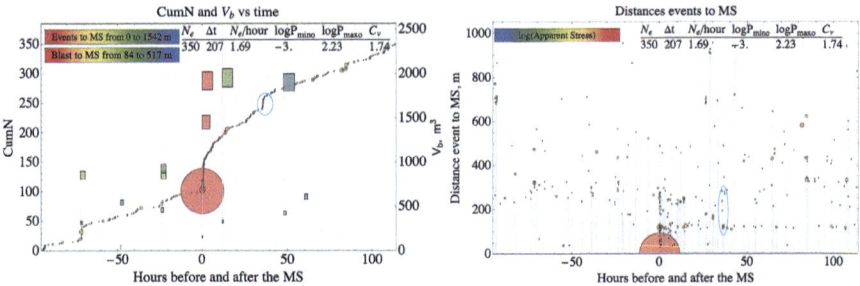

Fig. 5.40 CumN and volumes of rock blasted *(left)* and distances *(right)* of 100 events before and 250 after the MS

Seismic Response to Blasting Before and After the Main Shock Figure 5.40 left shows the cumulative number of 100 events before and 250 after the MS as well as the time and volumes of rock blasted. Size of the event scales with the radius of source volume, and the colour indicates the distance of that event from the MS. In the figure, the size of the rectangle scales with the characteristic size of the blast, $(V_b)^{1/3}$, and colour scales with the distance of that blast from MS.

Note that not every blast or blast sequence triggered a seismic response. Let us look at the seven blast sequences from 72 hours before to 60 hours after the MS:

1. Two blasts 72 hours before the MS. $V_{b1} = 1216$ ^{3} at distance 313 m away from the future MS and $V_{b2} = 452$ m^3 at distance of 86 m. Rock mass responded seismically to these two blasts.
2. Three blasts 48 hours before the MS. $V_{b1} = 84$ m^3 at distance 517 m, $V_{b2} = 49$ m^3 at 495 m, and $V_{b3} = 778$ m^3 at 458 m. No seismic response.
3. Three blasts 24 hours before the MS. $V_{b1} = 1216$ m^3 at distance 313 m, $V_{b2} = 656$ m^3 at 345 m, and $V_{b3} = 1320$ m^3 at 409 m. A very light seismic response.
4. Three blasts 6 minutes before the MS. $V_{b1} = 226$ m^3 at distance 440 m, $V_{b2} = 2733$ m^3 at 86 m, and $V_{b3} = 2066$ m^3 at 84 m. Strong seismic response that triggered $\log P = 2.23$ event and 95 aftershocks within the first 12 hours.
5. Two blasts 12 hours after the MS. $V_{b1} = 469$ m^3 at distance 517 m, $V_{b2} = 2778$ m^3 at 345 m. Relatively weak seismic response, considering the volume of the second blast and the buoyancy of the rock mass after the MS. The response was delayed by 1.5 hour and delivered 12 events during 1.2 hours. After these 12 events, the activity rate dropped.
6. Two blasts 48 hours after the MS. $V_{b1} = 606$ m^3 at distance 409 m, $V_{b2} = 2703$ m^3 at 440 m. Practically, no seismic response, just three very small events within 40 minutes.
7. One blast 60 hours after the MS. $V_{b1} = 867$ m^3 at distance 517 m. Practically, no seismic response, just a few very small events within 7 hours.

The fact that the rock mass responded seismically to blasts close to the future MS may suggest that this was an area of higher stress. Indeed, the apparent stress of

5.7 Seismic Rock Mass Response to Blasting Sequence

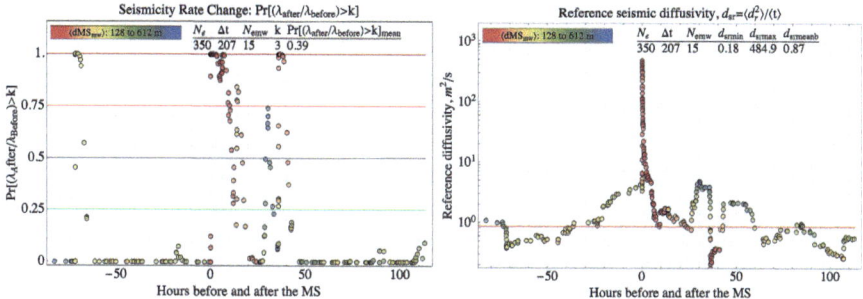

Fig. 5.41 Probability of seismicity rate change *(left)* and reference seismic diffusivity *(right)* in a 15 event moving window

the MS was 0.46 MPa, which is the highest in the data set. If correct, this would support the concept of tap-testing of the criticality of a system, i.e. how close is a given system to the critical point at which its behaviour would change qualitatively. Systems close to criticality are more excitable. The concept of tap-tests is that the immediate seismic response to blasting contains information about the general levels of stress.

There was a seismic flurry that started 35.5 hours after the MS and delivered 21 events within 1.5 hours. The bulk of these events located 120 metres from the MS, see Fig. 5.40 right that shows the distances of these events from the MS with colour scaling with the logarithm of apparent stress. It may be considered a spontaneous activity, unless it was the response to a blast that was not logged.

Figure 5.41 left shows a 15 event moving window of the probability that the rate of seismicity increased by at least 3 times with respect to the activity before the MS. The colour scales with the mean distance of events in each moving window to the MS that vary from 128 m in red to 612 m in blue. For details, see subsection 5.6.2 in this chapter. The probability jumped above the significant 0.75 level for 3 hours in reaction to seismic activity associated with two blasts 72 hours before the MS. These events were on average over 300 metres from the MS. It jumped to 1.0 at the time of the MS and stayed above 0.75 level for 10 hours. Then it dropped briefly and started to rise again 13.5 hours after the MS reacting to the 12 events after blasting. Practically, after the MS, the activity rate was 3 times higher than before the MS for 15 hours. The probability of activity rate change jumped above the 0.75 level again in response to what looks like a spontaneous flurry of small events that started 35.5 hours after the MS and stayed there for 2.8 hours.

Reference seismic diffusivity can be defined as the ratio of the mean distance squared between the reference location, e.g. the main shock or the blast or the point of injection in case of hydrofracturing, and seismic events to the mean time between these events, $d_{sr} = \langle d^2 \rangle / \langle t \rangle$. An increase in the seismic reference diffusivity signifies development of spatial correlation within the system. Figure 5.41 right shows a 15 event moving window of seismic diffusivity over the time of interest where colour scales with the mean distance of events in each moving window to the MS that vary from 128 m in red to 612 m in blue. The horizontal red line marks the mean reference diffusivity before the MS, $d_{srmeanb} = 0.87$. It ignored the light seismic response to the three blasts 24 hours after the MS mainly because they were relatively close to the MS. It started to increase 21.5 hours before the MS in reaction to both, a modest increase in the activity rate and larger distances of these events from the MS. It jumped at the time of the MS but dropped down to the pre-MS level 7 hours later. It increased modestly again for almost 10 hours 26 hours after the MS due to an increase in distances and in activity. The reference diffusivity dropped significantly during the flurry of activity 35 hours after the MS mainly due to the smaller distances from the MS and dropping in activity rate.

Final Comments It is important to schedule blasting sequences to limit the overlap of seismic relaxation. If larger blasts are too frequent and/or too close to each other, i.e. if the next blast is still during the excitation phase of the previous one, it may push a larger and larger part of the system into the sub-critical stage where it is more sensitive to small stress perturbations. Such systems are characterised by periods of low seismic activity followed by small bursts of activity, and they are less time predictable. There is also the possibility that such blasting may induce a larger seismic event that would not have happened otherwise, as opposed to advancing the clock for events that are almost ready to be triggered.

It is recommended to use seismic tap-testing to guesstimate how close the system is to the critical point, at which its behaviour could change qualitatively. Systems close to criticality are more excitable. There are three ways to tap-test the system: (1) Self tap-testing, where one can analyse the immediate response to small event(s) within the system. (2) Self tap-testing, where one can analyse the immediate response to remote larger events. (3) Active tap-testing where one can analyse the immediate seismic response to small blasts carried out in selected areas.

References

Aki, K. (1981). A probabilistic synthesis of precursory phenomena. In D. W. Simpson, & P. G. Richards (Eds.), *Earthquake prediction: An international review*, Maurice Eving (pp. 566–574). American Geophysical Union, Washington D.C.

Amitrano, D. (2003). Brittle-ductile transition and associated seismicity: Experimental and numerical studies and relationship with the b-value. *Journal of Geophysical Research, 108 (B1)*(2044), 1–15. http://doi.org/10.1029/2001JB000680.

Barnsley, M. F. (1988). *Fractals everywhere*. Academic Press.

Birch, D., Rigby, A., Simanjuntak, K., Mardiansyah, F., & Bayar, E. (2024). Establishing relations between seismic activity and production. In *9th International Conference on Mass Mining, Kiruna Sweden* (pp. 1634–1643).

Chen, C.-C., Wang, W.-C., Chang, Y.-F., Wu, Y.-M., & Lee, Y.-H. (2006). A correlation between the b-value and the fractal dimension from the aftershock sequence of the 1999 Chi-Chi, Taiwan, earthquake. *Geophysical Journal International, 167*(3), 1215–1219.

Davis, P. M., Jackson, D. D., & Kagan, Y. Y. (1989). The longer it has been since the last earthquake, the longer the expected time till the next? *Bulletin of the Seismological Society of America, 79*(5), 1439–1456.

de Beer, W. (2022). Quantitative correlation of production and seismic response in block caves. In Y. Potvin (Ed.), *5th International Conference on Block and Sublevel Caving, Perth* (pp. 473–486).

Epstein, B., & Lomnitz, C. (1966). A model for the occurrence of large earthquakes. *Nature, 211,* 954–956. http://doi.org/10.1038/211954b0.
Haldane, J. B. S. (1955). The measurement of variation. *Evolution, 9*(4), 484. http://doi.org/10.2307/2405484.
Heath, J. P. (2006). Quantifying temporal variability in population abundances. *Oikos, 115,* 573–581. http://doi.org/10.1111/j.2006.0030-1299.15067.x.
Henderson, J. R., Barton, D. J., & Foulger, G. R. (1999). Fractal clustering of induced seismicity in The Geysers geothermal area, California. *Geophysical Journal International, 139*(2), 317–324. http://doi.org/10.1046/j.1365-246x.1999.00939.x.
Hirano, R. (1924). Study of the aftershocks of the Kwanto earthquake. *Journal of the Meteorological Society of Japan, 77*(2).
Hirata, T. (1989). A correlation between the b value and the fractal dimension of earthquakes. *Journal of Geophysical Research, 94*(B6), 7507–7514.
Kijko, A. (1982). A modified form of the first Gumbel distribution: model for the occurrence of large earthquakes. Part 1 Derivation of distribution. *Acta Geophysica Polonica, 30*(4), 333–340.
King, G. (1983). The accommodation of large strains in the upper lithosphere of the earth and other solids by self-similar fault systems: The geometrical origin of b-value. *Pure and Applied Geophysics, 121*(5-6), 761–815. http://doi.org/10.1007/BF02590182.
Kisslinger, K. (1993). The stretched exponential function as an alternative model for aftershocks decay rate. *Journal of Geophysical Research, 98*(B2), 1913–1921.
Kohlrausch, v. R. (1854). Theorie des elektrischen Ruckstandes in der Leidener Flasche. *Annalem der Physik und Chemie (ed. J. C. Poggendorff), Leipzig, 91/1*(56), 179–214, p. 198.
Kvalseth, T. O. (2016). Coefficient of variation: the second-order alternative. *Journal of Applied Statistics.* http://doi.org/10.1080/02664763.2016.1174195.
Mandelbrot, B. B. (1967). How long is the coast of Britain? Statistical self-similarity and fractional dimension. *Science, 156,* 636–638.
Marsan, D. (2003). Triggering of seismicity at short timescales following Californian earthquakes. *Journal of Geophysical Research, 108*(B5), 2266.
Martinsson, J., & Torrman, W. (2020). Modelling the dynamic relationship between mining induced seismic activity and production rates, depth and size: A mine-wide hierarchical model. *Pure and Applied Geophysics, 177,* 2619–2639. http://doi.org/10.1007/s00024-019-02378-y.
McGarr, A. (1976a). Upper limit to earthquake size. *Nature, 262*(5567), 378–379.
McGarr, A. (1976b). Seismic moments and volume changes. *Journal of Geophysical Research, 81*(8), 1487–1494.
McGarr, A. (1984). Some applications of seismic source mechanism studies to assessing underground hazard. In N. C. Gay, & E. H. Wainwright (Eds.), *Proceedings 1st International Symposium on Rockbursts and Seismicity in Mines, Johannesburg, South Africa* (pp. 199–208). South African Institute of Mining and Metallurgy.
McGuire, R. K. (1977). Effects of uncertainty in seismicity on estimates of seismic hazard for the east coast of the United States. *Bulletin of the Seismological Society of America, 67*(3), 827–848.
Mendecki, A. J. (2001). Data-driven understanding of seismic rock mass response to mining: Keynote Address. In G. van Aswegen, R. J. Durrheim, & W. D. Ortlepp (Eds.), *Proceedings 5th International Symposium on Rockbursts and Seismicity in Mines, Johannesburg, South Africa* (pp. 1–9). South African Institute of Mining and Metallurgy.
Mendecki, A. J. (2005). Persistence of seismic rock mass response to mining. In Y. Potvin, & M. R. Hudyma (Eds.), *Proceedings 6th International Symposium on Rockburst and Seismicity in Mines, Perth, Australia* (pp. 97–105). Australian Centre for Geomechanics.
Mendecki, A. J. (2008). Forecasting seismic hazard in mines. In Y. Potvin, J. Carter, A. Diskin, & R. Jeffrey (Eds.), *Proceedings 1st Southern Hemisphere International Rock Mechanics Symposium, Perth, Australia* (pp. 55–69). Australian Centre for Geomechanics.
Molnar, P. (1979). Earthquake recurrence intervals and plate tectonics. *Bulletin of the Seismological Society of America, 69*(1), 115–133.
Omori, F. (1893). On after-shocks, *Tech. rep.*, College of Science Imperial University of Tokyo.

Omori, F. (1894). On the aftershocks of earthquakes. *Journal of College Science, Imperial University of Tokyo, 7*, 111–200.

Pearson, K. (1896). Mathematical contributions to the theory of evolution: Regression, heredity, and panmixia. *The Philosophical Transactions of the Royal Society of London, A, 253*–318.

Reasenberg, P. A., & Jones, L. M. (1989). Earthquake hazard after a mainshock in California. *Science, 243*, 1173–1176.

Savage, J. C. (1994). Empirical earthquake probabilities from observed recurrence intervals. *Bulletin of the Seismological Society of America, 84*(1), 219–221.

Shakal, A. F., & Willis, D. E. (1972). Estimated earthquake probabilities in the north Circum-Pacific area. *Bulletin of the Seismological Society of America, 62*(6), 1397–1410.

Smith, S. W. (1976). Determination of maximum earthquake magnitude. *Geophysical Research Letters, 3*(6), 351–354.

Sornette, D., & Knopoff, L. (1997). The paradox of the expected time until the next earthquake. *Bulletin of the Seismological Society of America, 87*(4), 789–798.

Utsu, T. (1957). Magnitudes of earthquakes and occurrence of their aftershocks (in Japanese). *Zisin, 2*(10), 35–45.

Vieira, F. M. C. C., Diering, D. H., & Durrheim, R. J. (2001). Methods to mine the ultra-deep tabular gold-bearing reefs of the Witwatersrand Basin, South Africa. In W. A. Hustrulid, & R. L. Bullock (Eds.), *Underground Mining Methods: Engineering Fundamentals and International Case Studies* (pp. 691–704). Society for Mining, Metallurgy and Exploration, Inc (SME).

Wang, J. (1994). On the correlation of observed Gutennberg-Richter's b-value and Omori's p-value for aftershocks. *Bulletin of the Seismological Society of America, 84*(6), 2008–2011.

Williams, G., & Watts, D. C. (1970). Non-symmetrical dielectric relaxation behavior arising from a simple empirical decay function. *Transactions of the Faraday Society, 66*, 80–85. http://doi.org/10.1039/TF9706600080.

Wyss, M. (1973). Towards a physical understanding of the earthquake frequency distribution. *Geophysical Journal of the Royal Astronomical Society, 31*(4), 341–359.

Wyss, M., Sammis, C. G., Nadeau, R. M., & Wiemer, S. (2004). Fractal dimension and b-value on creeping and locked patches of the San Andreas fault near Parkfield, California. *Bulletin of the Seismological Society of America, 94*(2), 410–421.

Open Access This chapter is licensed under the terms of the Creative Commons Attribution 4.0 International License (http://creativecommons.org/licenses/by/4.0/), which permits use, sharing, adaptation, distribution and reproduction in any medium or format, as long as you give appropriate credit to the original author(s) and the source, provide a link to the Creative Commons license and indicate if changes were made.

The images or other third party material in this chapter are included in the chapter's Creative Commons license, unless indicated otherwise in a credit line to the material. If material is not included in the chapter's Creative Commons license and your intended use is not permitted by statutory regulation or exceeds the permitted use, you will need to obtain permission directly from the copyright holder.

Chapter 6
Ground Motion Hazard

Abstract This chapter starts with the general description of transient strains and stresses and the dynamic stress concentration factors for a plane harmonic incident P- and S-wave hitting a tunnel. Then it gives a rudimentary explanation of the ground motion (GM) at seismic source, describes parameters that measure the intensity of GM, and presents a case study on amplification of GM at the skin of an excavation. Understanding, quantifying, and forecasting GM motion in the near and intermediate field of seismic radiation is the most important issue in mine induced seismicity since most damage caused by seismic events is observed in excavations very close to their sources. Since the maximum ground velocity at source is controlled by the strength of the rock mass, small and large events produce similar ground velocity at source, but large events affect a substantial volume of rock, and hence, the probability of hitting a vulnerable structure is considerably higher. In smaller mines, the strong ground velocities and displacements associated with larger events may affect the entire infrastructure.

The next section is dedicated to the ground motion prediction equation (GMPE) and its applications, e.g. plotting the expected GM at strategic locations in real time, seismic fragility curves that show the probability that tunnel support may be damaged under different seismic loads given its remaining deformation capacity, and the GM alert program, called GMAP. The next section describes the real-time version of GMAP, called GMAS, that does not require the GMPE and estimated source parameters to issue an alert.

The next section describes mapping seismic ground motion hazard, which in earthquake seismology is called the probabilistic seismic hazard assessment and its limitations, the major being the uncertainty in the spatial distribution of the expected seismicity. The last section, modelling seismic hazard, describes what is called the deterministic hazard, i.e. kinematic modelling of GM produced by potential seismic sources defined by their expected maximum potency or energy, placed at the most likely locations that can produce the highest intensity of GM motion at a given site. Examples of modelling GM produced by extended and complex sources are given.

6.1 Seismic Waves, Transient Strains, and Stresses

Introduction A wave is a disturbance that transports energy from its source through the rock without the transport of matter. It is the energy of the wave, not the particles of the medium, that travels through the medium. Wave motion can be transient, periodic, or random. Transient motion is the response of the medium to a sudden pulse-like excitation. Periodic motion is repetitive, recurring in the same form at fixed time increments, e.g. harmonic motion. In random motion, the instantaneous amplitude can be predicted only on a probabilistic basis (e.g. random noise).

The maximum disturbance in each cycle is known as the amplitude of the wave and is determined by its source. The amplitude of a wave is equal to the maximum displacement, velocity, or acceleration of ground motion from the equilibrium position. The number of cycles that pass by a fixed point per second is called frequency, f, and is expressed in hertz [Hz]. The frequency of a wave is determined by its source. The time in seconds required for a complete cycle to be produced or to pass a given point is the period of the wave, T, and it is the reciprocal of its frequency, $T = 1/f$ or $f = 1/T$. The length corresponding to one complete cycle of the wave is called the wavelength, Λ. It can be measured from crest to crest, trough to trough, or between corresponding points on adjacent pulses. Wavelength is affected by both the source and the medium.

Phase velocity quantifies how fast crests or troughs travel. The wavefront associated with any particular phase advances a distance Λ in time T, and therefore, the phase velocity $v_\pi = \Lambda/T = \omega/k$, where $k = 2\pi/\Lambda$ is the wavenumber and $\omega = 2\pi f$ is the circular frequency. Note that for a given frequency, the wavelength increases with increasing wave propagation velocity, $\Lambda = v_\pi/f$. In perfectly elastic homogeneous media, all frequencies travel with the same velocity. Attenuating media are dispersive and allow waves of different frequencies to travel at different velocities.

Wave Speed The speed of any wave depends upon the properties of the medium through which the wave is travelling. Typically, there are two essential types of properties which affect wave speed: inertial and elastic properties. The greater the inertia, or the mass density, of individual particles of the medium, the less responsive they will be to the interactions between neighbouring particles and the slower the wave. If all other factors would be equal, a sound wave would travel faster in a less dense material than in a more dense material. Elastic properties relate to the strength of interaction between particles measured by the tendency of a material to resist deformation upon applied force, which is called stiffness. The stiffer the material the faster the wave. Even though the inertial factor may favour gases, the elastic factor has a greater influence on the wave speed, and therefore: $v_{solids} > v_{liquids} > v_{gases}$. In general, the velocity of a seismic wave increases with rock mass stiffness and decreases with its density, $v \propto \sqrt{\text{stiffness/density}}$.

6.1 Seismic Waves, Transient Strains, and Stresses

P-Wave The dilatational or longitudinal or primary waves, where the elastic medium expands and contracts and in which particles move in the direction of propagation. The wave velocity is

$$v_P = \sqrt{\frac{\kappa + (4/3)\mu}{\rho}} = \sqrt{\frac{2\mu(1-\nu)}{\rho(1-2\nu)}} = \sqrt{\frac{Y(1-\nu)}{\rho(1-2\nu)(1+\nu)}}, \qquad (6.1)$$

where κ is the bulk modulus, μ the shear modulus, ρ is density, Y is the Young modulus, and ν the Poisson ratio. Note the dependence of v_P on both, the bulk and the shear modulus. The reason is that during propagation of dilatation the medium is subjected to a combination of compression and shear. The cross-sectional area of a small cube element normal to the direction of propagation will not be changed, but its dimension along the propagation will be altered. There is thus a change in the shape of the element as well as in its volume, and the resistance of the medium to shear as well as its compressibility comes into play (Kolsky, 1963).

S-Wave The transverse or shear or secondary waves, where the medium changes in shape, but not in volume. Particles move perpendicular to the direction of propagation, which occurs with velocity

$$v_S = \sqrt{\frac{\mu}{\rho}} = \sqrt{\frac{Y}{2\rho(1+\nu)}}.$$

Given a reference plane, the S-wave particle motion can arbitrarily be decomposed into horizontal SH and vertical SV components which, in homogeneous isotropic media, travel with the same speed. The polarisation, i.e. the direction of particle motion relative to the direction of wave propagation, of both P- and S-waves, is linear. In an anisotropic medium, where properties vary with direction, the S-wave splits into a fast and slow component. These split waves propagate with different velocities that cause a time delay and related phase shift. Accordingly, the two split S-wave components superimpose on an elliptical polarisation. The orientation of the main axis and the degree of ellipticity are controlled by the fast and slow velocity directions of the medium with respect to the direction of wave propagation and the degree of anisotropy. An independent propagation of the P- and S-waves is only guaranteed for sufficiently high frequencies where spatial variations in elastic properties occur over much larger distances than the wavelength of the waves involved. Fluids have no shear strength, $\mu = 0$, and thus do not propagate shear waves.

Ratio v_P/v_S The ratio of the P- to S-wave velocity depends only on the Poisson ratio ν, see Fig. 6.1, and since $-1 < \nu \leq 0.5$, we have

$$\frac{2}{\sqrt{3}} < \frac{v_P}{v_S} = \sqrt{\frac{2-2\nu}{1-2\nu}} < \infty. \qquad (6.2)$$

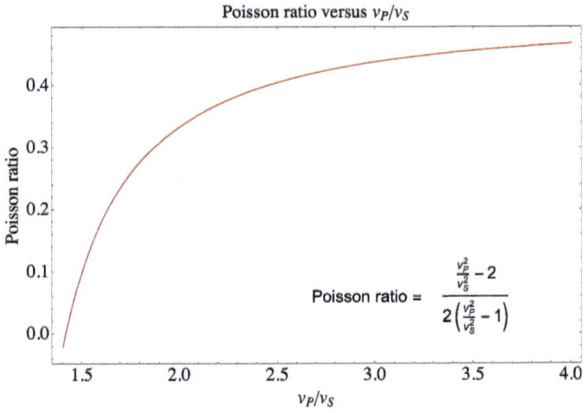

Fig. 6.1 v_P/v_S and Poisson ratio

Table 6.1 Relationship between elastic moduli

$Y =$	$2\mu(1+\nu)$	$3\kappa(1-2\nu)$	$\frac{9\kappa\mu}{3\kappa+\mu}$		$\frac{\lambda(1+\nu)(1-2\nu)}{\nu}$	$\frac{\mu(3\lambda+2\mu)}{\lambda+\mu}$
$\nu =$		$\frac{\lambda}{2(\lambda+\mu)}$	$\frac{Y}{2\mu}-1$	$\frac{3\kappa-2\mu}{2(3\kappa+\mu)}$	$\frac{\lambda}{3\kappa-\lambda}$	$\frac{3\kappa-Y}{6\kappa}$
$\mu =$		$\frac{Y}{2(1+\nu)}$	$\frac{\lambda(1-2\nu)}{2\nu}$	$\frac{3(\kappa-\lambda)}{2}$	$\frac{3\kappa(1-2\nu)}{2+2\nu}$	$\frac{3\kappa Y}{9\kappa-Y}$
$\kappa =$		$\frac{Y}{3(1-2\nu)}$	$\frac{2\mu(1+\nu)}{3(1-2\nu)}$	$\frac{\mu Y}{3(3\mu-Y)}$	$\frac{\lambda(1+\nu)}{3\nu}$	$\lambda+\frac{2}{3}\mu$
$\lambda =$		$\frac{2\mu\nu}{1-2\nu}$	$\frac{\mu(2\mu-Y)}{Y-3\mu}$	$\frac{\nu Y}{(1+\nu)(1-2\nu)}$	$\frac{3\kappa\nu}{1+\nu}$	$\kappa-\frac{2}{3}\mu$

Thus in a given medium, the P-wave is always faster than the S-wave. The ratio v_P/v_S increases with increasing ν, and for $\nu = 0.15$, the $v_P/v_S = 1.558$, for $\nu = 0.2$ $v_P/v_S = 1.633$. For a Poissonian solid, where the Poisson ratio $\nu = 0.25$, $v_P/v_S = \sqrt{3}$ and $v_S = v_P/\sqrt{3}$. For most consolidated rocks, the v_P/v_S ratio is between 1.5 and 2.0.

Elastic Constants Since hydrostatic pressure cannot cause an increase in volume and the volume decrease must remain finite, the bulk modulus κ can only assume positive values. The shear caused by a state of simple shearing has the direction of the stress, and therefore, the shear modulus μ must be positive. For an incompressible material, κ, and hence, λ (Lamé constant) must become infinite. These conditions ensure that the strain energy is positive, and they impose the following fundamental inequalities for the elastic moduli of isotropic materials: $\infty \geq \lambda > \frac{2}{3}\mu$, $\infty > \mu > 0$, $0 < Y \leq 3\mu$, $-1 < \nu \leq \frac{1}{2}$, where Y is the Young's modulus (Table 6.1). Abstract distance between two elastic media can be measured by $d = \sqrt{[\log(\kappa_1/\kappa_2)]^2 + [\log(\mu_1/\mu_2)]^2}$.

One can reduce the number of parameters to one under certain simplifying assumptions. The most frequent is the so-called Poisson's relation, $\lambda = \mu$, from

6.1 Seismic Waves, Transient Strains, and Stresses

which it follows: $\nu = 0.25$, $Y = 5\mu/2$, and $\kappa = 5\mu/3$. Another simplification is to assume that material is incompressible, and then $\kappa = \lambda = \infty$, $\nu = 0.5$, and $\mu = Y/3$.

Surface Waves Seismic surface waves are an example of guided waves that propagate along free surfaces, internal discontinuities, or other waveguides. Seismic surface waves are created by the interference of seismic body waves. There are two major kinds of surface waves: Love waves, which are shear waves trapped near the surface, and Rayleigh waves, which have rock particle motions that are very similar to the motions of water particles in ocean waves.

Love waves are generated when an SH ray hits a reflecting horizon near surface at post-critical angle, and all the energy is trapped within the wave guide (Love, 1911). They propagate by multiple reflections between the top and bottom surfaces of the low speed layer near the surface. Love waves are similar to S-waves with no vertical displacement. They move the ground from side to side horizontally parallel to the Earth's surface, at right angles to the direction of propagation, and produce horizontal shaking. The Love wave velocity is equal to that of shear waves in the upper layer for very short wavelengths, and to the velocity of shear waves in the lower layer for very long wavelengths.

Rayleigh waves, also known as a "ground roll", are the result of incident P and SV plane waves interacting at the free surface and traveling parallel to that surface. The particles oscillate in a vertical plane along the direction of propagation. There are two components to their oscillations: vertical up and down motion and the horizontal forward and back motion. At the free surface, the initial vertical motion is up, but the initial horizontal movement is backward. Thus the particle motion at the free surface is elliptical retrograde (i.e. the particle moves opposite to the direction of propagation at the top of its elliptical path), and the vertical displacement is about 1.5 times the horizontal displacement. The penetration into the ground increases with wavelength Λ. The vertical amplitude of particle motion decreases exponentially with depth. The horizontal amplitude becomes zero at a depth of $0.19 \cdot \Lambda$, and below that depth, it reverses its direction and the particle motion becomes forward elliptical. The Rayleigh wave velocity, v_R, varies from $0.9 v_S$, for the media with Poisson ratio $\nu = 0.1$, to $0.93 v_S$, for $\nu = 0.4$. For a Poissonian solid, $\nu = 0.25$, Rayleigh waves travel with a phase velocity $v_R \simeq 0.92 v_S$. In mines, surface waves are developed as the wavefront reaches the tabular excavations.

Transient Strains and Stresses Any function $f(x - v_\pi t)$ represents a wave propagating in the x direction, and the function $u(x, t) = u_0 \sin(2\pi f x/v_\pi - 2\pi f t)$ represents a simple harmonic wave with an amplitude u_0, frequency f propagating in the x direction with velocity v_π. The ground (particle) velocity at x then is $v = \partial u/\partial t = -2\pi f u_0 \cos(2\pi f t - 2\pi f x/v_\pi)$ and acceleration $a = \partial v/\partial t = (2\pi f)^2 u_0 \sin(2\pi f t - 2\pi f x/v_\pi)$. The ground velocity v is shifted in phase by $\pi/2$ with respect to u and by π with respect to a. The dynamic transient strain at point x is $\epsilon_d(x, t) = \partial u(x, t)/\partial x = 2\pi f (u_0/v_\pi) \cos(2\pi f t - 2\pi f x/v_\pi)$, and it shows that the particle velocity and the dynamic strain in a harmonic motion are related by $\epsilon_d = \text{v}/v_\pi$. In summary, for harmonic motion,

$$a = \omega \text{v} = \omega^2 u \quad \text{and} \quad \epsilon_d = \text{v}/v_\pi. \tag{6.3}$$

Therefore, ground displacement of 1 mm at a frequency of 20 Hz would result in a ground velocity of 0.125 m/s and an acceleration of 15.8 m/s^2, and the observed PGV of 10 mm/s associated with a wave propagating in the rock mass at 3300 m/s would cause $3 \cdot 10^{-6}$ strains, which is close to the upper limit for the elastic behaviour of rock. Larger ground velocities in this medium would result in inelastic deformation. If we assume that hard rock ruptures at shear strain $\epsilon_d = 5 \cdot 10^{-4}$ and that $v_\pi = v_S = 3250$ m/s, then the maximum expected ground velocity at source, $\text{v} = 1.625$ m/s, see Sect. 6.3. For two waves of the same frequency and different displacement amplitudes, $u_1/u_2 = \text{v}_1/\text{v}_2 = a_1/a_2$. These results are accurate for the harmonic pulse; otherwise, they are only approximations.

Assuming a linear elastic isotropic material in plane strain, the maximum normal stress is

$$\sigma_{max} = \frac{Y(1-\nu)}{(1+\nu)(1-2\nu)} \epsilon_{max} = \pm \frac{Y(1-\nu)}{(1+\nu)(1-2\nu)} \frac{|\text{v}_P|}{v_P} = \pm \rho v_P |PGV_P|, \tag{6.4}$$

where Y is the Young's modulus, ν is the Poisson ratio, ϵ_{max} is the maximum axial strain, PGV_P is the peak ground velocity in the direction of P-wave propagation, and v_P is the apparent P-wave velocity. A similar expression can be derived for the shear stress due to the propagating S-wave, $\tau_{max} = \pm \rho v_S |PGV_S|$, where v_S is the apparent S-wave velocity and PGV_S is the peak ground velocity in the direction of S-wave polarisation. These approximations are more accurate at lower frequencies and quite sensitive to the value of apparent propagation velocity which may be highly variable in the fractured rock surrounding excavations. Indeed, the selection of the appropriate apparent propagation velocity is not simple, and the associated uncertainty limits the accuracy of strain estimates from recorded data. At higher frequencies, there may be additional differential displacements and strains caused by spatial variability of ground motions, i.e. changes in the amplitudes and phases of the motions as well as arrival time perturbations of the waveforms at the various locations in the ground near the surface.

6.2 Static and Dynamic Stresses Around a Circular Tunnel

Underground structures, specifically those embedded in hard rock, are far more resilient to shaking than surface ones. Therefore, with the exception of large events, most damage caused by seismic events is observed in excavations very close to their sources. Since the maximum ground velocity at source is controlled by the strength of the rock mass, small and large events produce similar ground motion at source,

6.2 Static and Dynamic Stresses Around a Circular Tunnel

but large events affect a substantial volume of rock, and hence, the probability of hitting a vulnerable structure is considerably higher. In smaller mines, the strong ground velocities associated with larger events may affect the entire infrastructure. The maximum ground motion that can be experienced at a given site is controlled by the maximum ground velocity at source, the interaction of radiation from different parts of the source and from different travel paths, and by site effects.

The response of tunnels to seismic motion may be understood in terms of three principal types of inflicted deformation: axial, curvature, and hoop (tangential). Axial and curvature deformations develop when the direction of the incident wave is not exactly normal to the axis of the tunnel. Axial deformations are represented by alternating regions of compressive and tensile strain that travel as a wave-train on the surface of the tunnel along its axis. Curvature deformations also travel along the length of the tunnel trying to bend it in alternating directions. Hoop deformations result from waves of normal or nearly normal incidence to the tunnel axis. Three effects of the hoop deformations can be observed: (1) a distortion of the cross section, (2) dynamic stress concentration near the free surface of the tunnel, and (3) a circulation of seismic wave energy around the tunnel (which is possible only for wavelengths which are shorter than the radius of the tunnel).

An analytical treatment of the dynamic stress field near a tunnel of arbitrary shape and dimensions is not possible, yet valuable intuition can be obtained from case studies formulated for tunnels of simple geometry exposed to plane harmonic seismic waves. A typical example is the solution of the dynamic equations of motion for a plane SH wave at normal incidence to an infinite cylindrical tunnel of circular cross section surrounded by a perfectly elastic material. The same case can be reformulated for incident P-wave or SV-wave. The stationary case of a plane harmonic wave can be used as a stepping stone to the description of the dynamic effect on a tunnel due to a seismic pulse of complex frequency content. This would simply involve a direct and an inverse Fourier transform of the input data. This approach is of practical importance when the effect from an actual seismic wave on an existing tunnel needs to be evaluated. In this case, the velocity seismogram needs to be transformed into the frequency domain, and after computing the effect from each frequency, the obtained stress needs to be transformed back from the frequency domain to the time domain in order to compute the stress concentration factors and the related seismic response spectra.

At the other extreme of reformulating the problem, one can study the stress field around a tunnel due to a static remote loading. At a first glance, this static stress field should have very little in common with the picture of dynamic stress concentration. However, in many cases, the wavelength corresponding to the predominant frequency of a seismic wave can be one or two orders of magnitude greater than the radius of the tunnel, and therefore, the final results for the stress concentration factors in the dynamic and static cases turn out to be similar.

Of practical importance is the case of a plane wave at normal incidence to an infinite circular cylinder. The mathematical formulation of the problem is truly 2D and can be solved exactly. The corresponding static case was solved by Kirsch in the nineteenth century. The 2D-stress field near a circular opening under a constant remote stress $\sigma_{xx} = \sigma_1$, $\sigma_{yy} = \sigma_2$, and $\sigma_{xy} = \sigma_{12}$ is

$$\sigma_r = \frac{\sigma_1+\sigma_2}{2}\left[1-\left(\frac{a}{r}\right)^2\right] + \left[1 - 4\left(\frac{a}{r}\right)^2 + 3\left(\frac{a}{r}\right)^4\right]$$
$$\left[\left(\frac{\sigma_1-\sigma_2}{2}\right)\cos 2\theta + \sigma_{12}\sin 2\theta\right]$$

$$\sigma_\theta = \frac{\sigma_1+\sigma_2}{2}\left[1+\left(\frac{a}{r}\right)^2\right] - \left[1 + 3\left(\frac{a}{r}\right)^4\right]\left[\left(\frac{\sigma_1-\sigma_2}{2}\right)\cos 2\theta + \sigma_{12}\sin 2\theta\right]$$
(6.5)

$$\sigma_{r\theta} = \left[1 + 2\left(\frac{a}{r}\right)^2 - 3\left(\frac{a}{r}\right)^4\right]\left[\left(\frac{\sigma_2-\sigma_1}{2}\right)\sin 2\theta + \sigma_{12}\cos 2\theta\right],$$

where a is the radius of the circular hole (the cross section of the tunnel) and the components of the stress tensor are given in polar coordinates, $x = r\cos(\theta)$ and $y = r\sin(\theta)$, for $0 \leq r < \infty$ and $0 \leq \theta \leq 2\pi$.

The analytical static solution gives the radial stress component σ_r, the tangential (or hoop) stress σ_θ, and the shear stress $\sigma_{r\theta}$ at any point outside the hole, $r \geq a$ and $0 \leq \theta \leq 2\pi$. The circular hole in the 2D Kirsch case corresponds to the cross section of an infinite tunnel, and this is why the values for the stress components on the free surface $r = a$ are of special interest. One can verify that the components contributing to the normal stress are zero as required for the boundary of a cavity, $\sigma_r(r = a) = 0$ and $\sigma_{r\theta}(r = a) = 0$. The hoop stress on the surface of the tunnel, $r = a$, varies with the polar angle θ and can be expressed in terms of the corresponding value of the load σ_θ^∞—this would be the hoop stress if the hole was absent. The special case of the hoop stress at the edge of the hole for $r = a$ is

$$\sigma_\theta(r = a) = (\sigma_1 + \sigma_2) - 2(\sigma_1 - \sigma_2)\cos 2\theta - 4\sigma_{12}\sin 2\theta. \tag{6.6}$$

The stress concentration factor γ is defined as the ratio of the maximum hoop stress on the cavity wall ($r = a$) to the maximum remote stress. For the special case of biaxial loading, $\sigma_{12} = 0$ (or $\tau_{12} = 0$ in Fig. 6.2), $\sigma_1 > 0$, and $|\sigma_2| \leq \sigma_1$, the stress concentration factor is reached for $\theta = \pm\pi/2$ and has the value

$$\gamma = 1 + \frac{\sigma_2}{\sigma_1} + 2(1 - \frac{\sigma_2}{\sigma_1}) = 3 - \frac{\sigma_2}{\sigma_1}. \tag{6.7}$$

The biaxial static concentration factor ranges between 4 and 2. The maximum $\gamma = 4$ is reached for pure shear loading $\sigma_2 = -\sigma_1$, and the minimum $\gamma = 2$ corresponds to uniform, or hydrostatic, loading $\sigma_2 = \sigma_1$.

A special case of biaxial loading is $\sigma_2 = -\sigma_1 \nu/(1 - \nu)$, where ν is the Poisson ratio of the surrounding material. As it turns out, the static stress concentration factor for the above case can be very close to the similar dynamic quantity defined for a long tunnel exposed to a plane monochromatic P-wave.

6.2 Static and Dynamic Stresses Around a Circular Tunnel

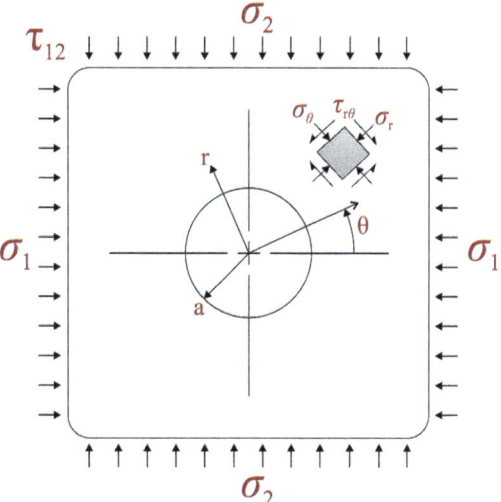

Fig. 6.2 2D elastic stresses around a hole

The general problem of evaluating the stress field at the boundary of an underground tunnel in the presence of a propagating seismic wave is quite complicated but, under some simplifying assumptions, a solution can be found (Mow & Pao, 1971). Usually, it is assumed that the tunnel is an infinitely long cylinder with circular cross section surrounded by a perfectly elastic material. Further, a plane harmonic incident wave is assumed to hit the tunnel at a right angle to the axis. In this formulation, the problem is two dimensional and admits an exact solution for the superposition of the incident and reflected waves. The dimension-less ratio of the maximum resultant tangent stress on the wall of the tunnel to the maximum stress in the incident wave defines the dynamic stress concentration factor. In the case of incident P- and SV-waves, the tangent stress is the hoop $\sigma_{\theta\theta}$, while for SH incidence $\sigma_{\theta z}$ is tangent to the boundary of the tunnel. All functions in the exact solutions are expressed in terms of the polar angle θ and the radial distance r to the axis of the tunnel. The polar angle is measured from the direction of the incident wave, and the radial variable, r, enters in the final solution through two dimensionless combinations: $\omega r/v_P$ and $\omega r/v_S$, where $\omega = 2\pi f$ is the circular frequency, v_P is the P-wave velocity, and v_S is the S-wave velocity. The exact evaluation of the stress components for any frequency, tunnel radius, and location in the surrounding rock can be computationally quite demanding, but obtaining asymptotic expressions for the dominant frequency of typical seismic waves and for a tunnel radius of just a few metres can be relatively easy. This would require taking the limit $\omega a/v_P \to 0$ and $\omega a/v_S \to 0$ in the expression for the tangent stress. The wavelength in the case of circular frequency ω and wave speed v is $\Lambda = 2\pi v/\omega$ so $\omega a/v = 2\pi a/\Lambda$ is usually a very small number. Indeed, the wavelength corresponding to the dominant frequency of a seismic event will be two orders of magnitude greater than the radius of the tunnel which justifies the limit $\omega a/v \propto a/\Lambda \to 0$. But this limit is also consistent with $\omega \to 0$ irrespective of the wave speed and the tunnel radius. The

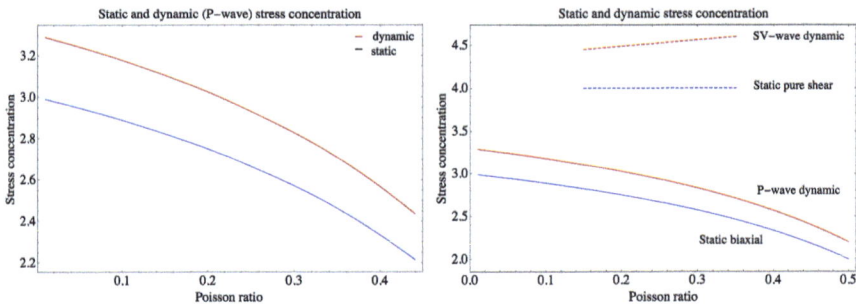

Fig. 6.3 Maximum dynamic stress concentration factor for P-wave *(left)* and SV-wave *(right)* on a cylindrical cavity

zero, or low, frequency regime is equivalent to the static case, hence, the relationship between the Kirsch solution and the dynamic stress concentration factor for P-wave incidence. The low frequency or quasi-static limits for the three types of plane harmonic waves at normal incidence to a tunnel are

$$\sigma_{\theta z} = \sigma_0 \left[1 + \left(\tfrac{a}{r}\right)^2\right] \sin(\theta) \quad \text{for SH-wave}$$
$$\sigma_{\theta\theta} = 4\sigma_0 \sin(2\theta) \quad \text{for SV-wave} \quad (6.8)$$
$$\sigma_{\theta\theta} = 2\tfrac{\sigma_0}{k^2}\left[(k^2 - 1) - 2\cos(2\theta)\right] \text{ for P-wave,}$$

where $k^2 = (v_P/v_S)^2 = (2 - 2\nu)/(1 - 2\nu)$. For practical purposes, these quasi-static limits can be very useful taking into account the fact that they are only 10% to 15% lower than the respective dynamic counterparts. However, this is not true for high frequencies when significant oscillations are observed in the stress concentration factors.

The properties of the rock surrounding the tunnel are also of importance when computing the stress concentration factors. This is shown in Fig. 6.3 where the dynamic and static stress concentration factors for incident P-wave and SV-wave are plotted as functions of the Poisson ratio of the rock.

The total stress is a superposition of the static and dynamic stresses. The dynamic stress concentration around the cavity for a P-wave for an isotropic, elastic medium was determined by Mow and Pao (1971), and it depends on ν and on the dimensionless frequency, Ω, of the wave

$$\Omega = \frac{2\pi f a}{v_P} = \frac{2\pi a}{\Lambda} \Rightarrow \frac{\Lambda}{a} = \frac{2\pi}{\Omega}, \quad (6.9)$$

where f is the frequency of the wave and Λ is the wavelength. Note that for a very long wavelength, i.e. very low frequency, Ω tends to zero and only the static stress concentration applies. The full expression for the dynamic stress concentration is rather complex, but its peaks are approximately 10% to 15% greater than the static one and occur at $\Omega \simeq 0.25$, and therefore, the maximum stress concentration occurs at $\Lambda/a \simeq 25$, i.e. at wavelengths approximately equal to 25 times the cavity radius. For SH-waves in which the particle motion is normal to the plane of the cross section, the dynamic stress concentration factor is 2.1 and corresponds to

$$\frac{2\pi f a}{v_S} = \frac{2\pi a}{\Lambda} = 0.4 \Rightarrow \frac{\Lambda}{a} = 16 \Rightarrow f = \frac{v_S}{16a}, \qquad (6.10)$$

and therefore, the maximum shear stress concentration occurs at wavelengths approximately equal to 16 times the cavity radius. For $v_\pi = v_S = 3000$ m/s and the characteristic radius of the cavity of $a = 4$ metres, the frequency of the maximum dynamic stress is at $f = 187.5/8 = 47$ Hz, and for $PGV_S = 0.1$ m/s, the maximum dynamic shear stress $\tau_{max} = 0.8$ MPa.

Mow and Pao (1971) selected the largest value of the dynamic stress concentration factor over the entire range of frequencies for a given ν and plotted it as a function of ν, for both P- and SV-wave, see Fig. 6.3.

6.3 Ground Motion at Source

Rupture is a propagating pulse that precedes slip at a seismic source. Its speed varies from $0.6v_S$ to $0.9v_S$ for sub-shear rupture and more than v_S for super-shear rupture. Slip follows rupture, and it is very fast at the tip of the rupture and slows dramatically past the rupture front. Slip velocity is the velocity of one side of the source with respect to the other. An average slip velocity varies from a few cm/s to a few m/s. Rupture may be unilateral, propagating in one direction across the source, bilateral, nucleating at the centre of the source and propagating in both directions, or it may be inhomogeneous. Seismic radiation and thus near-field motions strongly depend on rupture velocity: Slow ruptures radiate little seismic energy, while fast ruptures generate higher ground motion amplitudes. If all other factors are equal, the amplitudes of the radiated seismic waves increase with co-seismic deformation rate, rupture speed, rock strength, and with ambient stress.

The near-source ground velocity is equal to half of the slip velocity —the velocity of one side of the source with respect to the other. The ground velocity at source is controlled by the maximum stress at source which, in turn, is limited by the strength of the rock mass.

Consider a small piece of ground attached to an infinite source bounded on a plane by the extension of a rupture propagating with velocity $v_r \to \infty$ over a small increment of time Δt, and away from the source plane by the extension of the propagating S-wave with velocity v_S, see Fig. 6.4 left. If the applied effective stress σ_{eff} available to accelerate the two sides of the source is released instantaneously, then the rock mass acceleration, a, and velocity, v, can be derived from $a = F/m$, where here force $F = \sigma_{eff} (v_r \Delta t)^2$ and mass $m = \rho (v_r \Delta t)^2 v_S \Delta t$, and therefore

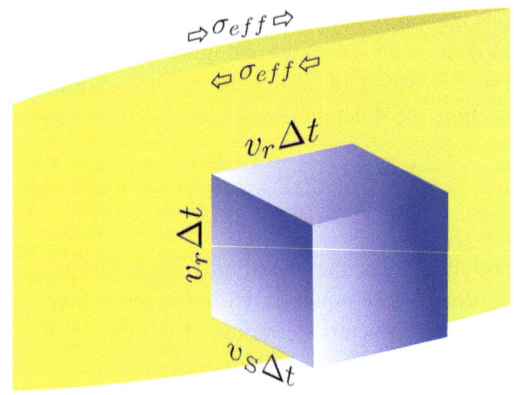

Fig. 6.4 Sketch of a volume of ground attached to a moving source

$$a = \frac{F}{m} = \frac{\sigma_{eff}}{\rho v_S \Delta t} = \frac{\sigma_{eff}}{\mu \Delta t} v_S \implies \dot{u} = \frac{F \Delta t}{m} = \frac{\sigma_{eff}}{\rho v_S} = \frac{\sigma_{eff}}{\mu} v_S = \epsilon_{eff} v_S, \quad (6.11)$$

where $\mu = \rho v_S^2$ is rigidity, ρv_S is the shear wave impedance, and ϵ_{eff} is the effective shear strain. If the effective stress $\sigma_{eff} = \mu$, then the ground velocity would be equal to S-wave velocity, $\dot{u} = v_S$. However, since $\sigma_{eff} \ll \mu$, the ground velocity $\dot{u} \ll v_S$.

According to the above equations: (1) Rock strength does not limit peak acceleration. For small Δt (at high frequencies, $1/\Delta t$), there is practically no limit on peak ground acceleration. A fracture may produce a steep change in ground velocity which results in high acceleration at high frequencies (Andrews et al., 2007). (2) The ground velocity will always be much smaller than the rupture velocity because the effective stress is much smaller than the shear modulus. (3) Rock strength limits ground velocity, which does not depend on frequency. (4) Ground velocity at source does not depend on the size of the event.

For a finite circular source of diameter $2r$ with instantaneous stress release, the effects of the edges of the crack will abate the ground velocity with time. For a simple taper, $\exp(-v_S t/r)$, given by Brune (1970), integration of equation (6.11) over the process time, r/v_S, (Kanamori, 1972) gives the average ground velocity,

$$\langle \dot{u} \rangle = 0.63 \sigma_{eff}/\mu. \quad (6.12)$$

If the effective strain, σ_{eff}/μ, at seismic source is between $5 \cdot 10^{-4}$ and 10^{-3} and $v_S = 3250$ m/s, the near-source ground velocity would vary between 1 and 2 m/s. For an inhomogeneous rupture, the effective strain may be considerably different at different parts of the source producing varying dynamic strain drops and varying ground motion. The highest ground motion would be produced by rupture of intact rock (Gay & Ortlepp, 1978; Ortlepp, 1997; van Aswegen, 2008).

The effective stress cannot be measured directly, but different approximations can be made. The best proxy would be the dynamic stress drop; however, it cannot be derived reliably from observations. One option is to assume that σ_{eff} is equal

6.3 Ground Motion at Source

Table 6.2 Models of near-source ground velocity as a function of rupture velocity according to Burridge (1969)[1], Ida (1973)[2], and McGarr and Fletcher (2001)[3], where $f(v_r)$ is a monotonic function that ranges from 0.11 to 0.4 as rupture velocity increases from $0.6v_S$ to $0.9v_S$

Model	$\langle \dot{u} \rangle$	For $v_r = 0.75v_S$
Bilateral rupture[1]	$\langle \dot{u} \rangle = \sigma_{eff} / [\rho v_S (1 + v_S/v_r)]$	$\langle \dot{u} \rangle = 0.43 \sigma_{eff} / (\rho v_S)$
Dynamic cohesive rupture[2]	$\langle \dot{u} \rangle = \sigma_{eff} v_r / (\rho v_S^2)$	$\langle \dot{u} \rangle = 0.75 \sigma_{eff} / (\rho v_S)$
Dynamic rupture scaling[3]	$\langle \dot{u} \rangle = 0.8 \sigma_{eff} / [\rho v_S f(v_r)]$	$\langle \dot{u} \rangle = 0.36 \sigma_{eff} / (\rho v_S)$

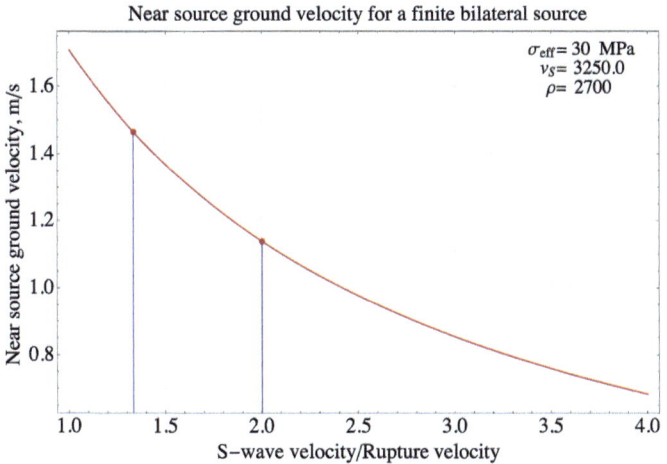

Fig. 6.5 Near-source ground velocity for a finite bilateral source and finite rupture velocity (Burridge, 1969)

to the bulk shear strength of the rock within the volume of interest, which averages for most hard rocks between 40 and 60 MPa. An intact rock may be considerably stronger. If a reliable database is available, one can, as a lower bound, also use the maximum stress drop derived from recorded waveforms, which in many hard rock mines averages between 2 and 4 MPa. Again the maximum values associated with failure of intact rock may go as high as 40 to 60 MPa.

The near-source ground velocity for a finite source and finite rupture velocity for different source models are quoted in Table 6.2. For $\sigma_{eff} = 30$ MPa, $\rho = 2700$ kg/m^3, $v_S = 3250$ m/s, assuming the rupture velocity v_r between $0.5v_S$ and $0.75v_S$, the estimates of the near-source ground motion would vary between 1.14 m/s and 1.47 m/s, see Fig. 6.5. In general, the faster the rupture the faster the slip and the higher the near-field ground motion.

Equations (6.11) and those in Table 6.2 are applicable when the rate of stress release with an increase in slip velocity is, at all times, less than the shear wave impedance, $d\sigma_{eff}/d\dot{u} < \rho v_S$, both in Pa·s/m. Slip rates given by these equations may underestimate strain rates at the edges of the moving source. If the rate of loading exceeds the rate at which energy can be removed by elastic waves, the system is no longer linear. To remove this excess energy, the large strains need to travel faster than small ones—the particle velocity exceeds the shock wave velocity. This is also what happens during super-shear rupture when the crack tip is moving faster than the *S*-wave velocity (e.g. Weertman, 1969; Burridge, 1973; Savage, 1971; Andrews, 1976; Archuleta, 1984; Spudich & Cranswick, 1984; Dunham, 2007; Lu et al., 2010; Andrews et al., 2007).

6.4 Ground Motion Characteristics

Near-source ground motions can be amplified by rupture **forward directivity**, which occurs when the rupture direction and slip direction are aligned and move towards the site. When the source ruptures towards the site at a speed close to shear wave velocity, most of the radiated energy arrives there in a short time interval, and the cumulation of these pulses results in a single large low frequency pulse observed at the beginning of the seismogram. Rupture directivity produces a long period pulse in the direction normal to the fault plane. In the case where the rupture propagates away from the site, the arrival of seismic waves is distributed in time. This condition, referred to as **backward directivity**, is characterised by ground motions with relatively long duration and lower amplitude. Neutral directivity occurs for sites located off to the side of the rupture area where the rupture is neither predominantly towards nor away from the site (Archuleta & Hartzell, 1981; Somerville et al., 1997).

Fling step is a result of a static ground displacement and is generally characterised by a unidirectional velocity pulse and a monotonic step in the displacement time history. Fling step displacements occur in the direction of slip and therefore are not strongly coupled with the rupture directivity pulse. In a strike-slip source, the directivity pulse occurs on the strike-normal component, while the fling step occurs on the strike parallel component. In a dip-slip source, both the fling step and the directivity pulse occur on the strike-normal component.

Secondary Sources Spall fractures are an example of secondary sources that occur when a high-intensity transient stress wave from the primary source reflects from a free surface. This phenomenon is the result of interference near a free surface between the portion of an oncoming incident compression wave which has not yet been reflected and the portion which has been reflected and transformed into a tensile wave. Usually, the amount of tension increases as the reflected wave moves back inward from the surface. The transient tensile stress resulting from the superposition is at the quarter wavelength depth and twice the maximum stress in the incident pulse. If the medium is not capable of withstanding these induced tensile stresses, it will break, creating a secondary source or sources of seismic radiation at that time and depth. The energy from the secondary sources arrives at the surface later in the strong motion, see Fig. 6.6.

They can be recognised by high frequency content in waveforms recorded by station(s) close to the spall but far from the primary source. Spalling occurs in rocks, soils, liquids and in cohesionless materials.

Fig. 6.6 Illustration of spall fracture

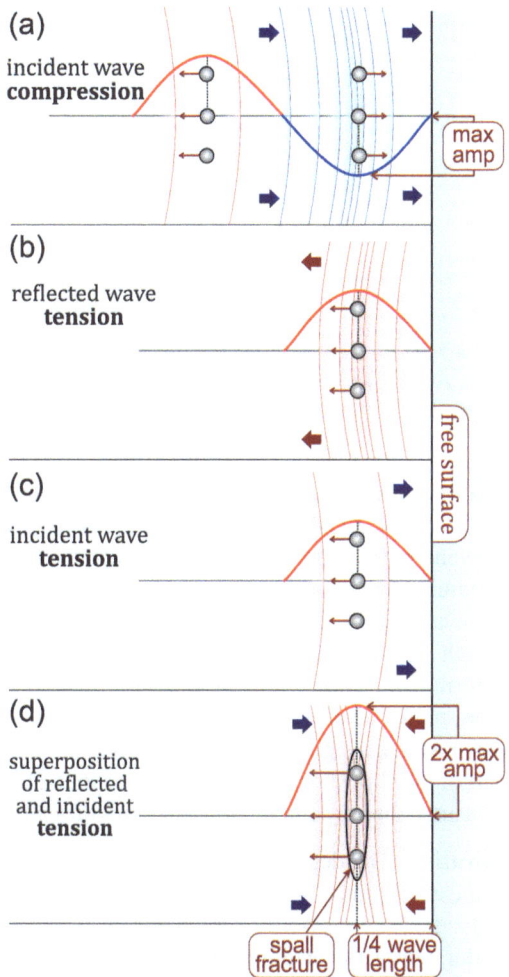

Figure 6.7 left shows 5.375 seconds of waveforms of the $\log P = 3.0$ event recorded in a mine 1230 metres from source with $PGV = 1.78$ cm/s at frequency approximately 5 Hz. Figure 6.7 right shows a zoomed section with a hidden secondary small high frequency event that was triggered close to the recording site.

6.4.1 Engineering Characteristics of Ground Motion

There is a number of parameters that measure the intensity of ground motion and its potential for damage. They are mainly based on measured maximum amplitude, energy, and duration.

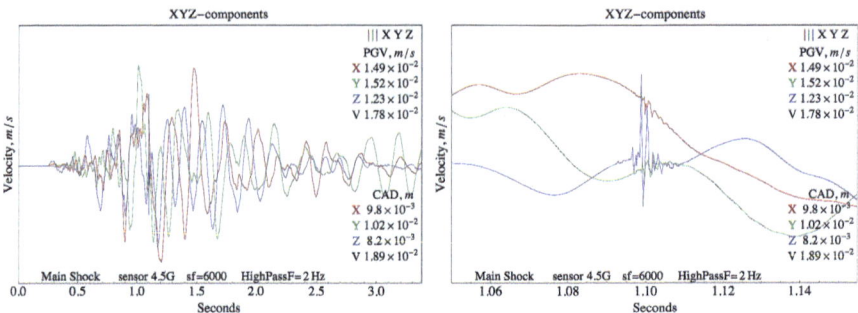

Fig. 6.7 Three component velocity waveforms of large event recorded 1230 metres away *(left)* and a zoomed section showing the hidden secondary event that happened close to the recording site *(right)*

Peak Ground Characteristics Ground motion characteristics include peak ground acceleration, PGA, velocity, PGV, and displacement, PGD. The PGA is most convenient for structural engineers, since the maximum force experienced by a rigid structure of mass m is $F_{max} = m \cdot PGA$. However, the PGA is a poor parameter for evaluating potential for damage. For example, a large PGA associated with a high frequency pulse may be absorbed by the inertia of the structure with little deformation, since $PGD \propto PGA/f^2$, where f is frequency. On the other hand, a more moderate acceleration associated with a long duration pulse of low frequency may result in a significant deformation of structures. The PGV, which at source is a half of the slip velocity, is less sensitive to the higher frequencies than PGA, can be measured directly and reliably, and provides a better indication of damage potential.

Duration The degradation of stiffness and strength of rock are sensitive not only to the amplitude of ground motion but also to its duration and an associated number of load or stress reversals above the elastic regime. There are three different types of duration: bracketed, uniform, and significant.

The **bracketed duration** measures the duration of the ground motion from the first to the last occurrence of amplitude exceeding a specified threshold. The **uniform duration** is defined as the sum of the time intervals during which the ground motion is greater than the threshold. A functional way to display it is to construct a histogram and the cumulative graph of the time that the amplitude of ground motion spent above a certain level. Figure 6.8 left shows waveforms with $PGA = 82.1$ m/s^2 produced by an event with $\log E = 6.5$ recorded 93 metres from the source, and Fig. 6.8 right the uniform duration plot. $PGA \geq 10$ m/s^2 lasted for 0.05 seconds and $PGA \geq 1$ m/s^2 lasted for 0.09 seconds.

The **significant duration** defines ground motion duration as the length of the time interval between the accumulations of two specified levels of ground motion energy at the site. For example, the amount of time in which the central 90% of the integral of the squared velocity or acceleration takes place, t_{90} (Trifunac & Brady, 1975). As distance increases, the bracketed and uniform durations tend to zero, but,

6.4 Ground Motion Characteristics

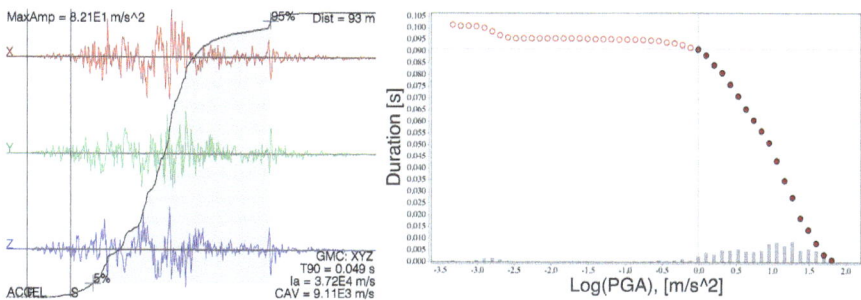

Fig. 6.8 An example of uniform duration associated with a complex event *(left)* measured on the cumulative graph of the time that the amplitude of ground motion spent above certain level *(right)*

as energy becomes dispersed with distance, the t_{90} tends to increase. It is useful then to calculate the central 90% of energy, $E_{90} = E(t_{95}) - E(t_5)$ and the power, $P_{90} = E_{90}/t_{90}$.

Arias intensity, $I_a = \pi/(2g) \int_0^{t_d} [a(t)]^2 dt$, where I_a is in m/s, $a(t)$ is the acceleration time history in units of acceleration due to gravity g, and t_d is the duration of ground motion (Arias, 1970).

The cumulative absolute velocity, $CAV = \int_0^{t_d} |a(t)| dt$, and has units of velocity, m/s (EPRI, 1988). In mines, however, the utility of I_a and CAV are limited for two reasons: (1) Mines mostly employ velocity transducers. (2) The highest accelerations recorded by piezoelectric accelerometers are associated with very high frequencies, where there is little displacement and little or no damage potential.

The **cumulative absolute displacement**, CAD, is defined as the integral of the absolute value of a velocity time series, $CAD = \int_0^{t_d} |v(t)| dt$, which has units of displacement. CAD is the area under the absolute velocity time history and is more sensitive to lower frequency ground motion, i.e. to larger displacements.

The gradient of the time integral is equal to the absolute velocity. Since $v(t) = du/dt$, the integral over velocity can be written as the summation of incremental peak-to-valley and valley-to-peak displacements, regardless of sign, in the displacement time series, $CAD = \sum_j^n |\Delta u_j|$, where Δu_j is the jth value of incremental displacement in the time series and n is the total number of incremental displacements. CAD can also be interpreted as the area under the plot of ground velocity, v, versus duration, $t_d (\geq v)$, see Fig. 6.9. Since CAD could be overly influenced by a time series of long duration that contained insignificant amplitudes, one can introduce an integration threshold, $CAD (\geq v_{min}) = \int_0^{t_d} |v(t)|_{\geq v_{min}} dt$. CAD is a better indicator of damage potential than a single PGV measurement and can replace PGV in the ground motion prediction equation.

Figure 6.10 top row shows the three-component velocity and displacement waveforms of a $\log P = 2.61$ event recorded by 4.5 Hz geophone and at the bottom row X, Y, and Z components of the absolute velocity that, after integration, deliver CAD. The PGV and CAD values are shown in the right margin of each graph.

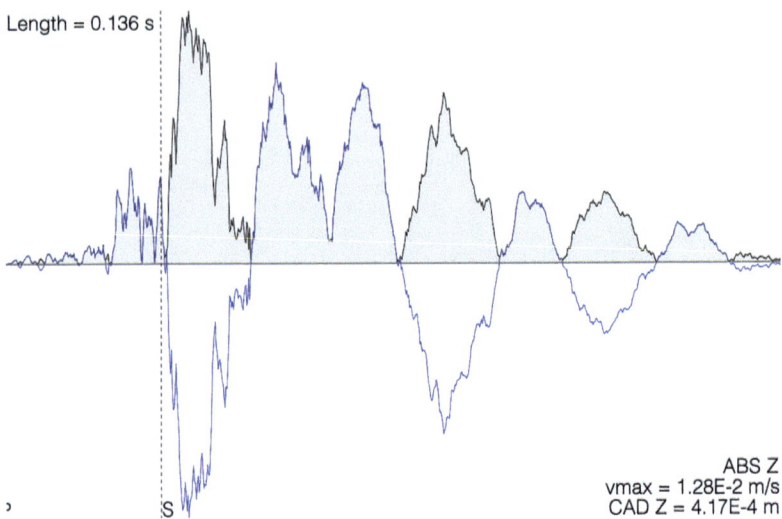

Fig. 6.9 An example of CAD measured on time series

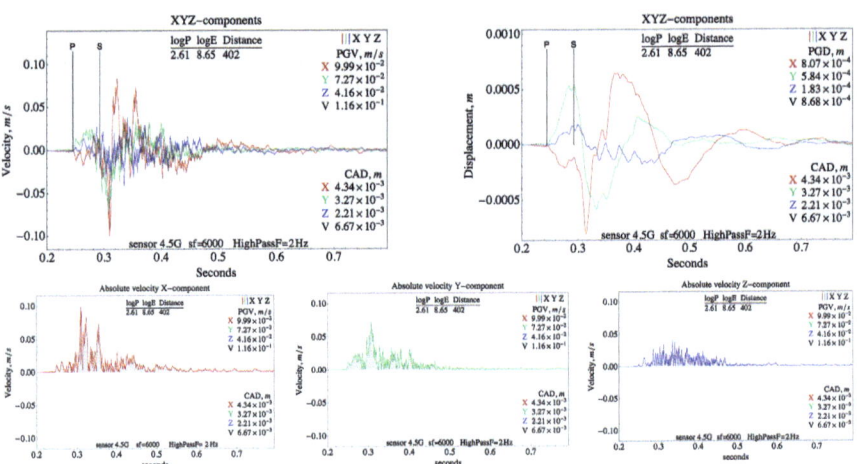

Fig. 6.10 Velocity waveforms and integrated displacements *(top)* of a $\log P = 2.61$ (m2.66) and energy $\log E = 8.65$ event recorded 402 metres from source. X, Y, and Z components of the absolute velocity that after integration delivers CAD *(bottom)*. At the bottom of each graph, there is the date and time of the event and information on sensor and filter applied to processing

Energy of Ground Motion The amount of S-wave (or P-wave) energy transmitted by the seismic wave across an area A normal to the direction of propagation during the time interval $t_2 - t_1$ is $E_{GM} = \rho A v_{P,S} \int_{t_1}^{t_2} \dot{u}^2 dt$, where ρ is the rock density, $v_{S,P}$ is the S- or P-wave velocity, and $\dot{u}$ is the ground velocity.

6.5 Ground Motion Amplification at the Skin of Excavation

The bulk of the damage caused by seismic events in mines occurs at the source region, i.e. within the volume $V = P/\Delta \epsilon$, with the strain change of the order of 10^{-4}, which for $\log P = 2$ gives $V = 10^6$ m^3. Complex, cascading sources may create considerably larger volumes of rock destruction. Damage to an excavation away from the source is related to the local geological site conditions that amplify ground motion, and to the capacity of support to sustain loading. Therefore, to select the appropriate support, it is advisable to estimate the expected site response. Site response can be estimated from theoretical models given that sufficient information about the near-surface geology is available, from weak ground motion or from the waveforms of seismic events recorded at the surface of the excavation and at a reference site.

Modelling the site response is very useful in understanding the physics of the problem; however, its success may be limited if the detailed 3D structural and geotechnical information is not accurate. The most reliable estimation of site response in mines, however, is obtained from waves recorded close to or at the skin of the excavation and the same waves recorded downhole deeper in the rock mass which is then taken as the reference site. The fundamental requirement then is that the distance between these two sites is small relative to the distance to the source and that the reference site should be free of site effect. One can then compare the ratio of the recorded ground motion characteristics and compute the spectral ratios of the S-wave group to estimate the frequency dependent site response.

Event1 $\log P = 1.0$ at Distance 55 m Figure 6.11 shows waveforms of a $\log P = 1.0$ seismic event recorded in an underground hard rock mine 55 metres away from the source by two three-component 14 Hz geophone sets sampled at 6 kHz, one installed at the surface of the 5 m by 6 m tunnel and the other 10 metres downhole.

The recorded average background noise level at the surface of the excavation on the x-component, vertical ↑, and y-component, parallel to the axis of the tunnel ||, is 16 times higher and on the z-component, orthogonal to the tunnel ⊥, 10 times higher than the respective components of the downhole sensor set. The maximum ground velocity at the surface is 1.7 times higher than downhole, with the x-component, $PGV_\uparrow$, amplified 1.72 times, the y-component, $PGV_{||}$, amplified 3.24 times, and the z-component, $PGV_\perp$, 1.24 times. The cumulative absolute displacement, CAD, is also amplified: $CAD_\uparrow$ 2.09 times, $CAD_{||}$ 2.7 times, and $CAD_\perp$ 1.16 times (Table 6.3).

Figure 6.12 left shows the spectra of the $\log P = 1.0$ event that includes 0.2 seconds of the pre-P-arrival noise and 0.2 seconds of coda, smoothed with the Konno-Ohmachi function

$$w(f, f_c) = \left\{ \mathrm{sinc} \left[b \log (f/f_c) \right] \right\}^4, \tag{6.13}$$

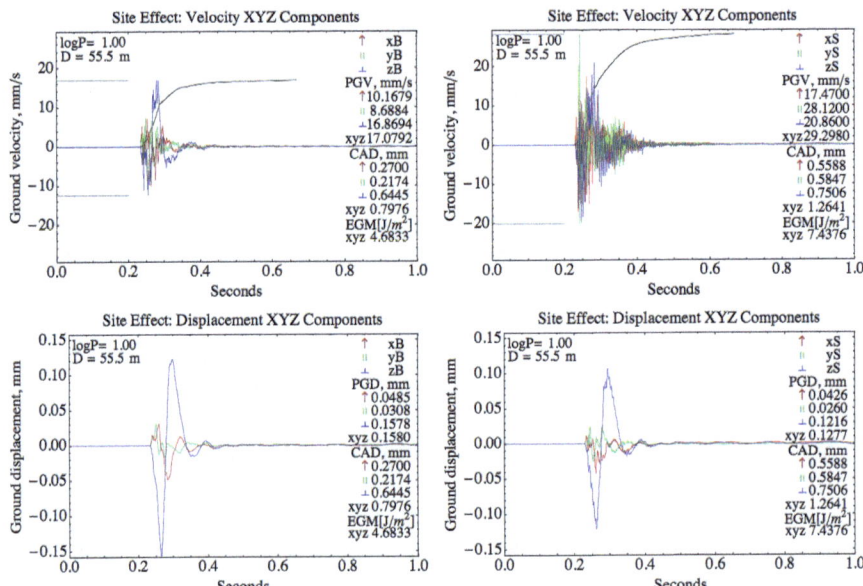

Fig. 6.11 Three-component velocity waveforms recorded at the end of the 10 metre borehole *(top left)*, at the surface of the u/g excavation *(top right)*, and integrated displacements *(bottom row)* of a log $P = 1.0$ event located 55.5 metres away. The grey area indicates the part of the waveform taken for spectral analysis of the event, and the grey lines at the top of the grey area show the part taken for spectral analysis of noise. The black line shows the cumulative energy of ground motion

Table 6.3 log $P = 1.0$ located 55 m away: PGV mm/s, PGD mm, and CAD mm, per component and their ratios

	$PGV_\uparrow$	$PGV_\parallel$	$PGV_\perp$	$PGD_\uparrow$	$PGD_\parallel$	$PGD_\perp$	$CAD_\uparrow$	$CAD_\parallel$	$CAD_\perp$
Surface	17.47	28.12	20.86	0.043	0.026	0.122	0.547	0.574	0.736
Borehole	10.17	8.69	16.87	0.049	0.031	0.158	0.261	0.213	0.633
Ratio	1.72	3.24	1.24	0.88	0.84	0.77	2.09	2.7	1.16

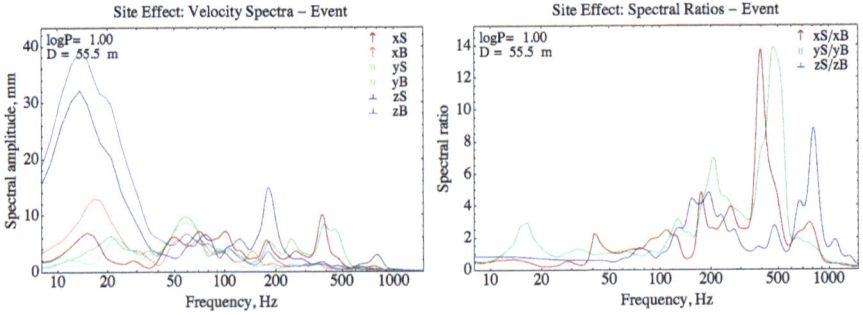

Fig. 6.12 Velocity spectra *(left)* and spectral ratios *(right)* of the log $P = 1.0$ event recorded 10 metres downhole and at the surface of the tunnel

6.5 Ground Motion Amplification at the Skin of Excavation

where $b = 40$ is the coefficient for bandwidth and f_c is the centre frequency (Konno & Ohmachi, 1998). The resulting spectral ratios, Fig. 6.12 right, show site amplifications between 2 and 3 times at 18, 40, and 120 Hz on the x- and y-components. Higher amplifications of 4 to 7 times are recorded at 150 to 250 Hz. Spectral ratios increase again past 300 Hz, most likely due to the interaction of higher frequencies with the tunnel, but at those high frequencies displacements are small, and therefore potential for damage is low.

Site amplification can also be quantified by taking PGV, or CAD, from continuous records every Δt and calculating the ratio between the records at the skin of excavation and in the borehole. While here frequency is disregarded, the method benefits greatly from the large number of measurements, e.g. taking sampling at 6 kHz, there are $5.184 \cdot 10^8$ samples per component per day to choose from. The method is also simple, and therefore, it can be implemented online to monitor changes in site effect at a particular site or between different sites as mining progresses.

Figure 6.13 left shows the smoothed probability density kernel of the observed logged ratio of the PGV for each component at the surface and the same measured downhole, taken every $\Delta t = 0.25$ second from the recorded continuous waveform over the 24 hour period that includes the $\log P = 1.0$ event shown in Fig. 6.11. To avoid spurious amplifications associated with weak ground motion, a threshold of 0.003 mm/s was applied to the data before the ratio was taken that reduced the number of ratios from 345600 to 669 on the x-component, to 621 on the y-component, and to 1084 on the z-component. It shows that the most probable amplification of ground motion on the x-component is $\log(PGV_S/PGV_B) = 0.2$, i.e. PGV_S on the x-component is most likely 1.6 times higher than PGV_B, on the y-component 2.2 times higher, and on the z-component 2.5 times higher.

Figure 6.13 right shows the survival function, i.e. the probability that the amplification ratio is greater than or equal to a given value. In this case, the

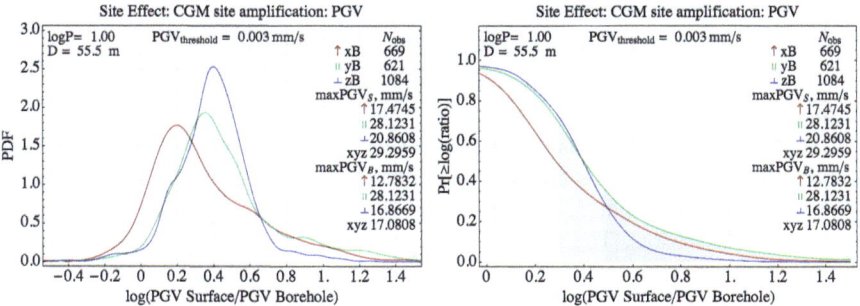

Fig. 6.13 Smooth probability density function of the logged ratio of the PGV per component measured at the surface and at 10 metres downhole taken every 0.25 second from the recorded continuous waveform over the 24 hour period that includes the $\log P = 1.0$ event *(left)*. The survival function for the same dataset, i.e. the probability that the amplification ratio is greater or equal to a given value is also shown *(right)*

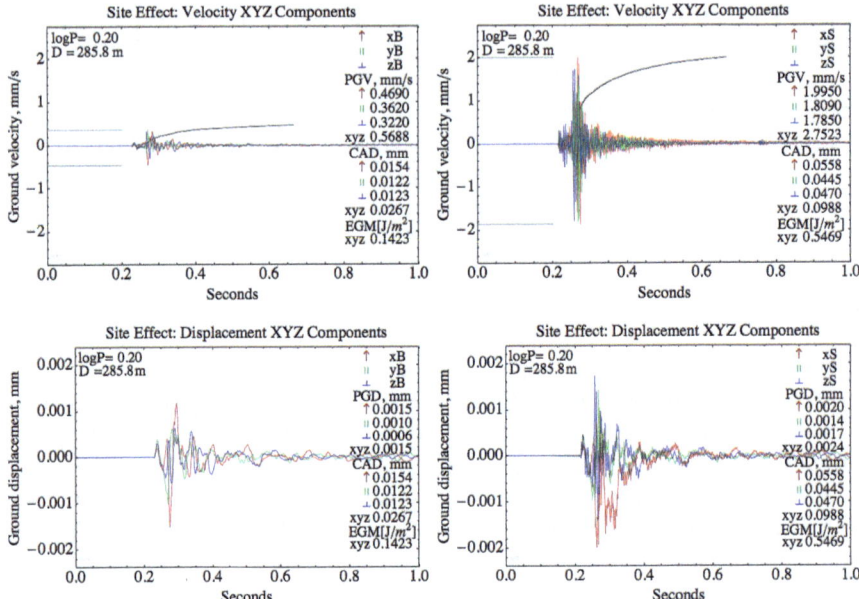

Fig. 6.14 Same as Fig. 6.11 but for $\log P = 0.2$ event

probability that the amplification of PGV on the x-component is greater than 2 is almost 50%, and on the y- and z-components it is 70%. The probability that the amplification of PGV on the x- and z-components is greater than 3 is 30%, and on the y-component 38%.

Note that logging the ratio solves the problem of lack of symmetry, i.e. if PGV_S is greater than PGV_B, the ratio can take theoretically any value greater than 1, but if PGV_S is less than PGV_B, the ratio is restricted to the range of 0 to 1. The logged ratio restores the symmetry, i.e. $\log(PGV_S/PGV_B) = -\log(PGV_B/PGV_S)$.

Event2 $\log P = 0.2$ **at Distance 286 m** Figure 6.14 shows waveforms of a smaller event with $\log P = 0.2$ located 286 metres away and recorded by the same set of sensors.

The recorded average background noise level at the surface of the excavation on the x-component, $\uparrow$, is 8 times higher, on the y-component $\|$, 11 times higher and on the z-component $\perp$, 5 times higher than at the respective components of the downhole sensors. The maximum ground velocity at the surface is 4.8 times higher than downhole, with the x-component, $PGV_\uparrow$, amplified 4.2 times, the y-component, $PGV_\|$, 5 times and the z-component, $PGV_\perp$, 5.5 times. The cumulative absolute displacement, CAD, is amplified 3.7 times, and the energy of ground motion, EGM, 3.8 times. The PGD is amplified less than PGV but more than in the previous case. The maximum amplification of the PGD is 2.7 times in the direction orthogonal to the tunnel, 1.4 times along the tunnel, and 1.3 times on the vertical component (Table 6.4).

6.5 Ground Motion Amplification at the Skin of Excavation

Table 6.4 $\log P = 0.2$ located 286 m away: PGV mm/s, PGD mm, and CAD mm, per component and their ratios

lsg	$PGV_\uparrow$	$PGV_\|\|$	$PGV_\perp$	$PGD_\uparrow$	$PGD_\|\|$	$PGD_\perp$	$CAD_\uparrow$	$CAD_\|\|$	$CAD_\perp$
Surface	1.995	1.809	1.785	0.00199	0.0014	0.00173	0.0515	0.042	0.0443
Borehole	0.469	0.362	0.322	0.00152	0.0010	0.00063	0.0141	0.0108	0.0113
Ratio	4.25	5.0	5.54	1.31	1.41	2.73	3.65	3.87	3.9

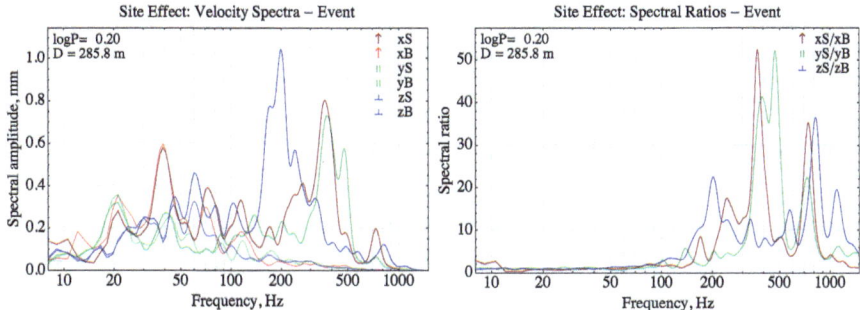

Fig. 6.15 Same as Fig. 6.12 but for $\log P = 0.2$ event

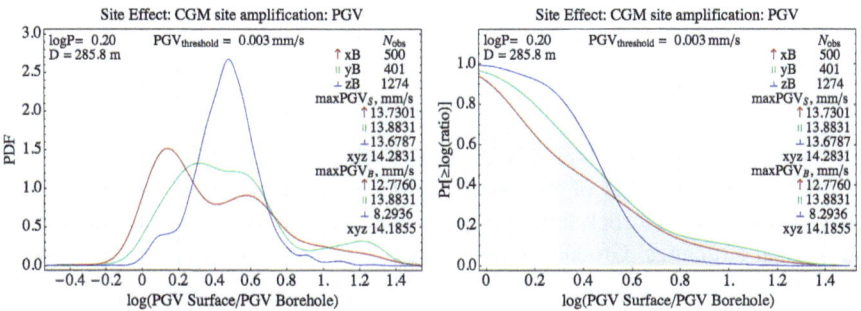

Fig. 6.16 Same as Fig. 6.13 but for $\log P = 0.2$ event

Figure 6.15 left shows the smoothed spectra of the $\log P = 0.2$ that also includes 0.2 seconds of the pre-P-arrival noise and 0.2 seconds of coda. The resulting spectral ratios, Fig. 6.15 right, show site amplifications up to 2 times in the frequency range up to 120 Hz on all components. Higher amplifications of 5 to over 20 times are recorded at 150 to 300 Hz, and then again up to 50 times past 300 Hz, most likely due to the interactions of shorter waves with the tunnel. Again, at such high frequencies, displacements are small, and therefore potential for damage is low.

Figure 6.16 left shows the smooth probability density kernel of the observed logged ratio of the PGV for each component at the surface and the same measured downhole.

The ratio is taken every $\Delta t = 0.25$ second from the recorded continuous waveform over the 24 hour period that includes the $\log P = 0.2$ event shown in Fig. 6.14. The same threshold of 0.003 mm/s was applied to avoid the spurious amplifications associated with weak ground motion, which reduced the number of ratios from 345600 to 500 on the x-component, 401 on the y-component, and 1274 on the z-component.

It shows that the distribution of amplification of ground motion on the x-component peaks at $\log(PGV_S/PGV_B) = 0.15$, i.e. where PGV_S on the x-component is 1.4 times higher than PGV_B. On the y-component, it is between 1.6 and 4 times higher and on the z-component 3 times higher. Figure 6.15 right shows the survival function, i.e. the probability that the amplification ratio is greater or equal to a given value. In this case, the probability that the amplification of PGV on any of the three components is greater than 2 is almost 85% and greater than 3 is 70%.

Summary (1) Larger seismic events produce strong ground motion at frequencies below 15 Hz, and at these frequencies both sets of sensors displace in tandem, and therefore, there is no, or very little, amplification. (2) The PGV is subjected to largest amplification, then CAD and least the PGD. The component of the largest amplification, being parallel, vertical or orthogonal, depends on the direction of the incoming wave.

6.5.1 Ejection Velocity

Ortlepp (1993) presented evidence of rock ejection velocities associated with seismic events in mines of the order of 10 m/s and greater and suggested that they may be due to a rock failure phenomenon different than the classical slip on geological structures. Ground motion can be modified at, or close to, the surface of an excavation during the buckling of slabs in the sidewalls inducing an ejection velocity that can well exceed that of ground motion at source.

Assuming buckling of excavation sidewalls subjected to high levels of compressive stress as a possible mechanism of failure (McGarr, 1997) estimated the ejection velocity, v_{ej}, as

$$v_{ej} = \sigma_c \sqrt{\frac{7-v^2}{2\rho Y}}, \tag{6.14}$$

where σ_c is the uniaxial compressive stress at failure in Pa, v is the Poisson ratio, Y is the Young modulus in Pa, and ρ is the rock density in kg/m^3 (Fig. 6.17). Indeed, taking reasonable values of σ_c, Y, v, and ρ for hard rocks, the ejection velocities could be well in excess of 10 m/s. The ejection velocity, v_{ej}, in Eq. (6.14) depends very weakly on the Poisson ratio.

6.5 Ground Motion Amplification at the Skin of Excavation

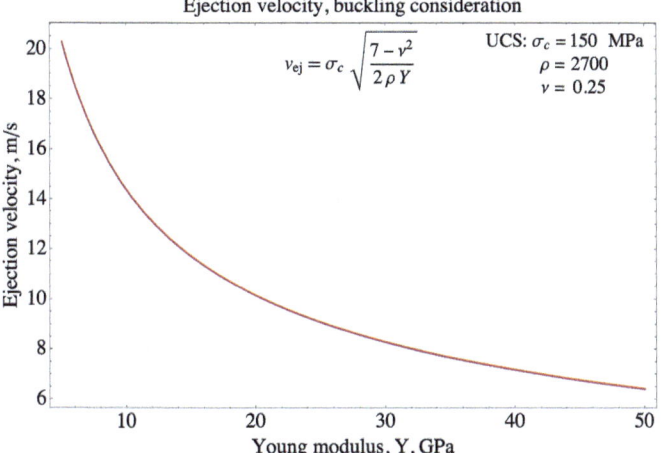

Fig. 6.17 Ejection velocity, Eq. (6.14)

For larger mass, m, such velocities can exert a considerable force, $F = mv_{ej}/\Delta t$, on a stiff support system, i.e. for short deformation time Δt. However, in case of face bursts driven by sudden loading where rock is shattered to small pieces, the average velocity of ejection is a decreasing function of m, i.e. $v_{ej} \sim 1/\sqrt{m}$.

If a block of rock is moving with velocity v_{ej} while being at a level where the gravitational potential energy is mgh_0 and eventually hits a wall at a point where its potential energy is mgh_1 and is stopped there, the work done, ΔW, will be numerically equal to the work for accelerating the body from rest to velocity v_{ej}, plus the work for moving the rock in the field of the Earth's gravity from the point of impact, level h_1, to the starting level h_0, and therefore,

$$\Delta W = \frac{1}{2} m v_{ej}^2 + mg(h_0 - h_1):$$

1. Ejection from the roof hits the floor. The depth of failure is d_f, and ρ is the density of the rock. The total mass of the ejected rock fragment is $m_f = \rho d_f A$, where A is the area of the face of the burden. The mass per unit area of the face is $m_A = \rho d_f$. The work done on the floor when it is hit by the ejected rock is $\Delta W = \frac{1}{2} m_f v_{ej}^2 + m_f gh$, in Jules, where h is the height of the tunnel. The work per face-unit area of the ejected rock is, $\Delta W_A = \frac{1}{2} m_A v_{ej}^2 + m_A gh$, in Jules/m^2.
2. Ejection from the floor. A block of rock is detached from the floor and pushed up by a dynamic event with an initial velocity v_{ej}. If the ejected rock hits the roof, the work done on the roof when it is hit by the ejected rock is $\Delta W = \frac{1}{2} m_f v_{ej}^2 - m_f gh$, and the work per face-unit area of the ejected rock is $\Delta W_A = \frac{1}{2} m_A v_{ej}^2 - m_A gh$. If the ejected rock does not reach as high as the roof, $\Delta W_A = \frac{1}{2} m_A v_{ej}^2$.

3. Ejection from the side wall. A block of rock is detached from one of the walls of the tunnel at height h_0 and pushed towards the opposite wall by a dynamic event with an initial velocity v_{ej}. The vector of the initial velocity subtends angle θ with the horizontal:

   (a) If the ejected rock hits the opposite wall at some height $h_1 < h_0$, then the work per face-unit area of the ejected rock is $\Delta W_A = \frac{1}{2} m_A v_{ej}^2 + m_A g (h_0 - h_1)$.

   (b) If the ejected rock drops on floor some distance d from the wall $\Delta W_A = \frac{1}{2} m_A v_{ej}^2 + m_A g (h_0 - h_1)$, then the work per face-unit area of the ejected rock is $\Delta W_A = \frac{1}{2} m_A v_{ej}^2 + m_A g h_0$. The distance d at which the ejected rock hits the floor is related to the initial velocity by the relation,

$$v_{ej} = d \sqrt{\frac{g}{2 h_0 \cos^2 \theta + d \sin^2 \theta}}, \tag{6.15}$$

which for $\theta = 0$ gives $v_{ej} = d \sqrt{g/(2h_0)}$.

6.6 Simple GMPE and Its Utility

6.6.1 Introduction

The ground motion prediction equation (GMPE) gives the expected value of a given ground motion parameter, e.g. PGV or CAD, as a function of seismic energy, potency or magnitude, and distance.

There is not much literature on GMPE for mining induced seismicity and even less for underground sites. McGarr et al. (1981) used ground motion data recorded 3 km underground in a South African gold mine to develop the relationship $\log(R \cdot PGV) = 3.95 + 0.57 m_L$, where both R is in cm and the peak ground velocity, PGV, is in cm/s, and m_L is local magnitude.

Kaiser and Maloney (1997) proposed a similar equation but with the exponent $a^* = 0.5$ as a scaling law for support design in rockburst conditions. It is written in the form, $PGV = C^* \cdot M^{a^*}/R$, where M is seismic moment expressed in GNm. The parameter C^* depends on the stress drop environment and based on data from the Creighton Mine, $C^* = 0.1$ to 0.3 for events with stress drops less than 2.5 MPa and with $C^* = 0.5$ to 1.0 for higher stress drop events. However, it is recommended to adjust C^* to a specific dataset at hand. Translating to the seismic potency domain gives $PGV = C^* P^{1/2}/R$, where PGV is in m/s, P in m^3, and R in metres. Taking into account that $M = \mu P$ and $\mu = 30$ GPa, the parameter $C^* = (1.1$ to $1.64)$ for events with stress drop less than 2.5 MPa and $C^* = (2.74$ to $5.48)$ for higher stress drop events. The relations by McGarr et al. (1981) and by Kaiser and Maloney (1997) do not cater for attenuation and near-source saturation.

McGarr and Fletcher (2005) developed GMPE for PGV and PGA based on the ground motion recorded on surface due to coal mining that generated events with $m \leq 2.2$, recorded at distances of 500 metres to 10 km.

Mendecki (2008) developed and compared the GMPE-PGV for four underground mines: two gold mines in South Africa, one in Australia, and an iron ore mine in Sweden, all based on data recorded by three -component geophones installed in boreholes drilled from underground excavations.

Atkinson (2015) used the NGA-West 2 database, containing horizontal component response spectra and PGV for events $3.0 \leq m \leq 6.0$ recorded on surface at distances up to 40 km to develop a GMPE that could be applied to induced seismicity. She concluded that ground motion from small to moderate induced events may be significantly larger than that predicted by most currently used GMPE.

Mendecki et al. (2018) developed GMPE for PGA due to induced seismicity based on 15541 observations recorded by 14 surface sites in the central area of the Main Syncline of the Upper Silesia Coal Basin in Poland.

The development of a GMPE for underground mines is in some respects different to that in earthquake seismology. There are very few accelerometers installed in mines, and in most cases, these are piezoelectric which are not strong ground motion instruments. They deliver high accelerations at high frequencies which are of little interest since at those frequencies there is not much ground velocity and even less ground deformation. Double integration of frequently noisy acceleration waveforms to displacement may also prove difficult. It is only recently that the semiconductor micro-electromechanical systems accelerometers, MEMS, are being used in mines. In addition, underground support designs are based on the expected demand in terms of PGV and deformation. Most mines install a mixture of 4.5 and 14 Hz geophones, and only recently have lower frequency sensors been deployed. For larger events, the higher frequency sensors filter lower frequencies and underestimate the ground motion parameters, and to some degree seismic potency and energy.

Figure 6.18 shows the three-component velocity and integrated displacement waveforms of a $\log P = 2.86$ event recorded at a distance 2252 metres from the source by a 1 Hz sensor and a 4.5 Hz sensor located next to each other in the same borehole. While the shape of the waveforms is similar, the 4.5 Hz sensor recorded significantly lower ground motions. PGV recorded by the 4.5 Hz sensors is 1.8 times lower, PGD is 3.5 times lower, and the cumulative absolute displacement, CAD, is 2.3 times lower. The 14 Hz geophone would record even lower ground motion. For smaller events, the differences are less significant. It is therefore advisable to select PGV and CAD recorded by the same type of sensors while developing GMPE.

One can increase the signal range and the travel limits of geophones by overdamping. With normal damping of 0.7, the frequency response is flat to ground velocity above the natural frequency, f_n, and is proportional to f^2 below, which is caused by a double pole at that frequency, see Fig. 6.19. As the damping increases beyond 1, the poles separate, in such a way that the product of the pole frequencies remains constant. Between these poles, the velocity response is proportional to frequency, effectively making it flat to acceleration over this frequency range. For

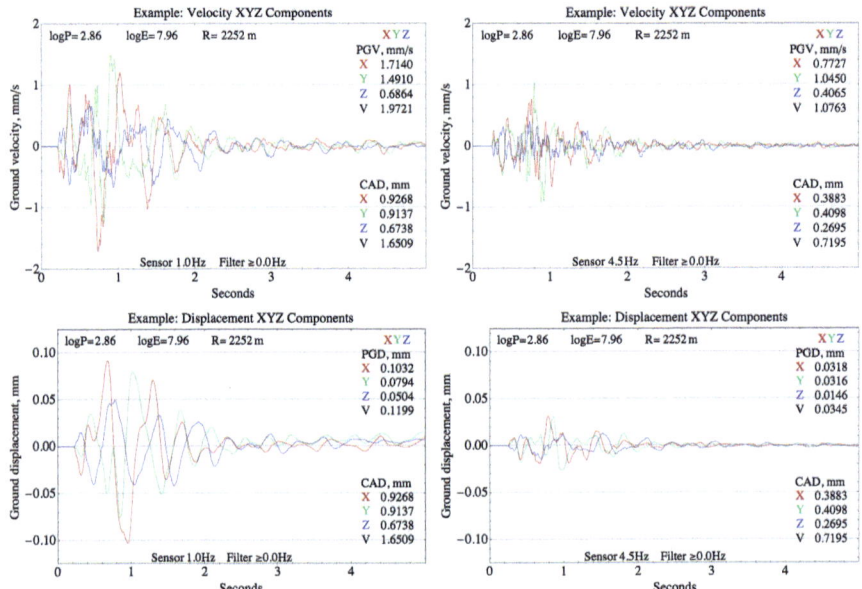

Fig. 6.18 Velocity and displacement waveforms of a $\log P = 2.86$ event recorded by 1 Hz *(left column)* and 4.5 Hz sensors *(right column)*, located at the same site

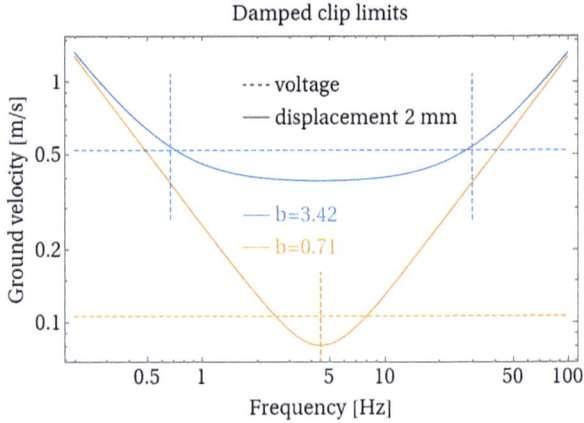

Fig. 6.19 The solid lines show the ground velocity required to produce 2 mm peak internal displacement in a 4.5 Hz geophone, for overdamped and maximally flat responses. The voltage limits for a typical audio ADC are marked by dashed horizontal lines, and pole frequencies by dashed vertical lines, after Mountfort and Mendecki (2019)

4.5 Hz geophones, the maximum damping which can reasonably be achieved is 3.4, which means the acceleration response covers the frequency band from 0.7 Hz to 30 Hz. In this configuration, the ADC voltage clip limit is raised by a factor of 5 to 0.5 m/s, which is then slightly greater than the minimum internal displacement clip limit.

Traditionally, the most important ground motion parameter in underground mines was the instrumental PGV which is used for support design. Recently, Mendecki (2018) developed a GMPE for the cumulative absolute displacement, CAD, to be applied to monitor the consumption of the deformation capacity of support due to seismicity. The main interest for mines are the ground motion parameters at distances between 50 and 1000 metres. Larger distances are of interest for surface structures, e.g. tailings dams or processing plants. For the sizes of events in mines, the PGV may drop by two orders of magnitude between 100 and 1000 metres from the source. CAD decays more slowly with distance than PGV, mainly because of the increased duration of waveforms with distance due to scattering. Although we caution that the GMPE is not the best tool to estimate the near-source ground motion (Mendecki, 2016), the geotechnical engineers, lacking other credible data, resort to such extrapolation when considering support specifications. Therefore, there is a need to constrain PGV at source to a physically acceptable level.

Seismic systems in mines are designed to locate events and to estimate their source parameters. For this reason, sensors are installed at least 6 to 10 metres into boreholes to avoid the very site effects that amplify ground motion at the skin of excavations. Since the GMPE derived from such measurements certainly underestimates seismic load, mines conduct separate site amplification measurements at selected locations (Milev & Spottiswood, 2005; Cichowicz, 2008; Mendecki, 2013; Dineva et al., 2016; Mendecki, 2016, 2017).

Unlike crustal seismology, in mines the bulk rock mass properties are changing due to rock extraction, specifically in caving and open pit mines, and, because new strong ground motion data is coming fast, the GMPE needs to be updated at least once a year. For the same reason, the GMPE developed for mines is characterised by large scatter. Moreover, many waveforms of larger events recorded at closer distances are displacement clipped or voltage saturated, which limits the number of observations in the near field. The distances to larger and intermediate size events are also uncertain because of the unknown orientation of sources and the complex nature of larger events.

6.6.2 Simple GMPE for PGV and CAD (Based on Mendecki, 2019)

Ground Motion Prediction Equation The ground motion prediction equation (GMPE) gives the expected value of a given ground motion parameter, GMP, e.g. peak ground velocity PGV or the cumulative absolute displacement CAD, as a function of seismic potency or energy or magnitude and distance. Its main utility in mines is to predict ground motion at selected sites in the near and intermediate fields of larger events that may occur in the future, on the basis of observations of ground motion caused by smaller events.

We start with the following prediction equation for a ground motion parameter, GMP, caused by an event of potency P at distance R,

$$\overline{GMP}(P, R) = c \cdot P^{c_P} \left(R + c_L \cdot P^{1/3}\right)^{-c_R}, \qquad (6.16)$$

where R for larger events should be measured orthogonal to the characteristic rupture plane, c_P is the potency dependence parameter, c_R controls the geometrical attenuation rate, and c is a free parameter (Esteva, 1970; Campbell, 1981). The term $c_L P^{1/3}$ modulates ground motion at small distances, where geometric attenuation is small, and to saturate them at source.

The amplitudes of GMP predicted by the GMPE above are positively correlated with c_P and c and negatively with c_R and c_L. The term $c_P \log P$ is consistent with the definition of earthquake magnitude as a logarithmic measure of the amplitude of ground motion. The term $-c_R \log R$ is consistent with the geometric spreading of the seismic wavefront as it propagates away from the source, and it also caters in part for the attenuation due to anelasticity and scattering.

The GMPE at source gives $GMP(R = 0) = cc_L^{-c_R} P^{c_P - c_R/3}$, i.e. at source $\log GMP$ is a linear function of $\log P$. For $c_P = c_R/3$ the GMP at source is independent of event size, $GMP(R = 0) = c/c_L^{c_R}$. For $c_P > c_R/3$, it delivers larger GMPs at source for events with larger potencies. The case $c_P < c_R/3$ predicts that lower potencies generate higher GMPs at source than larger potencies, which is rather unlikely. The case $c_R = 1.0$ with no attenuation and for $c_L = 0$ gives the familiar $GMP = cP^{c_P}/R$, McGarr et al. (1981); Kaiser and Maloney (1997).

From the GMPE given by Eq. (6.16), we can calculate $R = (cP^{c_P}/GMP)^{1/c_R} - c_L P^{1/3}$, i.e. the distance over which seismic source with potency P generates the ground motion parameter $\geq GMP$. We can also calculate the minimum potency, or $\log P$, that delivers a given level of GMP at a distance R. The case $c_P = c_R/3$ gives $P(R) = R^3 / \left[(c/GMP)^{1/c_R} - c_L\right]^3$, but in the general case the solution must be obtained numerically.

There are many forms of GMPE, and most of them developed for predicting surface ground motion resulting from earthquakes, see Douglas (2018) for a review. Some of them are complex and have more than five or even 10 coefficients to cater for magnitude, distance, site effects, source mechanisms (normal, strike slip, or reverse faulting), and, in some cases, even directivity. However, more complex models are more susceptible to the danger of overfitting, i.e. modelling spurious details of the data rather than the data generating process. The inversion procedure for parameters in more complex models should therefore be carried out in two stages to decouple potentially correlated variables, in this case, c_P and c_R (Joyner & Boore, 1993, 1994) or (Abrahamson & Youngs, 1992). However, such a process can only

6.6 Simple GMPE and Its Utility

alleviate the problem, and the real physical meaning of these parameters may be lost, a point well made by McGarr and Fletcher (2005).

Simple GMPE-PGV Equation (6.16) for PGV is $\overline{PGV}(P, R) = cP^{c_P}(R + c_L \cdot P^{1/3})^{-c_R}$, which at source, for $R = 0$, gives $\overline{PGV}(P, R = 0) = cc_L^{-c_R} P^{c_P - c_R/3}$. The average ground velocity at source is $PGV_0 = 0.63 v_S \epsilon_{eff}$. For $\sigma_{eff} = 25$ MPa, $\rho = 2700$ kg/m^3, $v_S = 3300$ m/s, and the rupture velocity $v_r = 0.75 v_S$, the estimates of the near-source ground motion vary between 1.0 m/s and 2.1 m/s.

The effective stress cannot be measured directly, but one can assume that it is equal to the bulk shear strength of the rock within the volume of interest, which for hard rock varies between 10 MPa, for an inhomogeneous rock to 100 MPa for an intact homogeneous hard rock. This, assuming the rigidity of the order of 10 GPa, translates to $10^{-4} \leq \epsilon_{eff} \leq 10^{-3}$. A more practical proxy for σ_{eff} or ϵ_{eff} is the upper limit of the static stress drop, $\Delta\sigma$, or strain change, $\Delta\epsilon$, derived from waveforms recorded in the area of interest.

Now, from $\overline{PGV}(P, R = 0) = 0.63 v_S \Delta\epsilon = cc_L^{-c_R} P^{c_P - c_R/3}$, we can derive $c = 0.63 v_S \Delta\epsilon c_L^{c_R} P^{-c_P + c_R/3}$. For $c_P = c_R/3$, parameter $c = 0.63 v_S \Delta\epsilon c_L^{c_R}$ is independent of potency P, and the GMPE for $\overline{PGV}$ is

$$\overline{PGV}(P, R) = 0.63 v_S \Delta\epsilon \left(\frac{c_L P^{1/3}}{R + c_L P^{1/3}} \right)^{c_R}. \tag{6.17}$$

While this expression has four parameters: v_S, $\Delta\epsilon$, c_L, and c_R, two of them, v_S and $\Delta\epsilon$, are constrained by the type of rock and can be assumed, and the other two, c_L and c_R, need to be inverted from data.

Simple GMPE-CAD Equation (6.16) for CAD is $\overline{CAD}(P, R) = c \cdot P^{c_P} (R + c_L \cdot P^{1/3})^{-c_R}$. For a circular crack with a uniform strain change $\Delta\epsilon$ over the source surface, the displacement profile is given by $u(x) = 24 \Delta\epsilon \sqrt{r^2 - x^2}/(7\pi)$, where x is the radial distance from the centre of the crack and r is the radius of the crack (Eshelby, 1957). The maximum displacement is in the middle of the crack, i.e. at $x = 0$, and therefore, $u_{max} = 24 r \Delta\epsilon/(7\pi)$. Integration over the crack length in polar coordinates, (x, φ), gives the mean displacement at source $\bar{u} = 24 \Delta\epsilon/(7\pi^2 r^2) \cdot \int_0^r x \, dx \int_0^{2\pi} d\varphi \sqrt{r^2 - x^2}$, which translates to $\bar{u} = 48 \Delta\epsilon/(7\pi r^2) \cdot \int_0^r \sqrt{r^2 - x^2} x \, dx$, and finally $\bar{u} = 16 r \Delta\epsilon/(7\pi)$. This gives seismic potency, $P = \bar{u} \pi r^2 = (16/7) r^3 \Delta\epsilon$, the source radius $r = (7/16)^{1/3} (P/\Delta\epsilon)^{1/3}$, and

$$u_{max} = 1.5 \bar{u} = q_0 \Delta\epsilon^{2/3} P^{1/3}, \tag{6.18}$$

where the constant $q_0 = (24/7\pi)(7/16)^{1/3} = 0.828494$. For the average strain change at source $\Delta\epsilon = 5 \cdot 10^{-4}$, the maximum displacement $u_{max} = 0.0053 \sqrt[3]{P}$, which is not far from $u_{max} = 0.0046 \sqrt[3]{P}$ given by McGarr and Fletcher (2003).

If we assume that the cumulative absolute displacement at the source is equal to the maximum source displacement, i.e. $CAD_0 = u_{max} = q_0 \Delta^{2/3} P^{1/3}$, then at source, $\overline{CAD}(P, R = 0) = cc_L^{-c_R} P^{c_P - c_R/3} = CAD_0 = q_0 \Delta \epsilon^{2/3} P^{1/3}$. For $c_P = (1 + c_R)/3$, parameter $c = q_0 \Delta \epsilon^{2/3} c_L^{c_R}$ is independent of potency P, and, after simple algebra, the GMPE for $\overline{CAD}$ can be written as

$$\overline{CAD}(P, R) = q_0 \Delta \epsilon^{2/3} P^{1/3} \left(\frac{c_L P^{1/3}}{R + c_L P^{1/3}} \right)^{c_R}. \tag{6.19}$$

Equation (6.19) has three parameters, $\Delta \epsilon$, c_L, and c_R, and since $\Delta \epsilon$ is constrained by the type of rock and can be assumed, the other two, c_L and c_R, need to be inverted from data.

6.6.3 SGMPE Example

We analysed ground motion data from 265 seismic events in the potency range $0.0 \leq \log P \leq 2.79$ recorded at distances between 87 and 996 metres from source that delivered 837 observations of PGV and CAD. The minimum PGV is 0.0218 cm/s, and the maximum 9.9 cm/s. The minimum CAD is 0.0028, cm and the maximum 0.512 cm.

Figure 6.20 shows $\log \Delta \epsilon$ vs. $\log P$ and $\log f_0$ vs. $\log P$ on the background of different constant strain changes at source. The upper limit of strain change at source is assumed to be $6 \cdot 10^{-4}$.

Figure 6.21 shows $\log P$ vs. PGV and $\log P$ vs. CAD of the final dataset accepted for fitting.

GMPE-PGV Assuming $v_S = 3950$ m/s and $\Delta \epsilon = 6 \cdot 10^{-4}$, the GMPE for PGV is obtained as

$$\overline{PGV}(P, R) = 1.4931 \left(\frac{3.5872 P^{1/3}}{R + 3.5872 P^{1/3}} \right)^{1.466}, \tag{6.20}$$

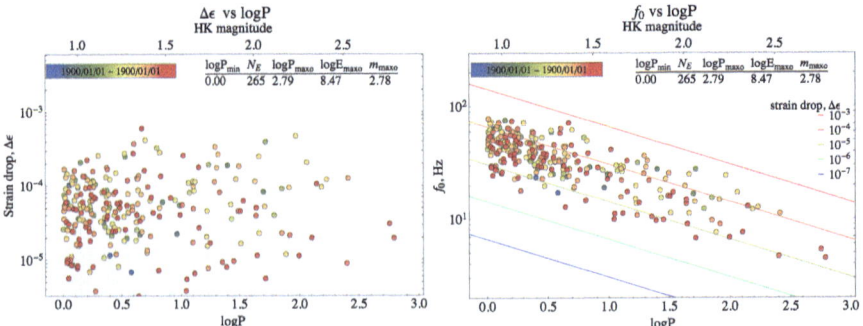

Fig. 6.20 $\log \Delta \epsilon$ vs. $\log P$ *(left)* and $\log f_0$ vs. $\log P$ *(right)* of events accepted for fitting

6.6 Simple GMPE and Its Utility

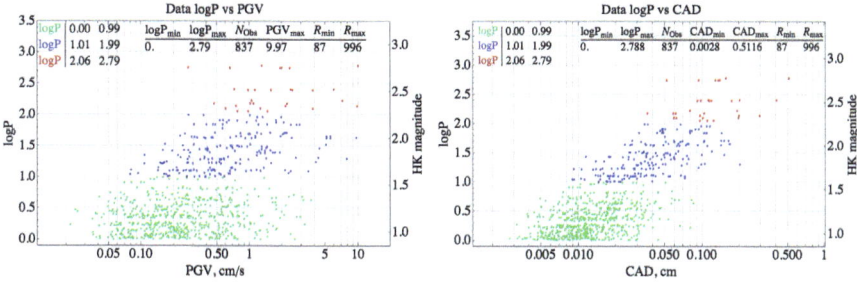

Fig. 6.21 $\log P$ vs. PGV (*left*) and $\log P$ vs. CAD (*right*) of the datasets accepted for fitting

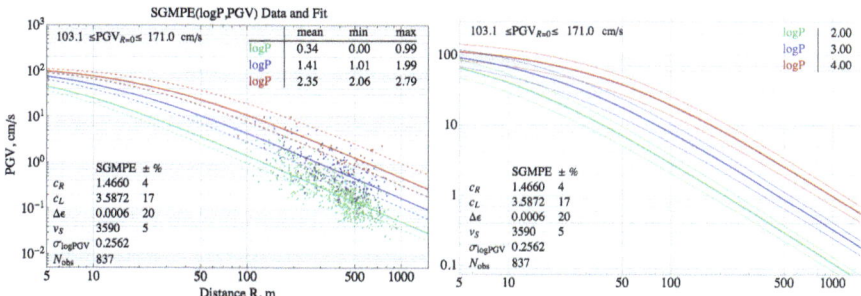

Fig. 6.22 The GMPE-PGV fit and data (*left*) and predictions for larger potencies (*right*)

where PGV is in m/s, P in m^3, and R in metres. The $c_R = 1.466$ with the standard errors of $sd_{cR} = \pm 0.03576$ and $c_L = 3.5872$ with the standard error $sd_{cL} = \pm 0.37495$, and $\sigma_{\log PGV} = 0.25622$.

If we assume 5% uncertainty in v_S and 20% uncertainty in $\Delta \epsilon$, then the expected peak ground velocity at source varies between $103.1 \leq PGV_0 \leq 171.0$ cm/s, irrespective of $\log P$. Figure 6.22 left shows the data selected for fitting with dots coloured by size range and the fitted model plotted in the middle of each PGV data range. Figure 6.22 right shows extrapolations for larger events.

Figure 6.23 shows the results of residual analysis: log (Obs/Pred) as a function of $\log P$ and $\log$ (Pred) vs. $\log$ (Obs).

GMPE-CAD Assuming $\Delta \epsilon = 6 \cdot 10^{-4}$, the GMPE for CAD is obtained as

$$\overline{CAD}(P, R) = 0.00589 \cdot P^{1/3} \left(\frac{0.3788 P^{1/3}}{R + 0.3788 P^{1/3}} \right)^{0.6083}, \quad (6.21)$$

where CAD is in m, P in m^3, and R in metres, see Fig. 6.24. The $c_R = 0.6083$ with the standard errors of $sd_{cR} = \pm 0.02395$ and $c_L = 0.3788$ with the standard errors of $sd_{cL} = \pm 0.0959$ and $\sigma_{\log CAD} = 0.1908$. If, in addition, we assume 20% uncertainty in $\Delta \epsilon$, the expected maximum seismic displacement at source for $\log P = 3.0$ varies between 4.5 and 5.89 cm.

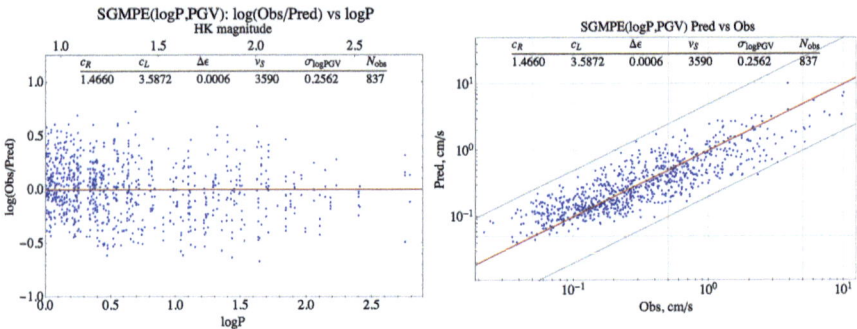

Fig. 6.23 $\log\left(PGV_{obs}/PGV_{pred}\right)$ vs. $\log P$ *(left)* and $\log PGV_{pred}$ vs. $\log PGV_{obs}$ *(right)*

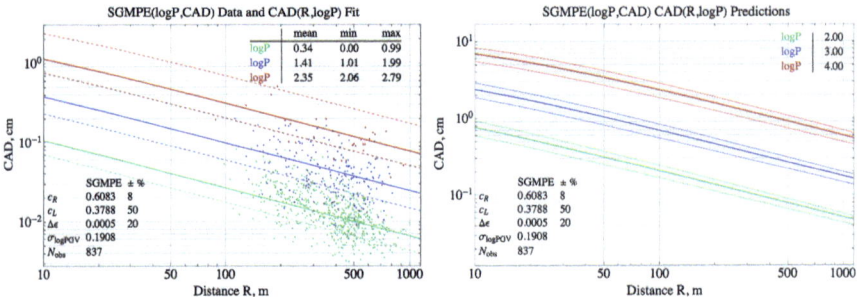

Fig. 6.24 The GMPE-CAD fit and data *(left)* and predictions for larger potencies *(right)*

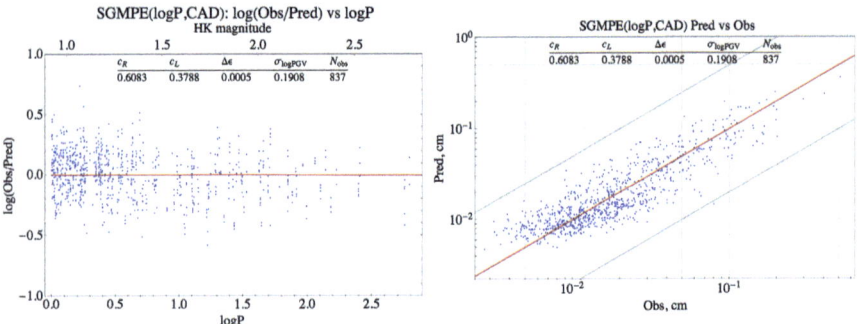

Fig. 6.25 $\log\left(CAD_{obs}/CAD_{pred}\right)$ vs. $\log P$ *(left)* and $\log CAD_{pred}$ vs. $\log CAD_{obs}$ *(right)*

Figure 6.24 left shows the data selected for fitting with dots coloured by size range and the fitted model plotted in the middle of each CAD data range. Figure 6.24 right shows extrapolations for larger events.

Figure 6.25 shows the results of residual analysis: log (Obs/Pred) as a function of $\log P$ and log (Pred) vs. log (Obs).

6.6.4 Perceptibility of Ground Motion

The velocity of ground motion at the source of a seismic event is mostly independent of its size, i.e. large and small events have similar velocities of ground deformation at source. However, small events affect smaller volumes of rock and radiate waves of higher frequencies that are attenuated faster than lower frequencies. Therefore, smaller events are perceptible over shorter distances.

Humans experience the effect of ground motion as a movement of the ground and as a sound. Large seismic events, e.g. with $\log P \geq 3.0$ ($m_{HK} = 2.95$), radiate most of their energy in a frequency band of 5 to 20 Hz and are easily felt as a movement over a distance of a few kilometres. Small events, e.g. with $\log P = -2.0$ ($m_{HK} \simeq -0.38$), which can be seen as a 15 metre size crack, radiate most of their energy at frequencies of 100 Hz and more and can be felt and/or heard underground at 100 metres away.

In general, ground motions lower than 1 mm/s are hardly perceptible as movement, but, by interacting with the environment, they may generate a perceptible noise. This is what happens underground when waves generated by a seismic event interact with the fracture zone around a tunnel and with support elements and generate an audible noise. Ground motion between 1 and 10 mm/s is perceptible as a movement and higher than 10 mm/s becomes unpleasant. Ground motion stronger than 10 cm/s can cause local falls of ground or strain bursts and above 50 cm/s can cause rock failure and damage to underground excavations. Since localised blasts generate PGV at higher frequencies, therefore producing less displacement, they will be felt as movement over shorter distances, but their high frequency content will generate more audible noise. In general, the same PGV generated at lower frequency will be felt as movement over larger distance.

Figure 6.26 illustrates the perceptibility and the expected damage potential to the underground structure by different levels of PGV with superimposed SGMPE for selected $\log P$.

6.6.5 Damage Inspection and Cumulative CAD Plots

The basic outcome of ground motion hazard analysis for a given site is a seismic hazard curve that shows the annual rate, or probability, at which a specific ground motion level will be exceeded. This is outside the scope of this chapter. It is expected that CAD, which includes the peak and the duration of ground motion, may be a better indicator of damage potential than the PGV alone, being a single measurement over the whole waveform. Below we present two simple applications: the potential damage inspection plot and the cumulative CAD plot.

From the simple GMPE-PGV given by Eq. (6.17), we can calculate the distance, R, over which a seismic source with potency P generates the velocity of ground motion $\geq PGV$,

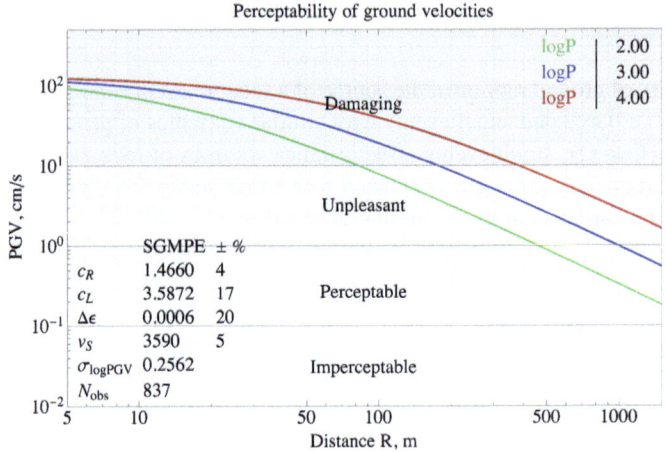

Fig. 6.26 Perceptibility of ground motion illustrated on the GMPE shown in Fig. 6.22

$$R (\geq PGV, P) = c_L P^{1/3} \left[(0.63 v_S \Delta\epsilon / PGV)^{1/c_R} - 1 \right]. \tag{6.22}$$

A similar equation can be derived from GMPE-CAD. We can also calculate the minimum potency, or $\log P$, that delivers a given level of PGV as a function of distance,

$$P (PGV, R) = (R/c_L)^3 \left[(0.63 v_S \Delta\epsilon / PGV)^{1/c_R} - 1 \right]^{-3}. \tag{6.23}$$

Now we can plot $\log P$ vs. distance R of seismic events, on the background of envelopes of a minimum $\log P$ that delivers a given level of PGV as a function of distance, for a number of strategic sites.

We analysed the last 60 days of seismic history before a $\log P = 2.61$ event at a mine here referred to as MineD, see the seismic hazard case study described in Chap. 5 and subsection Example in Chap. 3, and Sect. 6.6.8.3 below. The size of the event scales with the radius of source volume, and the colour indicates the distance of the event to the site.

Figure 6.27 left shows events with $\log P \geq -1.0$ vs. distance and the thresholds of ground motion, PGV_x, set as 1, 5, and 10 cm/s. The envelopes of a minimum $\log P$ stop where, according to the SGMPE, a $\log P_{max}$ event cannot deliver a given PGV beyond that distance. A seismic event that crosses the calibrated envelope for a given site may trigger damage inspection. Note that the developed GMPE does not take into account site effects, i.e. the amplification of ground motion at the skin of excavations, and therefore all these estimates are in solid rock and most likely underestimated.

6.6 Simple GMPE and Its Utility

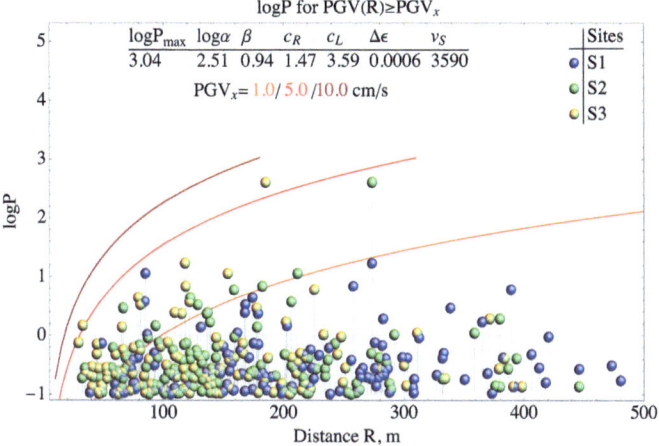

Fig. 6.27 897 events with $\log P \geq -1.0$ that occurred between 08 May and 07 July vs. distances to three sites at the mine. The orange, red, and darker red envelopes indicate a minimum $\log P$ that delivers $PGV = 1$, $PGV = 5$, $PGV = 10$ cm/s, respectively, as a function of distance

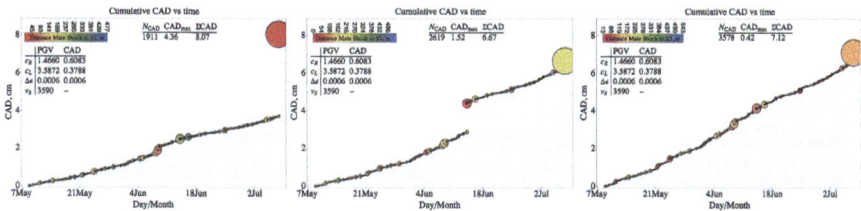

Fig. 6.28 Estimates of cumulative co-seismic displacement at the same sites, S1 *(left)*, S2 *(centre)*, and S3 *(right)*, over the same period of time

Another useful application is monitoring the consumption of the deformation capacity of the support due to seismicity. Figure 6.28 right shows the cumulative seismic deformation, $\mathrm{Cum}CAD$, due to seismic events with $\log P \geq -1.0$ that exceeded $PGV = 10^{-7}$ m/s, at the same three sites over the same period as in Fig. 6.27. It shows a big jump in CAD at site S1, which was deliberately located at the source of $\log P = 2.61$ event that delivered 4.36 cm of CAD.

It shows that Site1 was subjected to a relatively low level of seismic deformation before the main shock. Site 2 accumulated 6.67 cm and 1.52 cm of that was due to the $\log P = 1.24$ event that located very close to this site. Site S3 accumulated 7.12 cm, and the biggest jump was the $\log P = 2.61$ main shock that contributed 0.42 cm. Again, the observed CAD were recorded by sensors in boreholes, and therefore, they do not take into account the amplifying effect of the fracture zone close to excavations and the reaction of the support.

6.6.6 Probability Pr ($\geq PGV, R, \Delta T$)

To calculate the probability that ground motion may exceed a given threshold at distance R from a source, regardless of source location, within the time interval ΔT, we need to replace potency in Equation 6.20 for the probability of having an event $\geq P$ within the time interval ΔT,

$$\Pr(\geq P, \Delta T) = 1 - \exp\left[-\frac{\Delta T}{\Delta t}\alpha\left(P^{-\beta} - P_{max}^{-\beta}\right)\right], \tag{6.24}$$

with the potency P derived from the GMPE. The potency as a function of PGV and distance, $P(PGV, R)$, can be derived analytically from Equation 6.17 for SGMPE-PGV,

$$P(PGV, R) = (R/c_L)^3 \left[(PGV_0/PGV)^{1/c_R} - 1\right]^{-3}, \tag{6.25}$$

and, when inserted into Eq. (6.24), gives

$$\Pr(\geq PGV, R, \Delta T) = 1 - \exp\left\{-\alpha\frac{\Delta T}{\Delta t}\left[\left(\frac{R/c_L}{(PGV_0/PGV)^{1/c_R} - 1}\right)^{-3\beta} - P_{max}^{-\beta}\right]\right\}. \tag{6.26}$$

Figure 6.29 left shows the distance over which a given $\log P$ generates ground velocities $PGV \geq 15$, 10, and 5 cm/s. Figure 6.29 right shows the probabilities of having ground velocity $PGV \geq 15$, 10, and 5 cm/s within 180 days, as a function of the distance from the source.

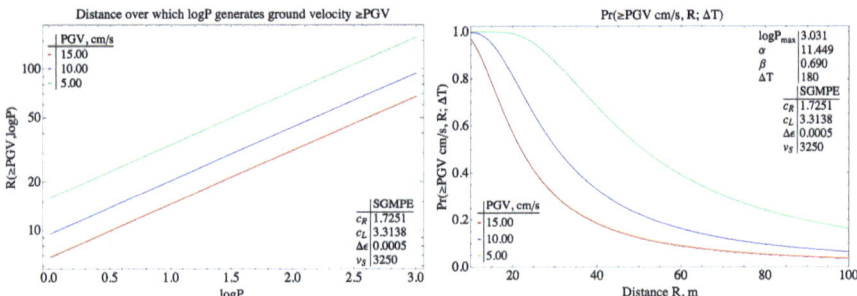

Fig. 6.29 Distance over which a given $\log P$ generates ground velocity $\geq PGV$ (*left*), and $\Pr(\geq PGV, R, \Delta T)$ as a function of distance from the source for three PGV thresholds (*right*)

6.6.7 Seismic Fragility Curves and Damage Potential

Fragility curves are functions that describe the conditional probability that the system may be subjected to a different degree of damage over the full range of loads to which that system might be exposed. The probability of damage is a function of both uncertainty in the capacity and uncertainty in the demand, and as demand increases relative to capacity, the probability of damage approaches one.

It is usually assumed that uncertainty in the capacity term follows a log-normal distribution, and therefore, the fragility curve also follows a log-normal distribution. Conforming to the accepted convention, the equation for the fragility function can be written as

$$\Pr(D \geq d | X = x) = F_d(x) = \Phi \left[\frac{1}{\beta_{cd}} \ln \frac{x}{\theta_{cd}} \right], \qquad (6.27)$$

where D is an uncertain damage state of the accepted damage classification, d is a particular value of D, $\Pr(D \geq d | X = x)$ is the conditional probability that $D \geq d$ is true given $X = x$ is true, X is the engineering demand parameter, x is a particular value of X, $F_d(x)$ is a fragility function for damage state d evaluated at x, Φ is the standard normal cumulative distribution function, θ_{cd} is the median capacity of the structure to resist damage state d, and β_{cd} is the standard deviation of the natural logarithm of the capacity of the structure to resist damage state d (Porter, 2018). In this formulation, only the median and standard deviation of the capacity are required to define the fragility function. For ground motion hazard, Eq. 6.27 can be written as

$$F_d(GMP) = \Phi \left[\frac{1}{\sigma_{\ln GMP_c}} \ln \frac{GMP}{\overline{GMP_c}} \right], \qquad (6.28)$$

where the engineering demand parameters GMP can be PGV, PGA, or CAD. The mean capacity of the structure to resist damage $\overline{GMP_c} = GMP_c \exp(\beta_c/2)$, and the standard deviation $\sigma_{GMP_c} = \overline{GMP_c} \sqrt{\exp(\sigma_{\ln GMP_c})^2 - 1}$.

The nature of the log-normal fragility function dictates that $F_d(GMP = GMP_c) = 0.5$, and the uncertainty in capacity $\sigma_{\ln GMP_c}$ dictates how steep is the fragility curve. If there is little uncertainty in capacity of the system, the fragility curve will be steep, i.e. there is great degree of certainty that the system will fail at that load. This situation applies to brittle or to well-understood systems. For complex inhomogeneous systems, e.g. the support of u/g tunnels in seismically active mines, the uncertainty in the capacity, GMP_c, is larger and the fragility curves are flatter.

Conventionally, the capability of geotechnical structures is evaluated by the design factor of safety, in this case, the ratio of the design capacity to the expected demand, $F_S = GMP_c/GMP_d$. The demand should be estimated for the maximum expected potency, $P = P_{max}$ or $\log P = \log P_{max}$. Obviously, all structures are designed to a factor of safety greater than one to provide an adequate margin of safety, $M_S = GMP_c - GMP_d$. The SGMPE allows not only to estimate the expected demand in terms of PGV or CAD, but also the expected linear extent of damage, $R_d(GMP_c, P)$, in metres in case the demand exceeds the in situ capacity, $GMP_d \geq GMP_c$,

$$R_d(GMP_c, P) = c_L P^{1/3} \left[(GMP_d/GMP_c)^{1/c_R} - 1 \right], \quad (6.29)$$

which may be of interest to underground tunnels. Equation (6.29) gives $R_d = 0$ when the demand is equal to capacity, i.e. $GMP_d = GMP_c$, which is the limiting case, without a margin of safety. Any further increase in demand leads to a damaged state, i.e. $R_d > 0$. Negative distances, $R_d < 0$, indicate a positive margin of safety.

Figure 6.30 shows an example of the fragility curves for four different capacities in terms of PGV_c (left) and CAD_c (right) assuming the structure is at or very close to a source of event with $\log P = 2.0, 2.5,$ and 3.0.

We assumed $0.5 \leq \sigma_{\ln PGV_c} \leq 0.6$ and $0.4 \leq \sigma_{\ln CAD_c} \leq 0.5$. The PGV graph also shows the corresponding dynamic co-seismic strains, $\epsilon_d = PGV/v_S$ as a reference, and the distances $R_d(PGV_c, P)$ and $R_d(CAD_c, P)$ within which the in situ demand exceeds, $R_d > 0$, or is below, $R_d < 0$ the capacity of the support. The parameters of the SGMPE are $v_S = 3590$ m/s, $\Delta \epsilon = 6 \cdot 10^{-4}$, and $c_{RPGV} = 1.466$, $c_{LPGV} = 3.587$, $c_{RCAD} = 0.6083$, $c_{LCAD} = 0.3788$. At source, the $PGV_0 = 0.63 v_S \Delta \epsilon = 1.357$ m/s is independent of $\log P$, and $CAD_0 = q_0 \Delta^{2/3} P^{1/3}$ is a function of $\log P$ and gives 0.0274, 0.04, and 0.0589 metres at source for $\log P = 2.0, 2.5,$ and 3.0, respectively.

The PGV fragility curves in Fig. 6.30 left show the probability of damage for four different in situ capacities, $PGV_c = 0.8, 1.0, 1.2,$ and 1.4 m/s. For the lowest capacity of 0.8 m/s, the distances within which the demand exceeds the capacity are 7.2, 10.6, and 15.6 m for $\log P = 2.0, 2.5,$ and 3.0, respectively. For $PGV_c = 1.0$ and 1.2, the probabilities of having damage are still over 50%. The minimum capacity to lower the probability of damage to below 50% by these events is 1.4

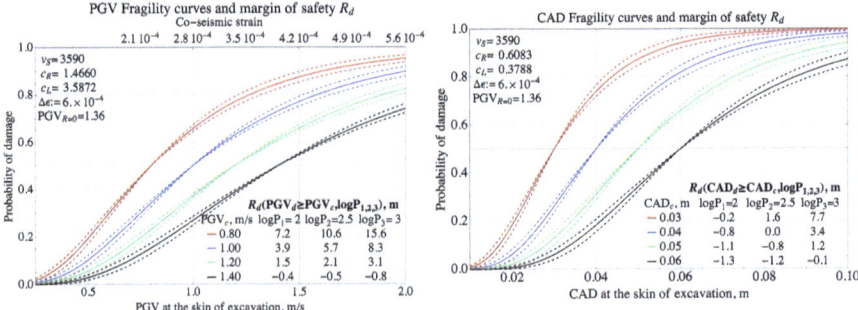

Fig. 6.30 Fragility curves for four different in situ capacities in terms of PGV_c (left) and CAD_c (right). Distances $R_d(PGV_c, \log P)$ and $R_d(CAD_c, \log P)$ within which the demand exceeds the capacity are calculated for three different $\log P$ using SGMPE. Dashed lines indicate the range of probabilities for assumed $\sigma_{\ln GMP_c}$

Table 6.5 Factors and margins of safety for limited CAD fragility curve given in Fig. 6.30

log P	Factors and margins of safety, $F_S\|M_S$			
	$CAD_c = 3$ cm	$CAD_c = 4$ cm	$CAD_c = 5$ cm	$CAD_c = 6$ cm
2.0	1.097 | 0.003	1.462 | 0.013	1.828 | 0.023	2.193 | 0.033
2.5	0.747 | −0.010	0.996 | −0.000	1.245 | 0.010	1.494 | 0.001
3.0	0.509 | −0.029	0.679 | −0.019	0.848 | −0.009	1.018 | 0.001

m/s. The PGV fragility curves are of limited practical value since it is difficult to establish the remaining in situ PGV, or the energy capacity, of the support.

The CAD fragility curves in Fig. 6.30 right are of more practical value because it is easier to assess the remaining in situ deformation capacity of support. The respective factors and margins of safety for CAD are given in Table 6.5.

6.6.8 Seismic Ground Motion Alert Program—GMAP

6.6.8.1 Introduction

This section is based on Mendecki (2023). Seismic risk is associated with the probability of a loss, which is the product of seismic hazard, the vulnerability of a site, and the exposure. Seismic hazard is the probability of having potentially damaging ground motion at a given site at a future time. The vulnerability of that site is its ability to sustain a certain level of ground velocity and ground displacement. The exposure is defined by the elements at risk which, in the case of safety, is the number of people that could be affected by a potential rockburst.

While the prediction of the time and the location of a potentially damaging seismic event is impossible, one should limit people's exposure at times of increased seismic activity and/or increased seismic loading by mid-size or larger events close to working places. The reason is very simple, as the rate of seismic activity increases so does the likelihood that one of these events may be larger and damaging.

One of the objectives of seismic monitoring in mines is "To detect strong and unexpected changes in the spatial and/or temporal behaviour of seismic parameters that could lead to rock mass instability and affect working places immediately or in the short term", see Mendecki (2016) section 1.2. By "strong", we understand changes exceeding a specified reference level, and the "unexpected" here means spontaneous, not associated with blasts or other controlled mining activities that usually trigger or induce such changes.

After blasting or after larger seismic events, the seismic activity rate is expected to increase, and over the years mine seismologists have developed different methods to monitor and quantify its decay in time and space to allow safe re-entry of personnel (e.g. Spottiswoode, 2000; Turner & Player, 2000; Malek & Leslie, 2006; Woodward et al., 2017; Gospodinov et al., 2022). While some methods use a combination of different seismic parameters, the fundamental one is always the seismic activity rate. One such method, which also includes guidelines for developing a re-entry protocol, is described in Vallejos (2010). Mendecki (2016), section 4.2, described a method to estimate the probability that the seismicity rates in two different time intervals are different by a given factor. Nordstrom et al. (2020) back analysed the utility of 14 different parameters as short term hazard indicators or as early warnings using data from Kiirunavaara mine in Sweden. They concluded that the most successful parameters are accelerated moment, accelerated apparent volume release, and increased activity rate. However, in many cases, the first two parameters are associated with the third one.

The short term hazard associated with aftershock activity quantified in terms of the probability of having an event above a certain size can be estimated by applying a rate function, e.g. the Omori or the stretched exponential, and the size distribution of seismicity for the given area to clusters of events, see Mendecki (2016) section 5.3.

An alert is raised if a given parameter, or a set of parameters, exceeds the imposed reference level(s). The reference level can be estimated by taking an average of a given parameter over times that were outside the influence of blasting and larger events and when there was normal safe working activity in a given area. In the case of seismic activity, it is expected that the coefficient of variation of the data selected to estimate the reference activity will be close to 1.0, i.e. not far from Poissonian. It is important that the reference level takes into account the current vulnerability of the site, or excavation. For example, an area with lower deformation capacity of support and/or a wider span of excavation should have a lower reference level.

Seismic exposure also needs to be defined in space; therefore, we need to be able to delineate the exclusion zone. In small mines, the exclusion zone can be the whole mine, but this is not practical for large mining operations. After an alert has been issued, there is a need to de-alert, and this again includes time and space. After some time, a part of the excluded area may be below the reference level, while others, specifically those close to the sources of larger events or blasts, may not. Typically, the exclusion zone volume increases as the size of the main shock increases. The exclusion time increases with the size of the main shock and decreases as the distance to the main shock increases.

All methods described above are the so-called polygon-based, where the polygon is defined as a seismogenic volume that generates seismicity affecting working places. Therefore, all relevant parameters are derived from the data selected from this polygon. This method has been widely applied in mines for many years. The most difficult, and also subjective, task here is the definition of the polygon. Different polygons select different datasets and therefore will produce different seismic characteristics.

In this section, we describe the polygon-less approach, where one takes into account the influence of all available seismic events, regardless of their location, on a particular working place. The preferred measures of influence are the rates of the following two ground motion parameters: (1) The cumulative absolute displacement, measured over a given minimum ground velocity, $CAD = \int_0^{t_d} |v(t)|_{\geq v_{min}} dt$. (2) The $ACAD$, which is the rate of CAD. The influence of CAD_{Rate} and $ACAD_{Rate}$

6.6 Simple GMPE and Its Utility

is moderated by the distance from the seismic source to the place of potential exposure. For example, a $\log P = 2.0$ event generates a similar level of PGV at 200 metres away as a $\log P = 1.0$ at 100 metres, and a $\log P = 2.0$ event generates a similar level of CAD at 200 metres away as a $\log P = 1.0$ at 50 metres.

6.6.8.2 GMAP Algorithm

GMAP is an influence based polygon-less two parameter method where one takes into account the influence of ground motion generated by all available seismic events, regardless of their location, on a particular working place. It is based on the rates of the cumulative absolute deformation, CAD calculated above a given ground velocity threshold, CAD_{Rate}, and on the activity rate of CAD events, $ACAD_{Rate}$. See Sect. 6.4.1 for more on CAD.

The following few steps describe the procedure to calculate GMAP ratings for each site defined by its coordinates (x, y, z), given the alert reference rates, CAD_{Rref1} and $ACAD_{Rref1}$, and alarm reference rates, CAD_{Rref2} and $ACAD_{Rref2}$. In this case study, alarm reference rates are set as 2 times the alert reference rates.

First define the moving time window, Δt, needed to calculate the rates. The window can be constant in time, or it may be defined by a number of events. For all events in the window:

1. Calculate the threshold seismic strain, $\epsilon_s = v_{min}/v_S$, and CAD using the relevant GMPE.
2. If for a given event seismic strain is greater than or equal to ϵ_s, accumulate CAD to get ΣCAD and calculate its activity, $ACAD$, defined by N_{CAD}, the number of times CAD is accumulated.
3. Calculate $CAD_{Rate} = \Sigma CAD/\Delta t$ and $ACAD_{Rate} = N_{CAD}/\Delta t$.
4. Normalise CAD_{Rate} and $ACAD_{Rate}$. They have different units, m/s and 1/s, but can be compared when normalised relative to their individual alert and alarm reference rates.

$$CAD_{RateN} = 1 + (CAD_{Rate} - CAD_{Rref1})/(CAD_{Rref2} - CAD_{Rref1})$$
$$ACAD_{RateN} = 1 + (ACAD_{Rate} - ACAD_{Rref1})/(ACAD_{Rref2} - ACAD_{Rref1}).$$

5. Take $\max [CAD_{RateN}, ACAD_{RateN}]$ to get the GMAP rating.
6. Repeat for each site and contour and/or move the window by an increment of time, or by one event, to create the time series of GMAP.

Note that any seismic activity or ground motion based alert or re-entry protocol is not a prediction. The method described here is also not a forecast, since it does not state the probabilities of occurrences. GMAP just responds to changes in the selected ground motion parameters.

CAD_{Rate} will react to small events located close to a given site and to larger events located up to a few hundred metres away. CAD_{Rate} is also sensitive to an increase in the frequency of mid-size events that are not associated with an overall increase in seismic activity. A single larger event located nearby should issue an instant alert/alarm at that site. $ACAD_{Rate}$ is sensitive to an increase in the activity rate influencing a given site.

Note that CAD excludes any time dependent aseismic deformation, e.g. bulking which is a function of time and may be delayed. Also the support system prevents part of that deformation from reaching the skin of excavation, so one cannot see or measure it, but this kind of seismic action contributes to the increase in the fracture zone around excavations and may contribute to damage when hit with a larger event.

CAD_{Rate} and $ACAD_{Rate}$ are calculated in a moving window. The moving window can be defined by a fixed number of events, in which case its duration varies with activity rate, or by a fixed time window. If after alert, the activity rate unexpectedly drops, the fixed number of events window will drop the GMAP rating very slowly, while with the fixed time window the rating will drop quickly. If the GMAP rating is dominated by one larger event with few aftershocks, then the rating will drop as soon as the large event drops from the moving window.

6.6.8.3 GMAP Example

GMAP Data We analysed the last 60 days of seismic history before a $\log P = 2.61$ event at a mine here referred to as MineD, see the seismic hazard case study described in Chap. 5 and subsection Example in Chap. 3. Figure 6.31 left shows the

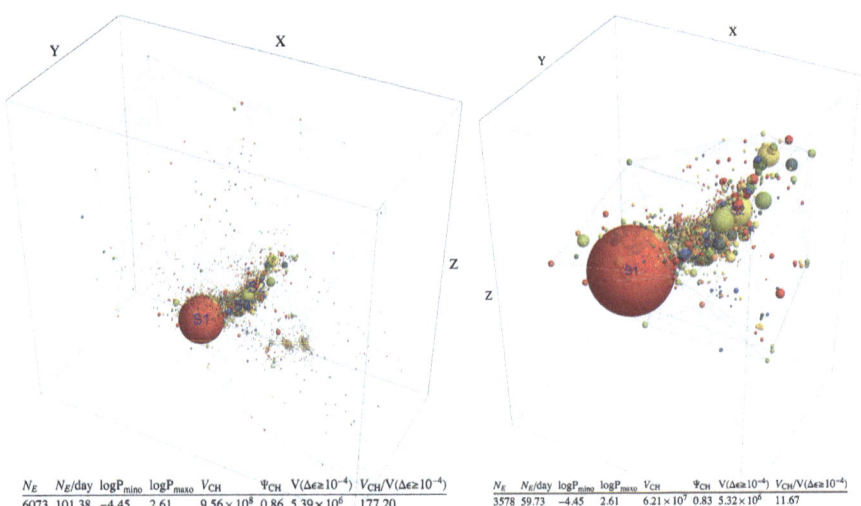

Fig. 6.31 Convex hull span over all available 6073 events *(left)* and over 3578 events that generated $PGV \geq 10^{-7}$ m/s at site S3 *(right)*

6.6 Simple GMPE and Its Utility

convex hull span over all 6073 events with $\log P \geq -4.5$ available for analysis. The size of the event represents the radius of the source volume taken as a sphere, $V = P/\Delta\epsilon$, where $\Delta\epsilon$ is the assumed strain change at the source, in this case $\Delta\epsilon = 10^{-4}$, and the colour indicates the time of the event, from the earliest in blue to the latest in red. In this section, we will test GMAP at site S3, which is the average location of all 6073 events. Figure 6.31 right shows the convex hull span over 3578 events that generated $PGV \geq 10^{-7}$ m/s at the centre of all 6073 events, shown as S3 in blue.

Distances of these events to site S3 range between 12 and 394 m. Tables at the bottom of these figures give the number of events, N_E, activity rate/day, the $\log P$ range, the volume of the convex hull, V_{CH}, and its sphericity index, Ψ_{CH}, the total volume of seismic sources with strain change ≥ 0.0001, $V\left(\Delta\epsilon \geq 10^{-4}\right)$, and the ratio of the convex hull to the volume of inelastic deformation of events $V_{CH}/V\left(\Delta\epsilon \geq 10^{-4}\right)$. After selection of the 3578 events, $V_{CH}/V\left(\Delta\epsilon \geq 10^{-4}\right)$ ratio dropped from 177.2 to 11.67, which indicates that the volume selected for stability is reasonably saturated with co-seismic inelastic deformation. The sphericity index, Ψ_{CH}, here dropped slightly from 0.86 to 0.83, however, while running moving windows through these 3578 events it varied between 0.65 and 0.86.

Figure 6.32 shows the cumulative number of events, $\text{Cum}N$, and the cumulative apparent volume, $\text{Cum}V_A$ in km^3, for all 6073 events. The size of the event represents the radius of the source volume and colour scales with distance to the site. There were 17 mid-size events with $\log P \geq 0.0$ which gives an activity rate 0.28/day, and the coefficient of variation of all events is $C_v = 1.32$, which indicates a degree of time clustering. The vertical red, blue, and green lines indicate times of the three largest events, respectively. The $\text{Cum}N$ vs. time plot is quite steady with an almost constant activity rate, while $\text{Cum}V_A$ shows a bit more structure and an increase in its rate before the main shock. However, the apparent volume does not enter into GMAP.

The largest event with $\log P = 2.61$ occurred on 07 July and located 130 m from site S3, the second largest with $\log P = 1.24$ on 14 June located 99 m away, and the third largest with $\log P = 1.1$ on 08 June 137 m away. The distance between the largest event and the second largest is 273 m, the largest and the third largest 85 m, and the second and the third largest 211 m. The distance between an event and a given site is taken as the Euclidean distance minus the radius of the event, $r = \sqrt[3]{3P/(4\pi\Delta\epsilon)}$, where the strain change, $\Delta\epsilon$, is the same used in the development of the SGMPE, in this case $6 \cdot 10^{-4}$.

Time History of GMAP Parameters Figure 6.33 top row shows the cumulative CAD and N_{CAD} vs. time. The vertical red, blue, and green lines on cumulative plots here indicate times of the three largest CAD events, respectively, i.e. events that generated the three largest seismic deformation at site S3. Figure 6.33 bottom row shows normalised CAD_{Rate} and $ACAD_{Rate}$ and the resulting GMAP alerts.

The reference rate CAD_{Rref1} was set as 0.072 cm/day and $ACAD_{Rref1}$ as 44 CAD events per day. Alarm levels were set as double the alert levels. Alert plots are

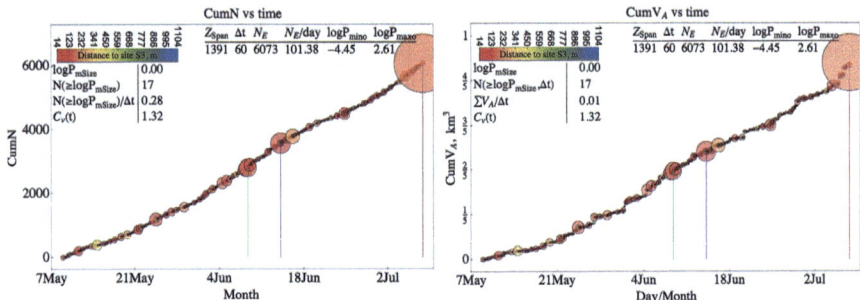

Fig. 6.32 CumN *(left)* and CumV_A *(right)* vs. time plotted for all 6073 events

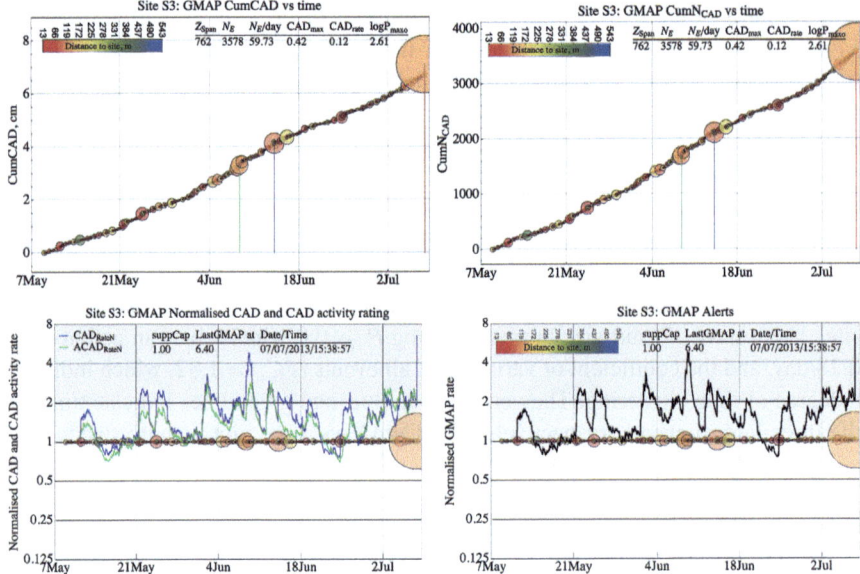

Fig. 6.33 Cumulative CAD and N_{CAD} vs. time, *(top row)*, and time histories of normalised CAD_{Rate}, $ACAD_{Rate}$, and resulting GMAP alerts *(bottom row)*

based on a fixed number 119 event moving window, which is equal to 2 times the average activity rate per day.

Of the total 6073 seismic events only 3578 classified as CAD events because most of them were too far and/or too small to induce $PGV \geq 10^{-7}$ m/s at the site. Most of these events located between 20 and 120 metres from site S3. The total CAD over that time was 7.12 cm, and the largest event with $\log P = 2.61$ that was 130 metres away imposed 0.42 cm of CAD. The second largest CAD was 0.09 cm associated with event with a $\log P = 1.24$ located 99 metres away.

The selected dataset is characterised by a reasonably stable activity rate which always makes alerting more difficult. However, the data has a significant number

6.6 Simple GMPE and Its Utility

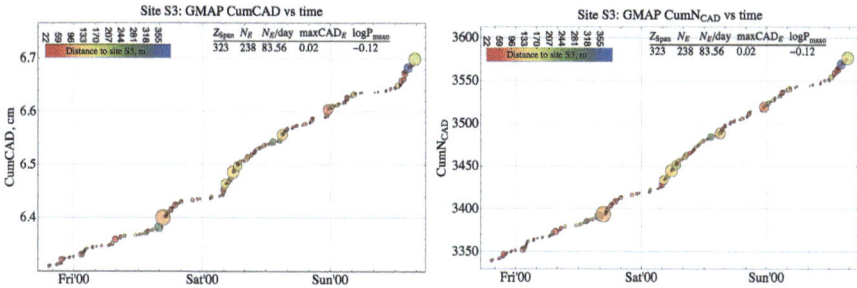

Fig. 6.34 The last 68.35 hours of $CumCAD$ *(left)* and $CumN_{CAD}$ before the main shock *(right)*

of mid-size events not far from the site of interest, therefore, with the exception of the last three days before the main shock, GMAP alert time history is dominated by CAD_{Rate} rather than $ACAD_{Rate}$.

The largest event with $\log P = 2.61$ occurred on 7 July at 15:38:57. The first alarm was on 02 July, then there were seven additional alarms, and the last one on 7 July at 15:30:40. There was also a spike type alarm exactly at the time of the event.

The second largest event with $\log P = 1.24$ and $CAD = 0.1$ cm occurred on 14 June at 04:36:38. There was only one Alarm before the event on 12 June at 09:52:48, and there was a spike type of alarm at the time of event.

The third largest event with $\log P = 1.1$ and $CAD = 0.06$ cm occurred on 8 June at 18:47:01 at distance 137 metres from site S3. At the time of this event GMAP was in a state of alert. There was the alarm issued on 06 June at 13:41:48, then three Alarms on June 07, and the last one on 08 June at 09:23:03.

Figure 6.34 shows the cumulative CAD and N_{CAD} for the last 68.35 hours before the largest event of $\log P = 2.61$, excluding the main shock.

During this time, there were 238 events giving the activity rate 83.56 events/day. The maximum event was with $\log P = -0.12$ that occurred 47 hours before the main shock at a distance of 97 metres and generated 0.17 mm of CAD. There was a clear acceleration in the cumulative CAD and N_{CAD} during the last 6 hours before the main shock. However, it was partially smoothed over by the 119 events moving window. The shorter window would have alerted earlier, but it would also have issued more false alarms.

Comparison with Stability Analysis Figure 6.35 left shows the time history of GMAP alerts, which is based on CAD_{Rate} and $ACAD_{Rate}$. Figure 6.35 right shows a simple version of the stability function for site S3, $\Psi_{CH} \cdot \text{Median}\left[\log \sigma_A\right]/\lambda$, where σ_A is apparent stress, λ the activity rate, and Ψ_{CH} sphericity of the convex hull span over the events in each moving window.

Both cases are based on the same dataset, the PGV threshold of 10^{-7} m/s, and the moving window of 119 events. See Sect. 3.5.2 in Chap. 3 for more on the examples of stability analysis. Note that the low level of the stability function indicates a less stable rock mass with higher potential for larger events or a swarm of events. In such cases, GMAP alerts go up. It is clear that these two plots are

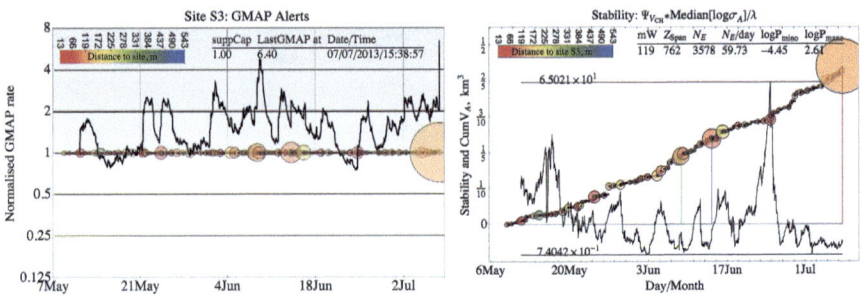

Fig. 6.35 GMAP *(left)* and stability *(right)*

qualitatively similar, or inversely correlated, i.e. both indicating increase potential for larger ground motion at the site at the same times. This is interesting because with the exception of the activity rate, λ, these two plots are based on different parameters. Let us consider the same three events we analysed above:

- The largest event with $\log P = 2.61$ on 7 July. From 02 July, GMAP is practically at alarm level, and the stability function is well below the 30 days mean.
- The second largest event with $\log P = 1.24$ on 14 June. GMAP is high and stability is low.
- The third largest event with $\log P = 1.1$ on 8 June. GMAP is high and stability is low.

6.6.8.4 General Comments

A prediction can be understood as a deterministic binary statement, true or false, about a future event that can be validated or falsified with a single observation. A forecast can be defined as a statement of probability about a future event. An individual forecast can never be validated by a single observation and requires multiple observations to establish a degree of confidence. As mentioned above, GMAP is neither a prediction nor a forecast, since a forecast requires stating the probabilities of occurrences.

The utility of GMAP is not only in issuing ground motion alerts or alarms after a sudden increase in GMAP rating, or in delineating the exclusion zones, but also in guiding control measures to mitigate seismic hazard. If the GMAP rating is systematically increasing or stays high in a given area, then the mine may change the spatial and temporal manner of rock extraction, e.g. change the sequence of blasting, scatter the production blasts, and/or slow down the rate of mining. There is a view that these control measures just delay the inevitable. However, experience shows that scattered rock extraction changes the nature of seismic release by producing more smaller or mid-size events and fewer large ones (van Aswegen & Mendecki,

1999; Handley et al., 2000; Vieira et al., 2001; Mendecki, 2001 Figure 8; Mendecki, 2005; Durrheim et al., 2005). Therefore, if guided by GMAP, the mine applies control measures to manage the seismic response, then the GMAP success or failure rate cannot be tested by the number of larger events that did or did not occur after alerts or alarms. In such cases, it would be more appropriate to measure it by the overall positive changes in the size distribution of seismic events. Paradoxically then, the case where there is no larger event after GMAP alerts would be considered a success. Like any other alerting method, GMAP has limitations:

1. GMAP may fail to alert for mid-size events located far, typically a few hundred metres from a given site, if there is no increase in seismic activity within that distance. During testing, GMAP never failed on a large event located within a few hundred metres from a given site; however, it may alert or alarm at the time of the event.
2. It needs to be calibrated, mainly the alert rate reference levels, CAD_{Rref1} and $ACAD_{Rref1}$, and the alarm multiplier.
3. GMAP requires a reasonably accurate GMPE that should be updated at least once a year.
4. It relies on seismic events being processed before the rating can be updated, and therefore, it depends on the speed and the quality of seismological processing, i.e. it is not a real-time system.

6.7 Seismic Ground Motion Alert System—GMAS

GMAS, like GMAP, is based on the two ground motion parameters, CAD_{Rate} and $ACAD_{Rate}$, but it computes these rates in real time from the waveforms provided by seismic sensors at a given location. Unlike GMAP, GMAS does not need a ground motion prediction equation, is not subject to association, and does not rely on the seismological processing that delays the process and introduces uncertainties, and therefore, it gives instantaneously local alerts or alarms when given threshold parameters are exceeded.

The following few steps describe the procedure to calculate GMAS ratings for an area around the GMAS sensor, given the alert reference rates, CAD_{Rref1} and $ACAD_{Rref1}$, and alarm reference rates, CAD_{Rref2} and $ACAD_{Rref2}$, which in this case study are set as 2 times the alert reference rates.

Define the moving time window, Δt, needed to calculate the rates. The window can be constant in time, or it may be defined by a number of events. For all events in the window:

1. Take a segment, say $\delta t = 0.25$ seconds long, of the continuous data stream as recorded by a seismic site and test if the PGV in this section exceeds a predefined threshold, say $PGV \geq 10^{-5}$ m/s. If so, declare a ground motion event.
2. Integrate that segment of the waveforms to get CAD for the GM event.

3. Accumulate CAD in the window to get ΣCAD and calculate its activity $ACAD$ defined by N_{CAD}, and the number of times CAD is accumulated.
4. Calculate $CAD_{Rate} = \Sigma CAD/\Delta t$ and CAD activity rate, $ACAD_{Rate} = N_{CAD}/\Delta t$.
5. Normalise CAD_{Rate} and $ACAD_{Rate}$. They have different units, m/s and 1/s, but can be compared when normalised relative to their individual alert and alarm reference rates.

$$CAD_{RateN} = 1 + (CAD_{Rate} - CAD_{Rref1})/(CAD_{Rref2} - CAD_{Rref1})$$
$$ACAD_{RateN} = 1 + (ACAD_{Rate} - ACAD_{Rref1})/(ACAD_{Ref2} - ACAD_{Ref1}).$$

6. Take max $[CAD_{RateN}, ACAD_{RateN}]$ to get GMAS rating.

GMAS Example We analysed 15.36 hours of continuous three-component waveforms recorded at Site10 by 4.5 Hz geophones installed at 8 metres in a borehole and sampled at 6 kHz,. The first sample was at 00:00:09 and the last at 15:21:32. A seismic event with $\log P = 1.1$ ($m = 1.65$) and $\log E = 6.4$ occurred at 10:21:43:608 and located 196 metres from the sensor. Figure 6.36 shows the recorded velocity and integrated displacement waveforms. The long tail on the displacement waveforms may indicate a permanent displacement.

Figure 6.37 shows the time history of PGV, CAD, and CumCAD of all 221087 segments of data with the red lines marking the mean values. The mean rate of cumulative CAD was 1.21 cm/day. There were a few bursts of CAD activity in the

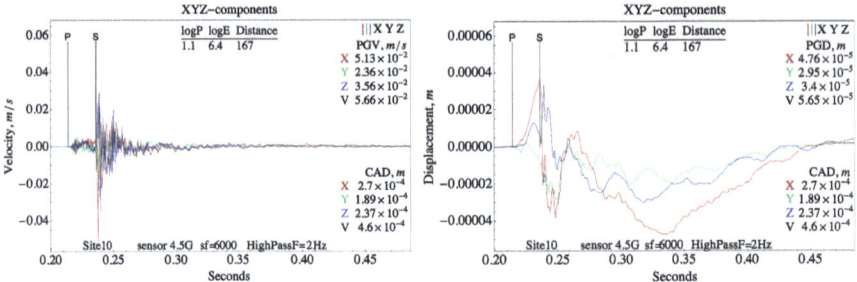

Fig. 6.36 Velocity and displacement waveforms of the $\log P = 1.1$ event

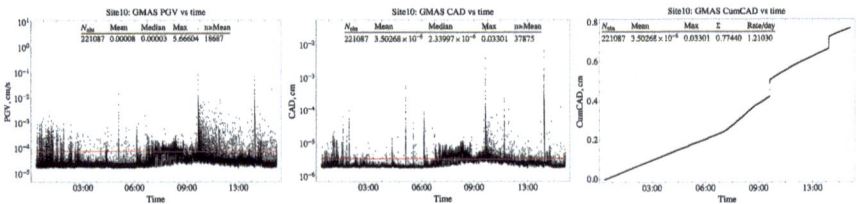

Fig. 6.37 PGV (left), CAD (centre), and CumCAD (right) time histories of all 221087 segments of data

6.7 Seismic Ground Motion Alert System—GMAS

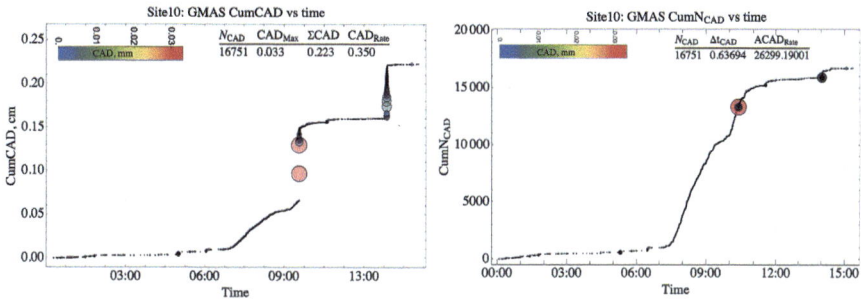

Fig. 6.38 CumCAD (left) and CumN$_{CAD}$ (right) of GM events

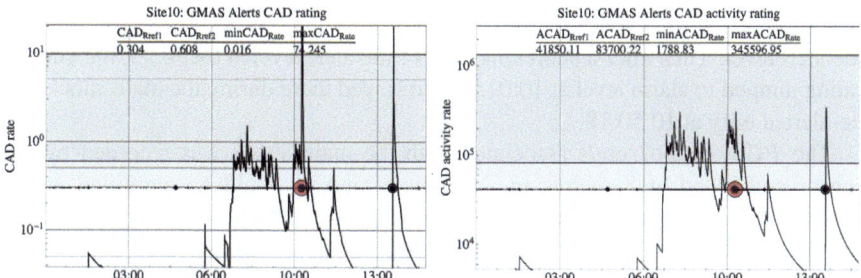

Fig. 6.39 Time histories of CAD_{rate} and $ACAD_{rate}$

morning, but the CumCAD rate plot remained structureless, it increased moderately at 07:30, but then remained steady until the main shock. Obviously, the cumulative number of all segments of data here would be just a straight line.

Figure 6.38 shows the CumCAD and CumN_{CAD} vs. time where we selected segments of data with $CAD \geq \overline{CAD} = 3.5 \cdot 10^{-6}$ cm and with predominant frequency less than 150 Hz. There are 16751 such GM events, and now the data show more interesting structure, including a significant increase in both, the rate of CumCAD, and the rate of CumN_{CAD} approximately 3 hours before the $\log P = 1.1$ event. The largest $PGV = 5.67$ cm/s, and the second largest $CAD = 0.031$ cm occurred at 10:21:43.358. The second largest $PGV = 2.24$ cm/s and the largest $CAD = 0.033$ mm occurred at 10:21:43.608, all of them associated with the $\log P = 1.1$ event.

Reference alert level for CAD_{Rate} was set at 0.3042 cm/day, and alarm level was set as 2 times the alert level. Reference alert level for $ACAD_{Rate}$ was set at 41850 GM events/day and alarm level at double the alert level. The moving window was set at 350 GM events, see Fig. 6.39.

Here we assumed the normalised GMAS alert level as 1 and alarm level as 2. Figure 6.40 shows time histories of normalised CAD_{Rate} and $ACAD_{Rate}$, in this case dominated by $ACAD_{Rate}$, and time histories of the final GMAS alerts.

The first GMAS alert was at 07:31:29 followed by an alarm at 07:38:16, and the GMAS stayed there practically till 09:24:54 when it dropped all the way to below

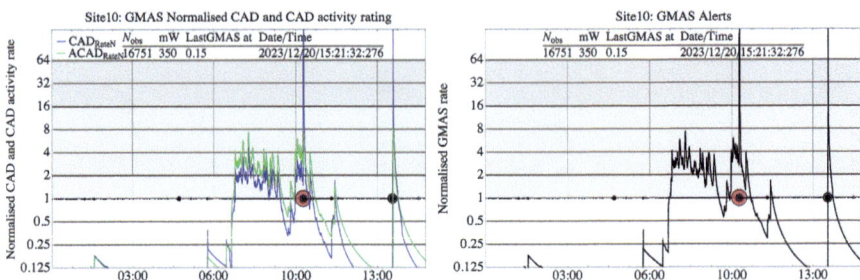

Fig. 6.40 Time histories of normalised CAD_{rate}, $ACAD_{rate}$ *(left)* and resulting GMAS alerts *(right)*

the alert level. Then after a short spike above the alert level at 09:50:53, the GMAS rating jumped to alarm level at 10:01:55 and stayed there during the main shock and de-alerted only at 10:50:18.

The $PGV = 5.67$ cm/s associated with the main shock was recorded by the sensor embedded at 8 metres in a borehole, and it would be amplified at the surface of the excavation. This level of ground motion may not be damaging, but it would certainly be felt strongly by people close to this site, and they would spontaneously evacuate the area. There was another GMAS alarm at 14:01:17 triggered by a flurry of smaller GM events with maximum $PGV = 0.08$ cm/s and max $CAD = 0.0073$ cm.

The area where the sensor is located is not well covered by the seismic system, and there is limited seismic data available; therefore, GMAP and stability analysis would not alert for this event at that site.

General Comments Like GMAP, GMAS is not a prediction, and it is also not a forecast since it does not state the probabilities of occurrences. GMAS just responds to changes in the local ground motion parameters, but, unlike GMAP, it works in real time. Like any other alerting method, GMAS has limitations:

1. GMAS is an inexpensive way to monitor and Alert/Alarm on stronger, potentially damaging ground motion. It does not need a ground motion prediction equation, is not subject to association, and does not rely on seismological processing, and therefore, it works in real time and, in cases of complex seismic events, may give a short notice to evacuate.
2. GMAS is local and in most cases will Alert/Alarm personnel to frequent mid-size events relatively close to the excavation. Its area of influence will scale positively with the recorded intensity of GM and its frequency.
3. If run on a number relatively closely spaced sites, it forms a real-time ground motion hazard monitoring system for the area.
4. One needs to define what constitutes a false alarm. This is an important question because it affects the calibration process. It would certainly be the case if an alarm goes off and people in the area do not feel anything. But it may not be

the case when GMAS alarms on GM that make people uncomfortable, and they would evacuate anyway.
5. GMAS needs to be calibrated, mainly the alert rate reference levels, CAD_{Rref1} and $ACAD_{Rref1}$. One can easily calibrate GMAS to Alert/Alarm at the time of a large GM event. However, it is difficult, if not impossible, to eliminate alarms on weaker but perceivable ground motion that may affect a small area. One can reduce the number of such alarms by increasing reference levels and/or the duration of the moving window. However, it is also advisable to define the number and the level of alerts and/or alarms in a given time span before advising personnel to evacuate. This should be a part of the calibration process that then feeds into the re-entry protocol.

6.8 Mapping Seismic Ground Motion Hazard

This section is based on the key note lecture delivered at the Rockburst and Seismicity in Mines Symposium in Santiago, Chile Mendecki, 2017.

The size distribution analysis disregards space; therefore, the given probabilities apply to the whole mine, while in many cases it is quite obvious that seismic hazard varies considerably in space, and these differences should be taken it into account when managing seismic risk. Subdividing space into sub-volumes, fitting a power law to the data extracted from each sub-volume, and calculating their probabilities are not the best strategy to estimate spatial hazard. There are two potential problems with this approach: (1) Seismic activity within these sub-volumes may not be independent, and therefore, it would be inappropriate to fit two different power laws to data. (2) There is a trade-off between the spatial resolution and the amount of data one can extract from these sub-volumes, so there may not be sufficient data for a reliable power law fit.

A better option to delineate seismic hazard in space is to estimate how frequently a given level of ground motion can be reached or exceeded at a given site, X, in future time ΔT, i.e. $\Pr[\geq GMP(X), \Delta T]$, where the ground motion parameter GMP can be PGV, PGA, or CAD. The ground motion hazard incorporates the size distribution analysis, the ground motion prediction equation, GMPE, and the distribution of distances from the relevant seismic events to a given site.

6.8.1 Methodology

The presented methodology to estimate ground motion hazard is based on the principles described in Cornell (1968), McGuire (2004), and Baker et al. (2021), with changes appropriate to accommodate data provided by mine seismic networks. For a given site, $X = (x, y, z)$, this methodology consists of the following steps:

1. For a given seismogenic volume of rock and time span of data, Δt, estimate the expected size of the next largest event, P_{max}, and obtain the empirical and theoretical probability density function of the size distribution, $f(P)$, and the expected activity rate, $\lambda (\geq P_{min}) = N (\geq P_{min})/\Delta t$, of events above a given potency threshold P_{min}.
2. Develop a simple GMPE and the respective survival function, $\Pr[\geq GMP(X); P, R]$, i.e. the probability that for a given seismic potency P and distance R the recorded ground motion at point X will exceed a given level of GMP.
3. For given site, define the empirical or theoretical probability density function of the distribution of distances, $f[R(X)]$, of events with co-seismic strain below a given threshold, say 10^{-7}. This will deliver a different number of events, and therefore, different activity rate for each site. The empirical distribution of distances is frequently based on more recent, shorter data set than the one used to derive the size distribution.
4. Combine the above information to compute the rate of exceedance,

$$\lambda[\geq GMP(X)] = \lambda(\geq P_{min}) \sum_{i=1}^{n_P} \sum_{j=1}^{n_R} \Pr$$
$$[\geq GMP(X); P_i, R_j] f(P_i) f[R_j(X)], \qquad (6.30)$$

where the ranges of seismic potency and distances are discretised into n_P and n_R intervals, respectively, and $f(P_i)$ and $f[R_j(X)]$ are the theoretical and/or the empirical probability density functions. The above equation assumes that the size distribution and the distance distribution are independent.
5. Compute the probability of observing at least one event in a period of time ΔT into the future. If the probability distribution of time between events is close to Poissonian, i.e. independent of time, where the coefficient of variation of the inter-event times is close to one, then the probability of observing at least one event in a period of time ΔT is $\Pr[\geq GMP(X), \Delta T] = 1 - \exp[-\lambda[\geq GMP(X)] \cdot \Delta T]$. In cases where there is a clear trend in seismic activity, one can apply the non-stationary Poisson process with a suitable intensity function.

6.8.2 Example: Data and Size Distribution

The data set starts on 07 September 2007 and ends on 07 July 2013 just after a $\log P = 2.61$ ($m = 2.66$) event. It spans 2130 days and includes 2818 events with $\log P \geq -1.0$ ($m \geq 0.25$), which gives the rate of 1.32 events/day. These events delivered $\Sigma P = 2526.06$ m^3 of seismic potency at the rate of 1.186 m^3/day. The largest event has $\log P = 2.61$ ($m = 2.66$), and there are 30 events with $\log P \geq 1.0$ ($m \geq 1.59$). The mean recurrence interval, $\bar{t}$ ($\log P \geq -1.0$) = 0.756 days

6.8 Mapping Seismic Ground Motion Hazard

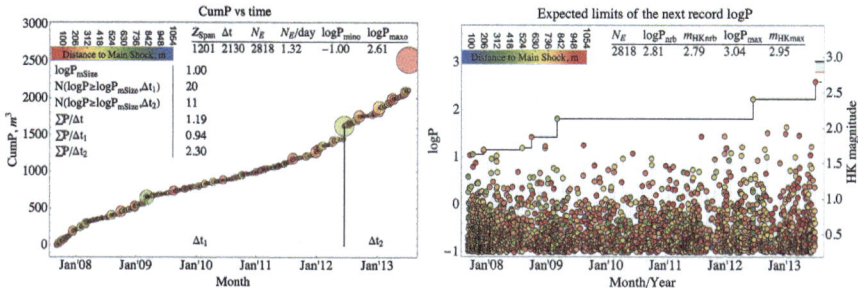

Fig. 6.41 Cum$P$ and the history of records vs. time

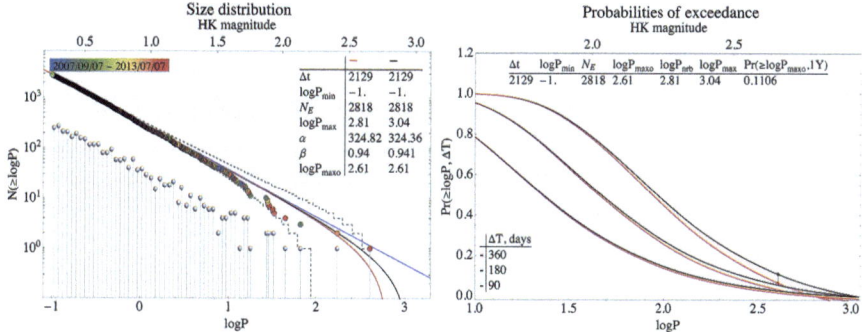

Fig. 6.42 Size distribution and the probabilities of exceedance

with the standard deviation of 1.125 days, which gives the coefficient of variation $C_v (\log P \geq -1.0) = 1.49$. The mean recurrence interval, $\bar{t} (\log P \geq 1.0) = 73.44$ days with the standard deviation of 94.76 days, which gives the coefficient of variation $C_v (\log P \geq 1.0) = 1.29$. The black vertical line is at the time of the second largest event with $\log P = 2.24$ that occurred on 13 June 2012 at the same depth but 500 m away from the main shock.

Figure 6.41 shows the cumulative potency, Cum$P$, and the history of records vs. time where the colour indicates the distance of the event from the main shock of $\log P = 2.61$.

Figure 6.42 left shows the size distribution of unbinned data where the colour indicates the time of the event, and two UT fits: the black with $\log P_{max} = 3.04$ and the red assuming that $\log P_{max} = \log P_{nrb} = 2.809$. The blue straight line is the open-ended fit with $\beta = 0.9415$ shown as a reference.

Figure 6.42 right shows the expected ranges of the probabilities of exceedance for the next 90, 180, and 360 days. The black line is for the upper range associated with $\log P_{max}$, and the red line shows the lower range associated with $\log P_{nrb}$. The probability of at least one event exceeding the maximum observed $\log P_{maxo} = 2.608$ within one year, $\Pr(\geq \log P_{maxo}; 1Y)$, lies between 0.066 and 0.111 and $\Pr(\geq \log P = 2.0; 1Y)$ is between 0.449 and 0.475.

6.8.3 GMPE and Survival Function, $\Pr(\geq GMP_x; P, R)$

The ground motion prediction model should provide a probability distribution instead of a single value of the ground motion parameter. In general, the ground motion prediction model can be described as $\ln \varphi = \overline{\ln \varphi} + \sigma$, where in this case $\varphi = PGV$ or $\varphi = CAD$, $\ln \varphi$ is a random variable, assumed to be well described by a normal distribution, and $\overline{\ln \varphi}$ and σ are the predicted mean and standard deviation of $\ln \varphi$, respectively. Note that if $\ln \varphi$ values are normally distributed, the non-logarithmic values are log-normally distributed.

Under such an assumption, the probability of exceeding a given level of ground motion parameter, GMP_x, is the survival function of the normal distribution,

$$\Pr(\geq \varphi; P, R) = 1 - \Phi\left[\left(\ln GMP_x - \overline{\ln \varphi}\right)/\sigma_{\ln \varphi}\right], \quad (6.31)$$

where Φ is the standard Gaussian cumulative distribution function. One can express it via the probability density function that can easily be evaluated numerically, $\Pr[\geq \varphi; P, R] = \int_{GMP_x}^{\infty} f(u)\, du$, where $f(u)$ is the probability density function of φ given P and R, which gives

$$\Pr[\geq \varphi; P, R] = \int_{PGV_x}^{\infty} \frac{1}{\sigma_{\ln \varphi}\sqrt{2\pi}} \exp\left[-\frac{1}{2}\left(\frac{\ln u - \overline{\ln \varphi}}{\sigma_{\ln \varphi}}\right)^2\right] du. \quad (6.32)$$

The scatter measured by $\sigma_{\ln \varphi}$ in data recorded in boreholes in solid rock in mines is expected to be lower than that recorded at the skin of underground excavations. Typical values of $\sigma_{\ln PGV}$ calculated to date in mines range from 0.4 to 0.65 and $\sigma_{\ln CAD}$ 0.3 to 0.55. Figure 6.43 shows the PGV and CAD survival functions for $\log P = 2.0$ at distances of 20, 50, and 100 metres for the SGMPE developed in Sect. 6.6.3.

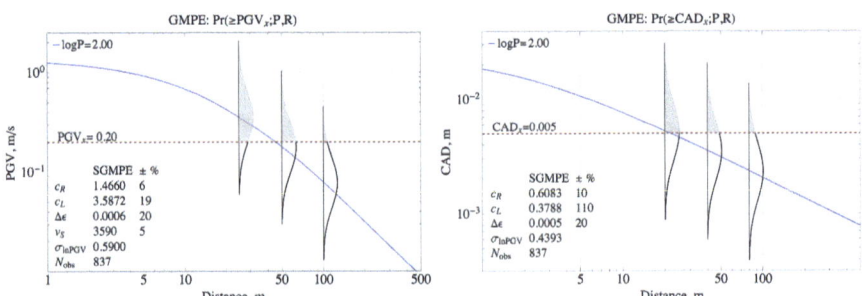

Fig. 6.43 Survival functions for $\log P = 2.0$: $\Pr[\geq PGV_x; P, R]$ at $PGV_x = 0.2$ m/s, $\sigma_{\ln PGV} = 0.59$ (left), and $\Pr[\geq CAD_x; P, R]$ at $CAD = 0.005$ m, $\sigma_{\ln CAD} = 0.4393$ $\sigma_{\log PGV} = 0.256$ (right), at distances 20, 50, and 100 metres from the seismic source

6.8.4 Distribution of Distances

Figure 6.44 left shows the convex hull span over all 6073 events available for analysis, and Fig. 6.31 right shows the convex hull span over 3578 events that generated $PGV \geq 10^{-7}$ m/s at the centre of all 6073 events, shown as S3 in blue. The three sites at which we will evaluate probabilities of exceedance are shown as S1, S2, and S3. This is the same data set we used in Sect. 6.6.8.3.

To estimate the expected ground motion at a given site, it is necessary to establish the distribution of distances from seismic sources to that site. The location of a seismic event is represented as a point which, in most cases, is assumed to be the rupture initiation. Earthquake seismologists use distance to the epicentre or hypocentre, distance to the closest point on the rupture surface, or distance to the closest point on the surface projection of the rupture.

However, in mines, seismicity follows rock extraction and clusters along geological features; therefore, the assumption of random spatial distribution frequently cannot be made. A better option is to construct an empirical probability distribution of distances to a given site. In most cases, these empirical probabilities cannot be fitted with any reasonable theoretical distribution.

Figure 6.45 shows the empirical probability density function of distances to the same three hypothetical sites S1, S2, and S3 as described in Sect. 6.6.8.3 and marked by blue dots. The log-normal distribution fit to the data defined by a mean and standard deviation is marked by solid red lines.

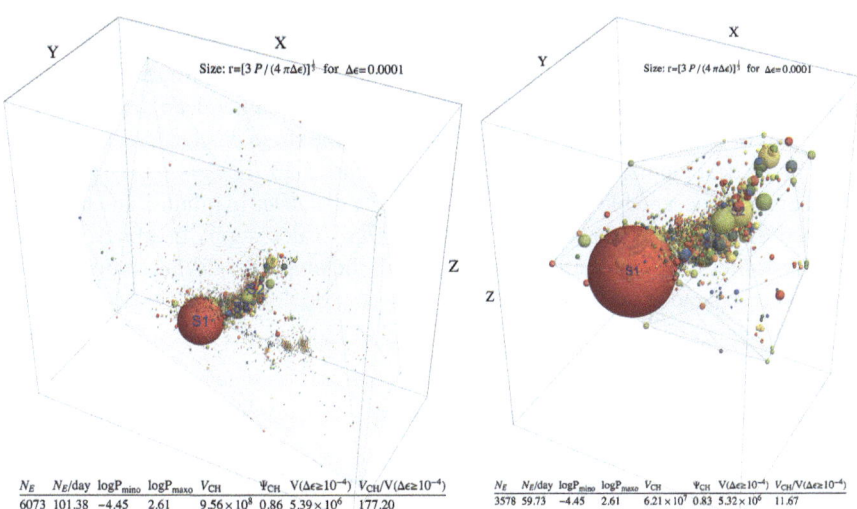

Fig. 6.44 Convex hull span over all available 6073 events *(left)* and over 3326 events that generated $PGV \geq 10^{-7}$ m/s at site S3 *(right)*. The three sites are shown as S1, S2, and S3 in blue

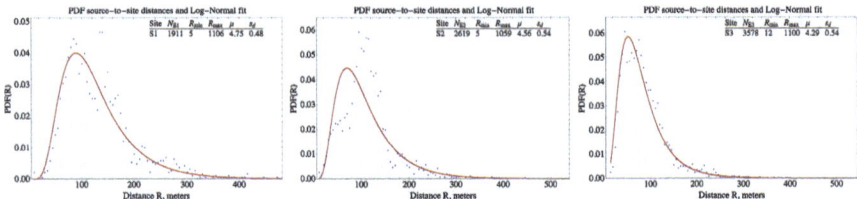

Fig. 6.45 The empirical probability density marked as blue dots and log-normal fit LGN (R) marked by red line for sites S1, S2, and S3

Table 6.6 Probabilities of exceedance of a given PGV and CAD over 90 and 360 days

Site	Pr $(PGV \geq$ 0.05, 90)	Pr $(PGV \geq$ 0.05, 360)	Pr $(CAD \geq$ 0.0005, 90)	Pr $(CAD \geq$ 0.0005, 360)
S1	0.032	0.1219	0.039	0.1471
S2	0.0851	0.2995	0.0665	0.2407
S3	0.128	0.4217	0.115	0.3866

The data set used to derive these distribution ranges from 08 May to 07 July, i.e. the last 60 days before the largest event, and includes 6073 events. The best log-normal distribution fit of distances is for site S3 with 3578 events that exceeded the strain threshold of 10^{-7}, and it has the mode at about 65 m and small standard deviation. The second best is for site S1 with 1106 events that has a mode at about 85 m, but the observed data also has a second mode at 130 m that cannot be reproduced by a log-normal distribution and consequently has larger standard deviation. The distribution for site S2 with 1059 events is the worst of the three, and it moved the observed mode at 100 m to almost 60 m, which would overestimate hazard. The distance between an event and a given site is taken as the Euclidean distance minus the radius of the event, $r = \sqrt[3]{3P/(4\pi \Delta \epsilon)}$, where the strain change, $\Delta \epsilon$, is the same used in the development of the SGMPE, in this case $6 \cdot 10^{-4}$.

There is no acceptable model for source to site distances in mining, and therefore we use the empirical distribution in GM hazard calculation. In mines, seismic activity follows rock extraction, and therefore, the empirical distances distribution should be updated at least every 3 months.

6.8.5 Probabilities and Hazard Maps

The basic outcome of ground motion hazard analysis for a given site is the probability of exceedance, i.e. the probability that there will be at least one event that will exceed a given GM parameter in time ΔT in the future. All plots below are based on the theoretical size distribution and empirical distances distributions.

Table 6.6 gives probabilities that there will be at least one event every 90 and 360 days with $PGV \geq 0.05$ m/s or $CAD \geq 0.0005$ m.

6.8 Mapping Seismic Ground Motion Hazard

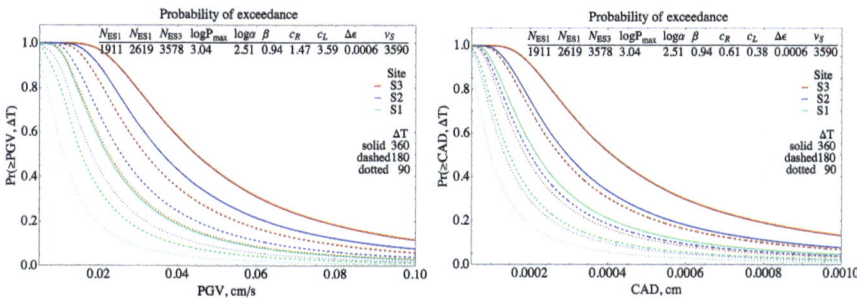

Fig. 6.46 Probabilities of observing at least one event greater than a given PGV *(left)* or CAD *(right)* within the period of 90, 180, and 360 days

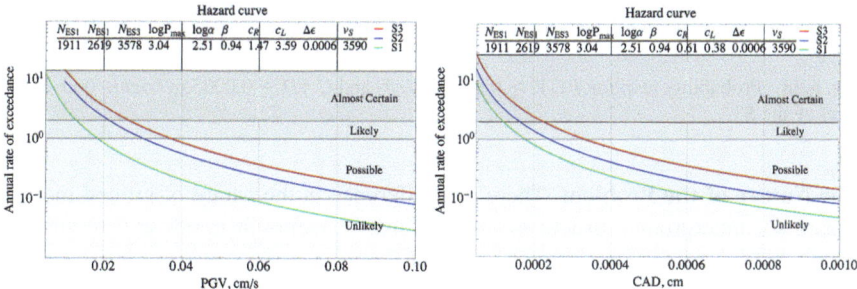

Fig. 6.47 Hazard curves for PGV *(left)* and CAD *(right)*

Figure 6.46 left shows the probabilities of exceedance of PGV over 90, 180, and 360 days for all three sites, and Fig. 6.46 right shows the same for CAD.

Ground motion hazard can also be presented as a seismic hazard curve that gives the annual rate at which a specific ground motion level may be exceeded. The hazard curve may be superimposed on the probability rating scheme, or likelihood scores, defined by the mine, see Fig. 6.47.

Figure 6.48 shows probability maps for $PGV \geq 0.05$ m/s and $CAD \geq 0.0005$ m calculated at the level of site S3.

6.8.6 Limitations

There is a wide spectrum of views in the literature on the utility of PSHA, from suggestions to drop it altogether (e.g. Mulargia et al., 2017) to the more pragmatic, stating that the shortcomings of the method do not invalidate the existence of the hazard curve, which comprises the basic assumption for PSHA (Anderson & Biasi, 2016).

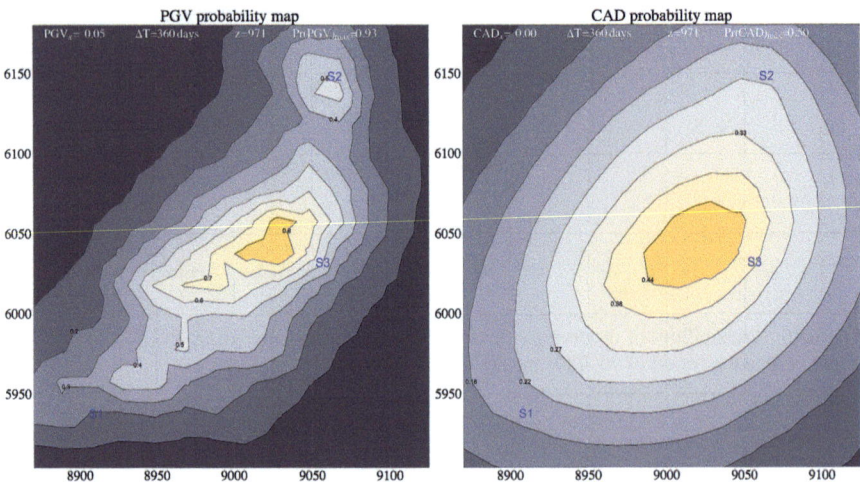

Fig. 6.48 Probability map for $PGV \geq 0.05$ m/s *(left)* and $CAD \geq 0.0005$ m *(right)*, both at the level of site S3

The Nature of the Problem There is a difference in the nature of ground motion hazard to underground structures due to seismic events in mines and to surface structures due to earthquakes. The main difference is the distances involved. As stated before, the near-source ground motion due to small and larger seismic events is similar, but the hypocentral distances to underground excavations in mines are very small, say from a few metres to a few hundred metres. Earthquakes are usually kilometres away. Therefore, earthquake engineers are mainly concerned with ground motions from larger earthquakes, but these earthquakes are less frequent and their activity rates are less certain.

Recurrence times for large events in mines are also uncertain, and therefore one can expect that the hazard maps for mines can better estimate the potential for less severe damage caused by smaller and medium size events than the infrequent large events.

This is why mines prone to larger events may wish to supplement probabilistic analysis with the deterministic one, i.e. ground motion simulation. Such a simulation involves kinematic modelling of ground motion produced by sources defined by their expected maximum potency, or magnitude, and placed at the most likely locations that can produce the strongest level of peak ground velocity at a given site or sites. For more details, see the next section in this book.

Probabilistic and deterministic methods for hazard assessment have advantages and disadvantages. Probabilistic methods can be viewed as inclusive of all deterministic events with a non-zero probability of occurrence. In this context, a proper deterministic method that models a particular larger event should ensure that that event is realistic, i.e. with a finite probability of occurrence. This points to the complementary nature of deterministic and probabilistic analyses: Deterministic events can be checked with a probabilistic analysis to ensure that the event is probable, and probabilistic analyses can be checked with deterministic events to see that rational, realistic hypotheses of concern have been included in the analyses (McGuire, 2001). However, the deterministic analysis can better account for specifics, i.e. the path and the site effects associated with a given strategic structure.

GMPE and Site Effect The reliability of a GMPE depends mainly on the selection of the appropriate data, and the main obstacle in selecting a good data set is the limited capability of the sensors employed to record stronger ground motion. This may limit the predicted level of ground motion at closer distances. It is important then for mines to deploy strong ground motion sensors. A more serious limitation though is the unknown amplification of ground motion at the skin of excavations.

Distribution of Distances There is no acceptable model for source to site distance distribution in mining, and therefore, in this example, we used the empirical distribution that reflects the recent past but does not extrapolate into the future. One option is to model the future spatial distribution of seismicity which for mines is very uncertain, or to update the analysis more frequently.

Size Distribution We assume that the magnitude distribution of future seismic events will be close to the one derived on the basis of existing data, i.e. that the predicted upper limit of the next record breaking potency, P_{max}, and parameters α and β of the assumed power law are reasonably accurate. This is not the most risky assumption since in most cases the future seismic rock mass response to mining is reasonably informed by its past size distribution. There are, of course, "black swan" events that considerably exceed the largest observed one, but they are not frequent enough to discourage forecasting. The experience of the author is that the bigger jumps occur more frequently when data follow the open-ended power law closer than the upper truncated one.

A Homogeneous Poisson Process In some cases, e.g. when rock extraction is interrupted, seismic activity in the time domain is far more clustered than in the volume mined domain. In such a case, it is recommended to perform the analysis in the volume mined domain. Alternatively, one can apply the non-stationary Poisson process with a suitable intensity function.

6.9 Modelling Ground Motion—Deterministic Hazard

6.9.1 Introduction

General Description Seismic waves in rock, propagating away from the source, interact with interfaces between different rock types, fractured zones, and underground excavations. In the process, they experience reflection, refraction, diffraction, scattering, and inelastic attenuation. Reflection, refraction, and diffraction are all manifestations of boundary behaviour of waves and are associated with some significant change of the wave vector leading to the bending of the wave path. Path bending can be observed only for waves in two or three dimensions. Reflection is a sudden change in the direction of wave propagation. It occurs when the wavefront reaches the interface between two different media. Refraction is caused by the change in speed when the wave travels through the interface between two different media. This change of wave speed leads to a change in the wavelength and in the direction of propagation. Diffraction is observed when a wave reaches some obstacle (an edge or an opening) of size co-measurable with the wavelength. It is a complex physical phenomenon involving not only bending of the wave path but also changes in the wave intensity due to interference. The said changes in the wave intensity can take the form of alternating minima and maxima when the peak value in a diffractive maximum is significantly larger than the intensity of the incident wave.

As an outgoing wave travels away from the source, its amplitude decreases due to geometrical spreading, attenuation, and scattering. Local effects include constructive and destructive interference of scattered waves, trapping in lower velocity layers and site effects. The combined effect from all these factors is that the wave field becomes more complex as the wavefront progresses.

The deterministic hazard, as it is discussed here, involves kinematic modelling of ground motion produced by specific seismic sources, i.e. sources, defined by their expected maximum potency or energy and placed at most likely locations that can produce the strongest level of peak ground velocity, PGV, or cumulative absolute displacement, CAD, at a given site. The likely locations of these sources may be inferred from the past data or, preferably, they can be determined by numerical modelling of the induced shear stresses on geological structures. In a kinematic model, the source process is defined by the spatial and temporal distribution of the slip vector, by the local slip velocity function, and by the rupture velocity, without taking into consideration the forces and stresses acting at the source.

There are three steps in the kinematic modelling process. Firstly, one needs to build a model of the mine including the geometry of any existing or planned excavations, geological structures, and rock mass properties such as rock density, wave velocities, and attenuation factors. Secondly, one needs to prescribe the location, the shape and the size of the source as well as the details of the slip rate that will generate seismic waves. The simplest example of a kinematic source model is one of zero size called a point source. A point source is defined by its location and the source time function, i.e. the local displacement time history. The seismic wave field created by a point source is relatively easy to reconstruct, but it can be compared with actual observations only at stations sufficiently far from the source. Since a true point source, that is one of zero volume, cannot exist, the modelling of the near and intermediate field effects requires the preparation of extended sources as input. The third element of the deterministic modelling of strong ground motion is the formulation of the initial and boundary conditions for the equations of motion.

Equations of Motion The equations of motion for elastic materials are the mathematical expression of Newton's second law: The rate of change in the momentum

6.9 Modelling Ground Motion—Deterministic Hazard

of a body is equal to the net applied force. The change of the momentum for the material in a small volume dV is $\rho \ddot{u} dV$, where $\ddot{u} = \partial^2 \mathbf{u}(x,t)/\partial t^2$ and $\mathbf{u}(\mathbf{x},t)$ is the displacement of the centre of mass. The net applied force comes from two contributions. One is due to interactions with the world outside the volume element. It is called body force. An example of body force is the weight or the force due to gravity. The other contribution to the net force applied on the volume element dV is a measure of the response of the surrounding medium to local deformations. The complete information for all response forces is contained in a single mathematical object called the stress tensor. For perfect elastic materials, the components of the stress tensor are linear functions of the components of the tensor of deformation which for small deformation is equal to the strain tensor. The forces due to the local deformation act on the surface dV, and their resultant is expressed as a combination of the spatial derivatives of the stress tensor $\nabla \cdot \sigma$. The components of the stress tensor are functions of the gradients of the displacement $\nabla \mathbf{u}$. These are the constitutive relations, and they can be either linear or nonlinear thus determining the overall properties of the equations of motion. The complete 3D equations of motion for a linear elastic solid are

$$\rho \ddot{\mathbf{u}} = \nabla \cdot \sigma + \mathbf{f} \quad \text{and} \quad \sigma_{ij} = c_{ijkl} \nabla_k \mathbf{u}_l, \tag{6.33}$$

where c_{ijkl} are constant coefficients. If the material is homogeneous, c_{ijkl} are symmetric: $c_{ijkl} = c_{klij} = c_{jikl} = c_{ijlk}$, which reduces the 81 components of the c_{ijkl} to 21 independent components. If the material is also isotropic, the number of independent components of c_{ijkl} is further reduced to just two Lame's parameters λ and μ. The equations of motion can now be written as the nine simultaneous equations: The first three are expressing Newton's second law

$$\begin{aligned}
\rho \partial_{tt} u_x &= \partial_x \sigma_{xx} + \partial_y \sigma_{xy} + \partial_z \sigma_{xz} + f_x \\
\rho \partial_{tt} u_y &= \partial_x \sigma_{xy} + \partial_y \sigma_{yy} + \partial_z \sigma_{yz} + f_y \\
\rho \partial_{tt} u_z &= \partial_x \sigma_{xz} + \partial_y \sigma_{yz} + \partial_z \sigma_{zz} + f_z
\end{aligned} \tag{6.34}$$

and the other six are the constitutive stress-strain relations

$$\begin{aligned}
\sigma_{xx} &= \lambda(\partial_y u_y + \partial_z u_z) + (\lambda + 2\mu) \partial_x u_x \\
\sigma_{yy} &= \lambda(\partial_x u_x + \partial_z u_z) + (\lambda + 2\mu) \partial_y u_y \\
\sigma_{zz} &= \lambda(\partial_x u_x + \partial_y u_y) + (\lambda + 2\mu) \partial_z u_z \\
\sigma_{xy} &= \mu \partial_y u_x \quad + \quad \mu \partial_x u_y \\
\sigma_{xz} &= \mu \partial_z u_x \quad + \quad \mu \partial_x u_z \\
\sigma_{yz} &= \mu \partial_z u_y \quad + \quad \mu \partial_y u_z.
\end{aligned} \tag{6.35}$$

In the above, ∂_{tt} is the operator of taking the double derivative with respect to time, $\partial_x, \partial_y, \partial_z$ are the operators of taking the first order spatial derivatives, u_x, u_y, u_z are the three components of the displacement vector, $\sigma_{xx}, \ldots, \sigma_{yz}$ are the nine components of the stress tensor, and f_x, f_y, f_z are the three components of the applied body force. The stress tensor is symmetric, so only six of its components are independent.

The spatial derivatives are a measure of the change in the function when moving from a given point to one of its neighbours. The governing differential equations of elastodynamics are the mathematical expression of the fact that a disturbance at one point in a continuum propagates by means of interactions between nearest neighbours. The result is a wave field which fills the volume of the elastic solid. The numerical methods for solving differential equations are simple in the case of a first order problems, that is, when only first order derivatives are present. The equations of elastodynamics are of second order, but they can easily be transformed into a system of first order differential equations by introducing the velocities $v_j(x, y, z, t) = \partial u_j(x, y, z, t)/\partial t$ as new unknown functions and differentiating the constitutive equations with respect to time,

$$\begin{aligned} \rho \partial_t v_x &= \partial_x \sigma_{xx} + \partial_y \sigma_{xy} + \partial_z \sigma_{xz} + f_x \\ \rho \partial_t v_y &= \partial_x \sigma_{xy} + \partial_y \sigma_{yy} + \partial_z \sigma_{yz} + f_y \\ \rho \partial_t v_z &= \partial_x \sigma_{xz} + \partial_y \sigma_{yz} + \partial_z \sigma_{zz} + f_z \end{aligned} \quad (6.36)$$

and

$$\begin{aligned} \partial_t \sigma_{xx} &= (\lambda + 2\mu) \partial_x v_x + \lambda \left(\partial_y v_y + \partial_z v_z \right) \\ \partial_t \sigma_{yy} &= (\lambda + 2\mu) \partial_y v_y + \lambda \left(\partial_x v_x + \partial_z v_z \right) \\ \partial_t \sigma_{zz} &= (\lambda + 2\mu) \partial_z v_z + \lambda \left(\partial_x v_x + \partial_y v_y \right) \\ \partial_t \sigma_{xy} &= \mu \partial_y v_x \quad + \quad \mu \partial_x v_y \\ \partial_t \sigma_{xz} &= \mu \partial_z v_x \quad + \quad \mu \partial_x v_z \\ \partial_t \sigma_{yz} &= \mu \partial_z v_y \quad + \quad \mu \partial_y v_z. \end{aligned} \quad (6.37)$$

These equations are meaningful only for a particular domain in space and in time, and therefore, the components of the source function, f_x, f_y, and f_z, need to be defined within the same domains. All differential equations have infinitely many solutions, and one needs to impose additional conditions on the components of the velocity and the stress to ensure the uniqueness of the elastic wave radiated by a given source.

Space-Time Grid The space-time grid is a set of discrete points separated in space by Δx, Δy, and Δz and in time by Δt. In most applications, the grid is regular: $\Delta x = \Delta y = \Delta z = h$. In a simple grid, all functions are approximated at the same grid points. In a partly staggered grid, displacement or particle velocity components are located at one set of grid points, whereas the stress tensor components are assigned to another set. A staggered grid of step h is just two regular grids of the same step but shifted by $h/2$ relative to each other in the three spatial directions. This procedure puts extra nodes at the middle of the ribs and the faces of the h-cubic cells and has the advantage that, when the components of the velocities and the stress are allocated to particular nodes, the spatial derivatives are approximated by central differences

6.9 Modelling Ground Motion—Deterministic Hazard

of step $h/2$, hence more accurately. The staggered distribution of quantities in space is related through the equations of motion to the staggered distribution of quantities in time.

The partial derivatives in the equations of motion need to be replaced by approximations which are related to the definition of derivatives, e.g. the first order partial derivative with respect to x for the function $u_z(x, y, z, t)$ at the point (x_0, y_0, z_0, t_0) can be approximated by the central difference

$$\partial_x u_z(x_0, y_0, z_0, t_0) \simeq [u_z(x_0 + h, y_0, z_0, t_0) - u_z(x_0 - h, y_0, z_0, t_0)] / (2h),$$

and for the time derivative by the forward difference at $(\mathbf{x}_0, t_0) = (x_0, y_0, z_0, t_0)$,

$$\partial_t \sigma_{xz}(\mathbf{x}_0, t_0) \simeq [\sigma_{xz}(\mathbf{x}_0, t_0 + \Delta t) - \sigma_{xz}(\mathbf{x}_0, t_0)] / \Delta t.$$

One and the same partial derivative of a given function can be approximated by infinitely many different finite difference expressions which can be obtained by combining different Taylor expansions of the same function. A finite system of equations for the desired approximation can be obtained from the truncated Taylor expansions. The lowest order term which was neglected in the truncation procedure defines the order of accuracy for the particular finite difference approximation. There are two meanings of "order" when the term is used to describe a finite differences approximation: One can approximate a derivative of a certain order, for instance, all derivatives in the nine differential equations for the velocity and the stress are of first order, and then one can describe a particular finite differences approximation as being of a given order of accuracy meaning the lowest order term in the truncated Taylor series which was neglected.

It can be shown that the forward difference is a first order approximation because the truncation error is proportional to Δt, and the central difference is a second order approximation because the truncation error is proportional to h^2. One example of a fourth order approximation formula is

$$\partial_x u_z(x_0) \simeq \frac{1}{h} \left\{ -\frac{1}{24} \left[u_z\left(x_0 + \frac{3}{2}h\right) - u_z\left(x_0 - \frac{3}{2}h\right) \right] \right.$$
$$\left. + \frac{9}{8} \left[u_z\left(x_0 + \frac{1}{2}h\right) - u_z\left(x_0 - \frac{1}{2}h\right) \right] \right\}.$$

The choice of h and Δt depends on the properties of the material and of the source time function. A good rule is $h = v_S/(k f_{max})$, where v_S is the velocity of the slower wave, in this case, S-wave f_{max} is the maximum frequency to be resolved and k represents the number of evaluations per wavelength, usually $k = 6$. The time step Δt depends on the shortest time taken by the P-wave to cross the distance h between two neighbouring nodes, and it is limited by the stability condition which, for the staggered grid, reads: $\Delta t < 0.5 h/v_P$. The finite difference representation of the partial derivatives applies only to points of the interior for which all nearest neighbours are within the model and for moments of time for which the time derivative can be approximated. This excludes the beginning, i.e. $t = t_0$.

The numerical solution of the dynamical equations of motion on a staggered grid is obtained by first imposing the initial conditions, that is, by giving the initial values of the velocity and the stress components at the respective grid nodes, and then solving for the nodal values at the next moment of time, evaluating the finite difference expressions on the right hand sides of the equations. In this way, the numerical solution is woven like a 3D carpet one time step at a time. There are two types of nodal values which require special attention: the nodes on the boundaries and the nodes on the seismic source. The boundaries and respectively the boundary nodes can be of two types: free surfaces and virtual boundaries. Free surfaces are characterised by zero normal stress, and the corresponding nodal boundary values must reflect this fact. Virtual surfaces are not related to any existing feature in the model domain. Their role is just to separate a finite volume out of the much bigger material world so that the numerical model would be of a reasonable size and could be solved. It is very difficult to impose boundary conditions on a virtual boundary because it must stay "invisible" for every incidence of waves on it. The nodes on the source are similar to the boundary nodes in the sense that the values of the physical fields at these points are not obtained by solving the equations but are taken from tabulated time functions. This is how a source point works: At the beginning, the source function is zero, and the source node behaves just like any other node in the model. That is, it obtains a new value after each iteration according to the solution of the finite differences equations. But when the time stepping procedure reaches the moment when the source function becomes non-zero, at that time step the velocity component in the source node takes its value from the source function. This value is not what the solution of the model would have given, and the difference acts as a perturbation of the wave field. In other words, it becomes a point source with a short, pulse-like velocity function. The time sequence of such pulses constitutes a point-like source which, together with others of the same type, represents the extended kinematic source model.

6.9.2 *Implementation and Examples*

This section is based on Mendecki and Lötter (2011). A numerical solution of the 3D wave equation can be achieved by finite difference modelling. In principle, the finite difference method does not have restrictions on the type of constitutive equation, boundary conditions, curved interfaces, and different source types and allows general material variability. The implementations of finite difference schemes for solving non-linear and non-isotropic problems, though, can be complicated and of little practical value.

As a first step, partial derivatives in Equation 6.33 are replaced by numerical first, second, or fourth order estimates obtained from Taylor expansions truncated after the desired number of terms. The decision on where on a finite difference

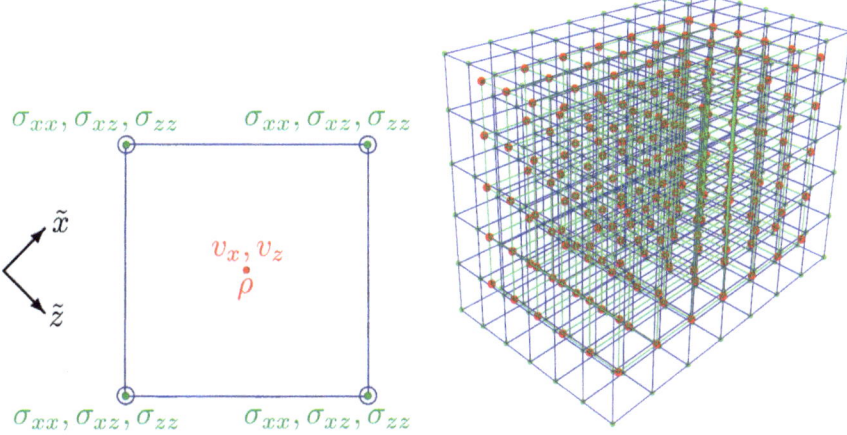

Fig. 6.49 Elementary 2D cell for velocities and stresses in a rotated staggered grid *(left)*, and its 3D view *(right)*

grid to place spatially dependent material properties, ρ, λ, and μ, and physical quantities, v_i, σ_{ij}, constitutes the choice between various staggered grid schemes. A fourth order in space scheme is described in Graves (1996). However, a staggered grid causes instability problems when the medium possesses high contrast discontinuities. Saenger et al. (2000) and Saenger and Bohlen (2004) proposed a new rotated staggered grid where all medium parameters are assigned to the centre of each elementary cell (Fig. 6.49). This modified finite differences scheme provides a more accurate way of including in the model underground excavations and interfaces between different materials. To reduce the influence of nonphysical wave reflections of any virtual boundaries, it is important to apply either absorbing boundary conditions (e.g. Clayton & Engquist, 1977) or an attenuating shell around the model that quenches the waves and minimises reflection.

The wave field from an extended source is modelled as the superposition of the wave fields from multiple point sources in a finite difference scheme. The seismic potency of a finite source is the sum of the potencies of the constituent point sources, and it is expected that these point sources will have similar principal axes but may vary in magnitude, as not all parts of the source will experience the same net displacement. Figure 6.50 demonstrates such a simple scheme with colours proportional to the final displacement experienced by each point. At the edge of this elliptical source, displacement tends towards zero (blue), while maximum displacement occurs at an off-centre point on the principal axis of the source ellipse.

When building a kinematic model of an extended source, the consistency of the set of source parameters and time histories for each point of the fracturing fault is very important. Two features of seismic sources that enforce some constraints on the possible distributions of these parameters are self-similarity of the final slip distribution and the expected characteristics of the high frequency part of the

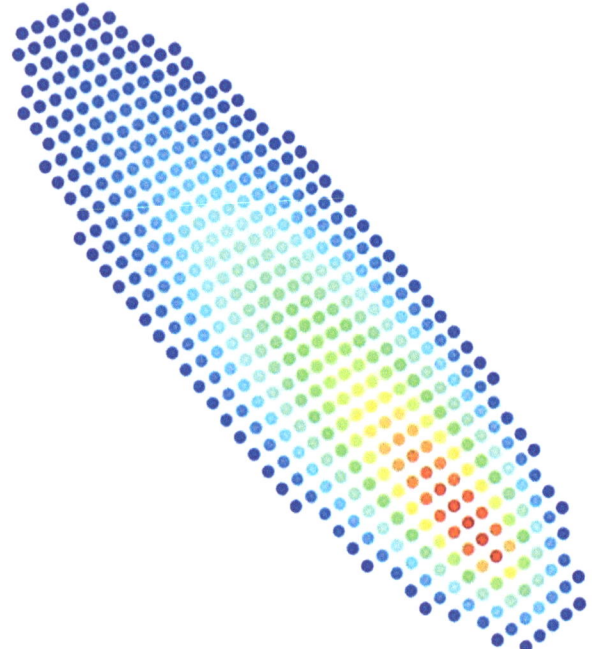

Fig. 6.50 An extended source built from a distribution of point sources

displacement spectrum as seen in the far field. These consistency constraints ensure smooth and monotone slip histories for all parts of the source, while remaining physically admissible.

To provide explicit constraints on the parameters controlling the rupture, the kinematic k-square earthquake source model of Herrero and Bernard (1994) can be used. The idea of the k-squared model and its generalisations is to prescribe the Fourier transform of the displacement in the 2D wavenumber space and then to obtain the spatial distribution in real space by performing the inverse Fourier transformation. One implementation of the k-square model proposes a final displacement at (x, y) on the fault as

$$u(x, y) = \int \int D(k_x, k_y) e^{i(xk_x + yk_y)} dk_x dk_y, \qquad (6.38)$$

where $D(k_x, k_y) = \exp[i\Phi(k_x, k_y)] / \sqrt{1 + \left[(k_x/k_c)^2 + (k_y/k_c)^2\right]^2}$, with the function $\Phi(k_x, k_y)$ in the Fourier transform being a random phase and $k_c \sim$ 1/(source size), is the corner wavenumber which, like the corner frequency in the Brune model, demarkates the low-wavenumber asymptote from the high-wavenumber behaviour in the Fourier transform of the slip distribution. Fourier transform decays as an inverse square. An extended source can then be constructed by distributing its total seismic potency over the sub-sources proportional to the modelled permanent displacement at their positions. For sub-source (i, j) at point (x_i, y_j) on the fault, we thus obtain $P_0^{ij} = u(x_i, y_j) \Delta A_{ij} = u_{ij} h^2$. Similarly, rise times over different parts of the source can be chosen proportionally to the final slip, $T_i = T^{max} u(\xi_i, u_{max})$, where T^{max} is a chosen rise time corresponding to the sub-source with maximum final displacement.

The modelled spatial distribution of the final displacement on the source cannot ensure on its own the correct high frequency behaviour of the far-field displacement spectra. A method of specifying the slip velocity time functions for the sub-sources needs to be added to the kinematic source model. Beresnev and Atkinson (1997) have given an example of a slip velocity time function which leads to a far-field displacement spectrum adhering to ω^{-2} decay. The parametrisation of their source time function $v(t; \tau, \zeta) = u^\infty \left(\frac{2}{\tau}\right)^2 t \cdot \exp(-2t/\tau)$ will reproduce the rise time for the sub-sources T_{ij} when $\tau = \frac{1}{2} T_{ij}$ with $T_{ij} \leq T^{max}$.

Every sub-source in an extended kinematic source model needs to be assigned its own initiation time t_0^{ij}. In solving a finite difference model, time is measured in time steps Δt starting from zero. The velocity time function of sub-source (i, j) is zero before the initiation time t_0^{ij} after which it follows the respective velocity time function, for instance the Bereznev expression. The choice of initiation times t_{ij}^0 for the sub-sources is equivalent to specifying the propagation of the rupture front over the fault. It is an important element of seismic source modelling because the spatial distribution of the initiation times is equivalent to the complete rupture scenario. At this stage, one can model acceleration of rupture, sub- or super-shear rupture velocity, and the stopping phase. For instance, one can make the simple assumption that rupture speed is faster when parallel to slip and slower when orthogonal to slip. In this case, one obtains an extended source in the shape of an ellipse, with eccentricity determined by the ratio of these two orthogonal rupture velocities. This is the scenario illustrated by Fig. 6.50.

Alternatively, one can choose a spatial distribution of the rupture velocity v_{ij}^{rupt} and compute the initiation time t_0^{ij} for a sub-source as $\sum (h/v_{lm}^{rupt})$ where the sum is taken over the fault nodes on the shortest path from the hypocentre to sub-source (i, j). Now, the ground motion experienced at a site is controlled by the maximum velocity of deformation at the seismic source, i.e. slip velocity, by the interaction of radiation from different sub-sources from different travel paths and by site effects.

6.9.2.1 Example 1: Extended Sources in Heterogeneous Media

Displacement on a fault originates at the focus, i.e. at the hypocentre, and propagates towards its edges, in a manner resembling the extended source described above. As the focus is not necessarily in the centre of the fault, much of the propagation is unidirectional.

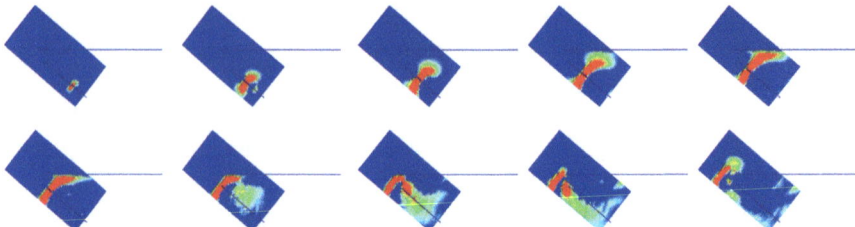

Fig. 6.51 Snapshots in section of time steps in a sub-shear rupture process, chronologically from left to right, top to bottom

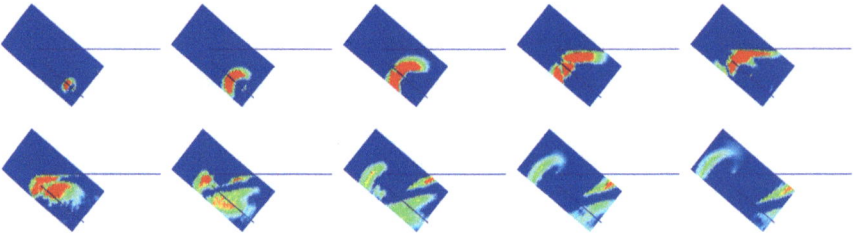

Fig. 6.52 Snapshots in section of time steps in a super-shear rupture process, chronologically from left to right, top to bottom

Traditionally, it has been assumed that the upper bound for rupture propagation velocity is the Rayleigh wave velocity (Broberg, 1996). However, field observation has shown several examples where the rupture velocity exceeds shear wave velocity, $v_r > v_S$, Dunham and Archuleta (2004).

To examine the extreme ground motions that can be caused at points placed close to the source in the direction of rupture propagation, we modelled two similar extended sources based on the same k-square slip distribution, but with different rupture scenarios.

In the first case, we took a sub-shear rupture speed $v_r = 0.9v_S$, while in the second case we let the rupture propagate with super-shear velocity $v_r = \sqrt{2}v_S$. Also present in this model was a tabular stope that caused reflections and partly obstructed the waves from directly traveling to the upper part of the rock mass. The displacement follows the rupture front from the focus towards its edges. As the focus is not necessarily in the centre of the fault, much of the propagation is unidirectional. From an inspection of the wave field in Figs. 6.51 and 6.52, for which corresponding frames refer to the same points in time, it can be seen that rupture progresses faster in the super-shear case and that a Mach cone evolves (see Fig. 6.52 frame 4).

6.9.2.2 Example 2: Complex Sources in Heterogeneous Media

We model a complex seismic source conceptualised by David Ortlepp and described in Ortlepp (1984) and Ortlepp (1997) page 63 caption (e), see Fig. 6.53 left. It is a

Fig. 6.53 Strike section through a stope showing a double source mechanism conceptualised by Ortlepp *(first)*. Rupture history of the Ortlepp source, with initiation time per point obtained from radial distance from the focus *(second)*. Slip history of the Ortlepp source, with final displacements determined by a k-square distribution *(third)*. Synthetic sensor placement around the fault-stope corner *(forth)*

system of two rectangular extended faults. A kinematic source model was created for each of the faults following the methodology outlined in the previous section. Rupture initiates at the fault which is further away from the stope. The wave radiated by this fault triggers rupture on the second fault, the one that interacts with the stope. Here we assume that the second rupture is induced by the first one almost immediately.

Due to the available freedom in placing faults and sub-sources in a finite difference grid, we are now able to independently choose both the orientations and the velocity of rupture, v_r, on these faults. In particular, we can chose for the initiating fault a sub-shear rupture velocity ($v_r < v_S$) and a super-shear rupture velocity ($v_r > v_S$) for the second fault. The rupture and slip histories of such a scenario are illustrated in Fig. 6.53.

One can see that rupture and slip do not need to be in the same direction at all source points. For our model, we introduce a fault with a dip of 60°, touching a horizontal stope. Some points of interest are marked around it. It is at these points where we record the modelled ground motion. In particular, as the rupture of the fault starts below and progresses upwards, we are especially interested in comparing ground motions of the footwall and the hanging wall.

After performing a sufficient number of iterations with the described kinematic source model, we investigated the recorded velocity seismograms at sensors 4, 7, and 9 (Fig. 6.54). It was found that particle velocities close to the fault exceed 10 m/s as one can see on the synthetics recorded at sensor 9. Both, sensors 4 and 7, which are opposite to each other on the footwall and hanging wall of the fault, experience about 2 m/s velocities, although a higher frequency content is observed at sensor 7 due to the interaction of the stope with the wave field.

Sensors 12 and 15 are placed opposite to each other on the hanging wall—sensor 12, and the footwall—sensor 15, of the stope. The corresponding synthetic velocity seismograms are shown in Fig. 6.55.

Clearly the footwall sensor records ground motion earlier than the hanging wall sensor. This is not so much due to the shorter straight line distance to source, but because of the longer path the elastic waves need to travel around the stope

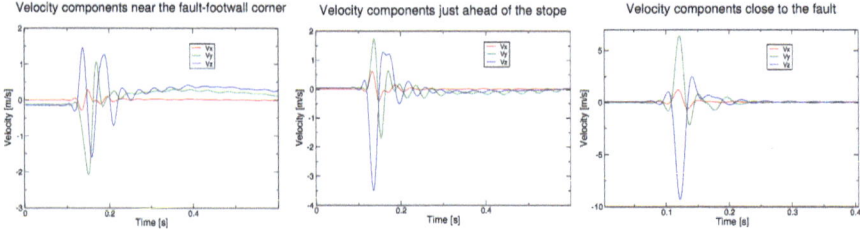

Fig. 6.54 Synthetic seismograms recorded around the fault. Sensor 4: Velocities near fault and footwall corner *(left)*, Sensor 7: Velocities just ahead of the stope *(centre)*. Sensor 9: Velocities close to the fault *(right)*

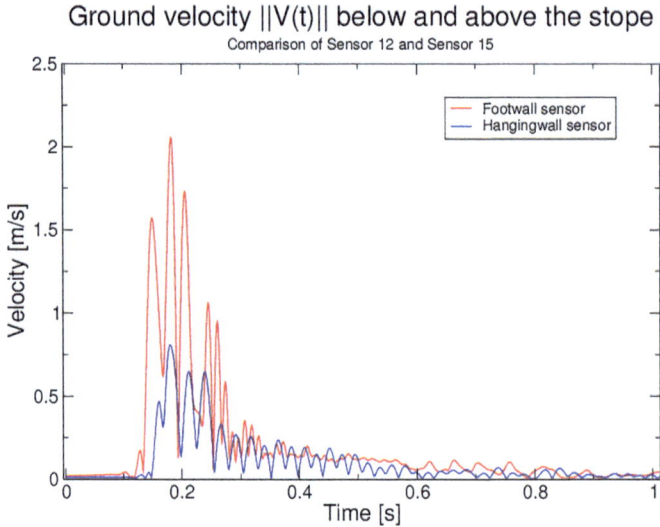

Fig. 6.55 Synthetic seismograms (as absolute velocity) comparing the hanging wall and footwall ground motions of sensors 12 and 15

which is modelled as a reflector. Evidently, the synthetic velocity seismograms recorded at sensor 12 have significantly lower amplitudes. This difference in the recorded velocity amplitudes from two stations so close to each other would not have been seen if the effect from the underground excavation was not a property of the numerical model.

Figure 6.56 shows successive snapshots of the seismic wave field from a two-fault Ortlepp source. In the first few frames, the radiation from the low-potency initial fault is visible, but upon initiation of the second, high-potency, fault, its contribution to the wave field becomes dominant. When the rupture on the second fault reaches the stope, interaction, reflection, and constructive interference lead to high ground motions.

6.9 Modelling Ground Motion—Deterministic Hazard

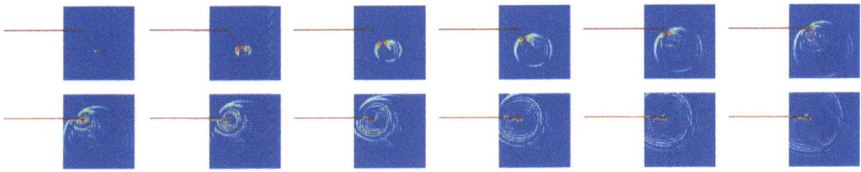

Fig. 6.56 Snapshots of a 2D section of the 3D wave field induced by the synthetic Ortlepp shear event

6.9.2.3 Example 3: Surface Ground Motions Induced by Mining Events

In this model, the event is 1.5 km below surface, at an epicentral distance of 5 km from a major city centre. The event is modelled as a rupture of an elliptical reverse fault with axes of 300 m and 100 m. The fault strikes at 0°, dips at 45°, and slips at a rake of 90°. The maximum slip velocity is chosen to be 2 m/s, and the average displacement over the whole source is set as $\bar{u} = 0.2$ m. We assumed $\log P = 3.5$ and a predominant frequency of 3 Hz.

The model domain is filled with a mostly homogeneous hard rock. There is a soil layer three-grid spacings thick just under the free surface (the grid is of spacing h =25m). For the hard rock, we choose $\rho = 2700$ kg/m³, $v_P = 5500$ m/s, and $v_S = 3500$ m/s, whereas for the 50 m soil layer, which is constructed over the top three-grid points in the rock below the air in the finite difference grid, we choose $\rho = 2000$ kg/m³, $v_P = 4000$ m/s, and $v_S = 2000$ m/s. To model the free surface effect and air above it, we choose $\rho = 2000$ kg/m³, $v_P = 300$ m/s, and $v_S = 0$ m/s with high density to avoid stability problems during the kinematic modelling phase on the finite difference grid.

We have computed waveforms in three points of interest indicated at Fig. 6.57. Forward modelling allows us to track velocities and dynamic stresses of the wave field not only at the points of interest, but in the entire model domain. Thus we can visualise the wave field on multiple planes of interest in subsequent time snapshots. This representation is shown in Fig. 6.58. With reference to Fig. 6.57, we have recorded synthetic seismograms at sensors 1, 2, and 3, and these are shown in Fig. 6.59.

We show only the vertical component (blue) and the horizontal component (green) because our source is symmetric relative to the plane of the sensors. The stronger horizontal ground motions at sensor 1 (Fig. 6.59 left) which are actually further from the hypocentre are expected to have a more damaging effect.

The predominant frequency at sensor 1 was calculated by computing a power spectrum over the two non-trivial components of the observed waveform. The predominant frequency of the ripple recorded by sensor 2 at about $t = 2$s is 10 Hz (estimated from the average over eight consecutive full periods).

While potentially a numerical effect, this can be compared to the expected horizontal S-wave resonance (20 Hz) and the expected vertical S-wave resonance (10 Hz). Both horizontal and vertical ground motions exceeding 15 mm/s are

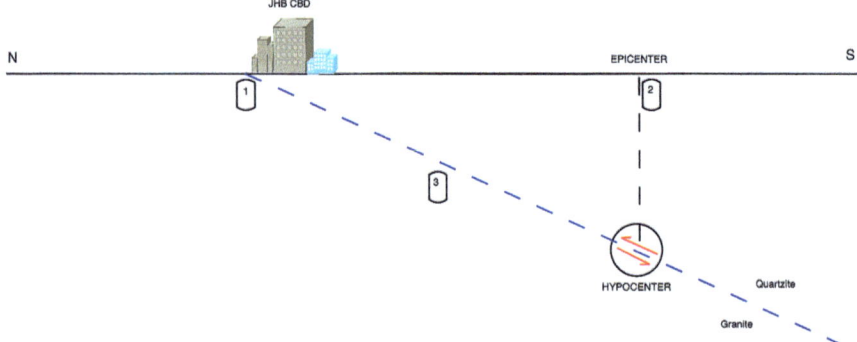

Fig. 6.57 Points of interest at which seismograms are recorded. Sensor 1 is just below the urban area of interest, sensor 2 at epicentre, and sensor 3 is halfway between the hypocentre and urban city centre

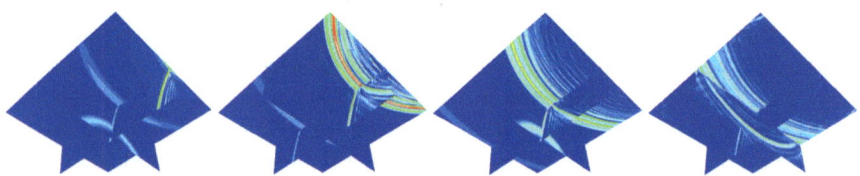

Fig. 6.58 Snapshots of velocity fields during subsequent stages of the kinematic model run of an underground event and observed surface ground motions. The intersection of the two vertical sections represents our point of interest, the area below the urban area where damage could potentially occur

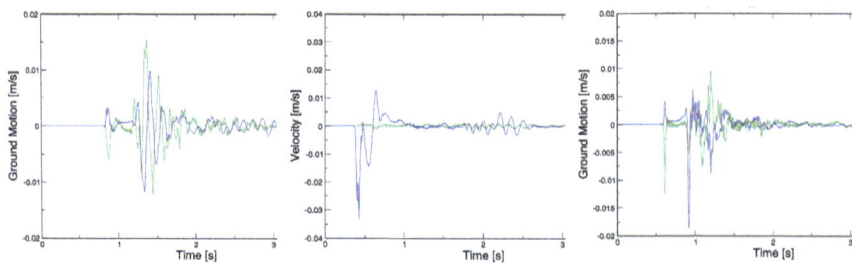

Fig. 6.59 Synthetic seismograms at sensors 1 *(left)*, 2 *(centre)*, and 3 *(right)*

observed on the surface and in the area of interest. The predominant frequency of this ground motion is 5.8 Hz. At sensor 2, a significant vertical particle velocity is recorded. After integrating to displacement, this leads to an upward permanent displacement of the surface equal to about 1 mm. It is to be expected of reverse faulting but unlikely to be particularly damaging to surface structures. At sensor

3, the initial arrivals and their immediate coda are clear of reflected waves, as this synthetic recording is underground, halfway between the hypocentre and the surface.

References

Abrahamson, N. A., & Youngs, R. R. (1992). A stable algorithm for regression analyses using the random effects model. *Bulletin of the Seismological Society of America, 82*(1), 505–510.

Anderson, J. G., & Biasi, G. P. (2016). What is the basic assumption for probabilistic seismic hazard assessment? *Seismological Research Letters, 87*(2A), 323–326. http://doi.org/10.1785/0220150232.

Andrews, D. J. (1976). Rupture velocity of plane strain shear cracks. *Journal of Geophysical Research, 81*(32), 5679–5687.

Andrews, D. J., Hanks, T. C., & Whitney, J. W. (2007). Physical limits on ground motion at Yucca Mountain. *Bulletin of the Seismological Society of America, 97*(6), 1771–1792.

Archuleta, R. J. (1984). A faulting model for the 1979 Imperial Valley earthquake. *Journal of Geophysical Research, 89*(B6), 4559–4585.

Archuleta, R. J., & Hartzell, S. H. (1981). Effects of fault finiteness on near-source ground motion. *Bulletin of the Seismological Society of America, 71*(4), 939–957.

Arias, A. (1970). A measure of earthquake intensity. In R. J. Hansen (Ed.), *Seismic Design for Nuclear Power Plants* (pp. 438–483). MIT Press.

Atkinson, G. M. (2015). Ground motion prediction equation for small to moderate events at short hypocentral distances, with application to induced seismicity hazards. *Bulletin of the Seismological Society of America, 105*(2A), 981–992. http://doi.org/10.1785/0120140142.

Baker, J. W., Bradley, B. A., & Stafford, P. J. (2021). *Seismic hazard and risk analysis*. Cambridge University Press.

Beresnev, I., & Atkinson, G. M. (1997). Modeling finite-fault radiation from the omega-n spectrum. *Bulletin of the Seismological Society of America, 87*(1), 67–84.

Broberg, K. (1996). How fast can a crack go? *Materials Science, 32*, 80–86. http://doi.org/10.1007/BF02538928.

Brune, J. N. (1970). Tectonic stress and the spectra of seismic shear waves from earthquakes. *Journal of Geophysical Research, 75*(26), 4997–5009.

Burridge, R. (1969). The numerical solution of certain integral equations with non-integrable kernels arising in the theory of crack propagation and elastic wave diffraction. *Philosophical Transactions of the Royal Society of London, A 265*, 353–381.

Burridge, R. (1973). Admissible speeds for plane-strain self-similar shear cracks with friction but lacking cohesion. *Geophysical Journal of the Royal Astronomical Society, 35*(4), 439–455.

Campbell, K. W. (1981). Near-source attenuation of peak horizontal acceleration. *Bulletin of the Seismological Society of America, 71*(6), 2039–2070.

Cichowicz, A. (2008). Near-field pulse-type motion of small events in deep gold mines: Observations, response spectra and drift spectra. In *14th World Conference on Earthquake Engineering, Beijing, China*.

Clayton, R., & Engquist, B. (1977). Absorbing boundary conditions for acoustic and elastic wave equations. *Bulletin of the Seismological Society of America, 67*(6), 1529–1540.

Cornell, C. A. (1968). Engineering seismic risk analysis. *Bulletin of the Seismological Society of America, 58*(5), 1583–1606.

Dineva, S., Mihaylov, D., Hansen-Haug, J., Nystrom, A., & Woldemedhin, B. (2016). Local seismic systems for study of the effect of seismic waves on rock mass and ground support in Swedish underground mines Zinkgruvan, Garpenberg, Kiruna). In *Ground Support 2016, Lulea, Sweden* (pp. 1–11).

Douglas, J. (2018). Ground motion prediction equations 1964–2018. *Review*. University of Strathclyde, Glasgow.
Dunham, E. M. (2007). Conditions governing the occurrence of supershear ruptures under slip-weakening friction. *Journal of Geophysical Research*, *112*(B07302), 1–24. http://doi.org/10.1029/2006JB004717.
Dunham, E. M., & Archuleta, R. J. (2004). Evidence for a supershear transient during the 2002 Denali fault earthquake. *Bulletin of the Seismological Society of America*, *94*(6B), S256–S26.
Durrheim, R. J., Spottiswoode, S. M., Roberts, M. K. C., & Brink, A. v. Z. (2005). Comparative seismology of the Witwatersrand basin and Bushveld Complex and emerging technologies to manage the risk of rockbursting. *The Journal of The South African Institute of Mining and Metallurgy*, *105*, 409–416.
EPRI (1988). A criterion for determining exceedance of the operating basis earthquake. *Tech. Rep. EPRI NP-5930*, Electrical Power Research Institute, Palo Alto, California.
Eshelby, J. D. (1957). The determination of the elastic field of an ellipsoidal inclusion and related problems. *Proceedings of the Royal Society of London, Series A, Mathematical and Physical Sciences*, *241*(1226), 376–396.
Esteva, L. (1970). Seismic risk and seismic design decisions. In R. J. Hansen (Ed.), *Seismic risk and seismic design criteria for nuclear power plants* (pp. 142–182). MIT Press.
Gay, N. C., & Ortlepp, W. D. (1978). Anatomy of a mining-induced fault zone. *Geological Society of America Bulletin*, *90*, 47–58.
Gospodinov, D., Dineva, S., & Dahner-Lindkvist, C. (2022). On the applicability of the RETAS model for forecasting aftershock probability in underground mines, Kiirunavaara Mine, Sweden. *Journal of Mine Seismology, Springer*, *4*(26). http://doi.org/10.1007/s10950-022-10108-6.
Graves, R. W. (1996). Simulating seismic wave propagation in 3D elastic media using staggered grid finite differences. *Bulletin of the Seismological Society of America*, *86*(4), 1091–1106.
Handley, M. F., de Lange, J. A. J., Essrich, F., & Banning, J. A. (2000). A review of the sequential grid mining method employed at Elandsrand Gold Mine. *The Journal of The Southern African Institute of Mining and Metallurgy*, *100*(3), 157–168.
Herrero, A., & Bernard, P. (1994). A kinematic self-similar rupture process for earthquakes. *Bulletin of the Seismological Society of America*, *84*(4), 1216–1228.
Ida, Y. (1973). The maximum acceleration of seismic ground motion. *Bulletin of the Seismological Society of America*, *63*(3), 959–968.
Joyner, W. B., & Boore, D. M. (1993). Methods for regression analysis of strong motion data. *Bulletin of the Seismological Society of America*, *83*(2), 469–487.
Joyner, W. B., & Boore, D. M. (1994). Methods for regression analysis of strong motion data - Errata. *Bulletin of the Seismological Society of America*, *84*(3), 955–956.
Kaiser, P. K., & Maloney, S. M. (1997). Scaling laws for the design of rock support. *Pure and Applied Geophysics*, *150*(3-4), 415–434.
Kanamori, H. (1972). Determination of effective tectonic stress associated with earthquake faulting. The Tottori earthquake of 1943. *Physics of the Earth and Planetary Interiors*, *5*, 426–434.
Kirsch, E. G. (1898). Die theorie der elastizitat und die bedurfnisse der festigkeitslehre. *Zeitschrift des Vereines deutscher Ingenieure*, *42*, 797–807.
Kolsky, H. (1963). *Stress waves in solids* (212 p.). Dover Publications.
Konno, K., & Ohmachi, T. (1998). Ground motion characteristics estimated from spectral ratio between horizontal and vertical components of microtremors. *Bulletin of the Seismological Society of America*, *88*(1), 228–241.
Love, A. E. H. (1911). *Some problems of geodynamics*. Cambridge University Press.
Lu, X., Rosakis, A. J., & Lapusta, N. (2010). Rupture modes in laboratory earthquakes: Effect of fault prestress and nucleation conditions. *Journal of Geophysical Research*, *115*(B12302), 1–25. http://doi.org/10.1029/2009JB006833.

References

Malek, F., & Leslie, I. (2006). Using seismic data for rockburst re-entry protocol at Inco's Copper Cliff North Mine. In *41st U.S. Symposium on Rock Mechanics, Golden, Colorado*, ARMA-06-1163.

McGarr, A. (1997). A mechanism for high-wall rock velocities in rockbursts. *Pure and Applied Geophysics*, *150*(3-4), 381–391.

McGarr, A., & Fletcher, J. B. (2001). A method for mapping apparent stress and energy radiation applied to the 1994 Northridge Earthquake Fault Zone-Revisited. *Geophysical Research Letters*, *28*(18), 3529–3532. http://doi.org/10.1029/2001GL013094.

McGarr, A., & Fletcher, J. B. (2003). Maximum slip in earthquake fault zones, apparent stress, and stick-slip friction. *Bulletin of the Seismological Society of America*, *93*(6), 2355–2362.

McGarr, A., & Fletcher, J. B. (2005). Development of ground-motion prediction equations relevant to shallow mining induced seismicity in the Trail Mountain area. *Bulletin of the Seismological Society of America*, *95*(1), 31–47. http://doi.org/10.1785/0120040046.

McGarr, A., Green, R. W. E., & Spottiswoode, S. M. (1981). Strong ground motion of mine tremors: Some implications for near-source ground motion parameters. *Bulletin of the Seismological Society of America*, *71*(1), 295–319.

McGuire, R. K. (2001). Deterministic vs probabilistic earthquake hazards risks. *Soil Dynamics and Earthquake Engineering*, *21*, 377–384.

McGuire, R. K. (2004). *Seismic hazard and risk analysis*. Earthquake Engineering Research Institute.

Mendecki, A. J. (2001). Data-driven understanding of seismic rock mass response to mining: Keynote Address. In G. van Aswegen, R. J. Durrheim, & W. D. Ortlepp (Eds.), *Proceedings 5th International Symposium on Rockbursts and Seismicity in Mines, Johannesburg, South Africa* (pp. 1–9). South African Institute of Mining and Metallurgy.

Mendecki, A. J. (2005). Persistence of seismic rock mass response to mining. In Y. Potvin, & M. R. Hudyma (Eds.), *Proceedings 6th International Symposium on Rockburst and Seismicity in Mines, Perth, Australia* (pp. 97–105), Australian Centre for Geomechanics.

Mendecki, A. J. (2008). Forecasting seismic hazard in mines. In Y. Potvin, J. Carter, A. Diskin, & R. Jeffrey (Eds.), *Proceedings 1st Southern Hemisphere International Rock Mechanics Symposium, Perth, Australia* (pp. 55–69). Australian Centre for Geomechanics.

Mendecki, A. J. (2013). Characteristics of seismic hazard in mines: Keynote Lecture. In A. Malovichko & D. A. Malovichko (Eds.), *Proceedings 8th International Symposium on Rockbursts and Seismicity in Mines, St Petersburg-Moscow, Russia* (pp. 275–292). ISBN:978-5-903258-28-4.

Mendecki, A. J. (2016). *Mine seismology reference book: seismic hazard* (1st edn.). Institute of Mine Seismology. ISBN:978-0-9942943-0-2. www.imseismology.org/msrb/.

Mendecki, A. J. (2017). Mapping seismic ground motion hazard: Keynote lecture. In J. A. Vallejos (Ed.), *Proceedings 9th International Symposium on Rockbursts and Seismicity in Mines, Santiago, Chile*.

Mendecki, A. J. (2018). Ground motion prediction equations for DMLZ. *Technical PTFI-REP-GMPE-201801-AJMv1*, Institute of Miner Seismology.

Mendecki, A. J. (2019). Simple GMPE for underground mines. *Acta Geophysica, Springer*, *67*(3), 837–847. http://doi.org/10.1007/s11600-019-00289-z.

Mendecki, A. J. (2023). Seismic ground motion alerts for mines. *Journal of Seismology, Springer*, *27*, 599–608. http://doi.org/10.1007/s10950-023-10147-7.

Mendecki, A. J., & Lötter, E. C. (2011). Modelling seismic hazard for mines. In *Australian Earthquake Engineering Society 2011 Conference, Barossa Valley*.

Mendecki, M. J., Duda, A., & Idziak, A. (2018). Ground-motion prediction equation and site effect characterization for the central area of the main syncline, Upper Silesia Coal Basin, Poland. *Open Geoscience*, *10*(1), 474–483. http://doi.org/10.1515/geo-2018-0037.

Milev, A. M., & Spottiswood, S. M. (2005). Strong ground motion and site response in deep South African mines. *The Journal of The South African Institute of Mining and Metallurgy*, *105*, 1–10.

Mountfort, I. P., & Mendecki, A. J. (2019). Measuring ground motion with geophones. *Acta Geophysica* (in submission).

Mow, C.-C., & Pao, Y.-H. (1971). The diffraction of elastic waves and dynamic stress concentration. *R-482-pr*, United States Air Force.

Mulargia, F., Stark, P. B., & Geller, R. J. (2017). Why is probabilistic seismic hazard analysis (PSHA) still used? *Physics of the Earth and Planetary Interiors, 264*, 63–75.

Nordstrom, E., Dineva, S., & Nordlund, E. (2020). Back analysis of short-term seismic hazard indicators of larger seismic events in deep underground mines (Lkab, Kiirunavaara mine, Sweden). *Pure and Applied Geophysics, 177*, 763–785, http://doi.org/10.1007/s00024-019-02352-8.

Ortlepp, W. D. (1984). Rockbusts in South Africa gold mines: A phenomenological view. In N. C. Gay, & E. H. Wainwright (Eds.), *Proceedings 1th International Congress on Rockbursts and Seismicity in Mines*, Symposium Series 6. The South African Institute of Mines and Metallurgy.

Ortlepp, W. D. (1993). High ground displacement velocities associated with rockburst damage. In R. P. Young (Ed.), *Proceedings 3rd International Symposium on Rockbursts and Seismicity in Mines, Kingston, Ontario, Canada* (pp. 101–106). Balkema, Rotterdam.

Ortlepp, W. D. (1997). *Rock fracture and rockbursts - An illustrative study.* Monograph Series M9 (126 p.). South African Institute of Mining and Metallurgy.

Porter, K. (2018). *A beginners guide to fragility, vulnerability, and risk.* University of Colorado Boulder.

Saenger, E., & Bohlen, T. (2004). Finite-difference modeling of viscoelastic and anisotropic wave propagation using the rotated staggered grid. *Geophysics, 69*, 583–591.

Saenger, E. H., Gold, N., & Shapiro, S. A. (2000). Modeling the propagation of elastic waves using a modified finite difference grid. *Wave Motion, 31*(1), 77–92. http://doi.org/10.1016/S0165-2125(99)00023-2.

Savage, J. C. (1971). Radiation from supersonic faulting. *Bulletin of the Seismological Society of America, 61*(4), 1009–1012.

Somerville, P. G., Smith, N. F., Graves, R. W., & Abrahamson, N. A. (1997). Modifications of empirical strong ground motion attenuation relations to include the amplitude and duration effects of rupture directivity. *Seismological Research Letters, 68*(1), 199–222.

Spottiswoode, S. M. (2000). Aftershocks and foreshocks of mine seismic events. In *3rd International Workshop on the Application of Geophysics to Rock and Soil Engineering, Melbourne* (pp. 1–6).

Spudich, P., & Cranswick, E. (1984). Direct observation of rupture propagation during the 1979 Imperial Valley earthquake using a short baseline accelerometer array. *Bulletin of the Seismological Society of America, 74*(6), 2083–2114.

Trifunac, M. D., & Brady, A. G. (1975). A study on the duration of strong earthquake ground motion. *Bulletin of the Seismological Society of America, 65*(3), 581–626.

Turner, M. H., & Player, J. R. (2000). Seismicity at Big Bell Mine. In *Proceedings Massmin 2000* (pp. 791–797). The Australasian Institute of Mining and Metallurgy.

Vallejos, J. A. (2010). Analysis of seismicity in mines and development of re-entry protocols, PhD thesis, Queen's University Kingston, Ontario, Canada.

van Aswegen, G. (2008). Ortlepp Shears - dynamic brittle shears of South African gold mines. In Y. Potvin (Ed.), *Proceedings 1st Southern Hemisphere International Rock Mechanics Symposium, Perth* (pp. 111–120).

van Aswegen, G., & Mendecki, A. J. (1999). Mine layout, geological features and seismic hazard. *Final report gap 303*, Safety in Mines Research Advisory Committee, South Africa (pp. 1–91).

Vieira, F. M. C. C., Diering, D. H., & Durrheim, R. J. (2001). Methods to mine the ultra-deep tabular gold-bearing reefs of the Witwatersrand Basin, South Africa. In W. A. Hustrulid, & R. L. Bullock (Eds.), *Underground Mining Methods: Engineering Fundamentals and International Case Studies* (pp. 691–704), Society for Mining, Metallurgy and Exploration, Inc (SME).

Weertman, J. (1969). Dislocation motion on an interface with friction that is dependent on sliding velocity. *Journal of Geophysical Research, 74*(27), 6617–6622.

Woodward, K., Wesseloo, J., & Potvin, Y. (2017). Temporal delineation and quantification of short term clustered mining seismicity. *Pure and Applied Geophysics, 174*, 2581–2599. http://doi.org/10.1007/s00024-017-1570-6.

Open Access This chapter is licensed under the terms of the Creative Commons Attribution 4.0 International License (http://creativecommons.org/licenses/by/4.0/), which permits use, sharing, adaptation, distribution and reproduction in any medium or format, as long as you give appropriate credit to the original author(s) and the source, provide a link to the Creative Commons license and indicate if changes were made.

The images or other third party material in this chapter are included in the chapter's Creative Commons license, unless indicated otherwise in a credit line to the material. If material is not included in the chapter's Creative Commons license and your intended use is not permitted by statutory regulation or exceeds the permitted use, you will need to obtain permission directly from the copyright holder.

Appendix A
Basic Concepts in Probability and Statistics

This chapter gives a simple overview of the subject with emphasis on statistical dependence. Having a number of parameters that may be relevant to the subject at hand, we would like to select a subset that is most informative, mainly those that are independent. However, some of the relevant parameters may not be totally independent, and then we need to select a set of parameters that are least dependent.

A.1 Basic Concepts in Probability

A.1.1 Random Variable

The frequentist view of probability is associated with the relative frequency of events in the long run, $\Pr(A) = n_A/n$, where n_A is the number of times event A occurred and n is the total number of trials. However, most processes are unique and do not occur repeatedly. The Bayesian approach allows one to update assessments of probability that integrate prior knowledge with observed events, thereby allowing better conclusions to be reached. Both the frequentist and the Bayesian approaches converge to the same results as increasingly more data, or information, is available. It is most frequently used to judge the relative validity of hypotheses in the face of sparse or uncertain data, or to adjust the parameters of a specific model. It is important, however, to distinguish between the Bayesian interpretation of probability and the calculus of Bayes' rule—which is nothing more than a formula for calculating conditional probability.

A random variable is a numerical description of the outcome of an experiment whose value depends on chance, i.e. whose outcome is not entirely predictable. There are two types of random variables: (1) discrete—that can take on only a finite number of values, (2) continuous—that may take on any value in an interval.

The basic concepts of the theory of probabilities are best explained by examples from combinatorics. The factorial $n!$ of a non-negative integer n is defined as equal to 1 for $n = 0$ and to the product of all positive integers less than or equal to n. The factorial function is defined by the product, $n! = \prod_{k=1}^{n} k$. A list of n different entries can be reordered in exactly $n!$ different ways called permutations. Permutation, $P(n, k) = n!/(n-k)!$, is the number of ways that k objects can be selected from n objects with the order being important, $P(n, n) = n!$. Combination, $C(n, k) = n!/[k!(n-k)!]$, is the number of ways that k objects can be selected from n objects with the order not being important.

A.1.2 Union and Conditional Probability

Union of two events A and B is the set of outcomes in either A or B or both and is denoted by $A \cup B$. Intersection of two events A and B is represented by the set of outcomes in both A and B simultaneously and is denoted by $A \cap B$. Mutually exclusive events are those which have no outcomes in common. Event B is inclusive in event A when all outcomes of B are contained in those of A, i.e. B is a subset of A. Probabilities of mutually exclusive events add up, $\Pr(A \cup B) = \Pr(A) + \Pr(B)$. The complement is everything else that could happen other than the proposition in question. The probability of the complement of event A, $\Pr(\bar{A}) = 1 - \Pr(A)$. Probability for either A or B, when they are not mutually exclusive, to occur is $\Pr(A) + \Pr(B) - \Pr(A \cap B)$. Marginal probability of an event A is the probability of A expressed via the joint probability, $\Pr(A) = \Pr(A \cap B) + \Pr(A \cap \bar{B})$. Conditional probability is the probability that event B occurs given that A has already occurred, $\Pr(B|A) = \Pr(A \cap B)/\Pr(A)$ or $\Pr(A|B) = (n_{AB}/n)/(n_B/n)$. Probabilities for events that are positively correlated tend to move in the same direction. If A and B are positively correlated, then: $\Pr(B|A) > \Pr(B)$ and $\Pr(A|B) > \Pr(A)$ and $\Pr(A \cap B) > \Pr(A) \cdot \Pr(B)$. If A and B are negatively correlated, then: $\Pr(B|A) < \Pr(B)$ and $\Pr(A|B) < \Pr(A)$ and $\Pr(A \cap B) < \Pr(A) \cdot \Pr(B)$. If A and B are uncorrelated or independent, then: $\Pr(B|A) = \Pr(B)$ and $\Pr(A|B) = \Pr(A)$ and $\Pr(A \cap B) = \Pr(A) \cdot \Pr(B)$. The Bayesian theorem allows going from $\Pr(B|A)$ to $\Pr(A|B)$ if we know the marginal probabilities of the outcomes of A and the probability of B, given the outcomes of A, $\Pr(A|B) = \Pr(B|A)\Pr(A)/\Pr(B)$.

The total probability theorem, $\Pr(B) = \sum_{j=1}^{n} \Pr(B|A_j)\Pr(A_j)$, gives the probability of B given probabilities of a set of mutually exclusive, collectively exhaustive events $A_1, A_2, \ldots, A_n$. For example, we can estimate the probability of having damage to one of the tunnels in a given area of a mine during the next 12 months given the following probabilities: $\Pr(0.0 \leq \log P \leq 1.0) = 0.827$, $\Pr(1.0 \leq \log P \leq 2.0) = 0.147$, and $\Pr(2.0 \leq \log P \leq 3.0) = 0.026$. In addition, we know from experience that $\Pr(\text{damage}|0.0 \leq \log P \leq 1.0) = 0,01$, $\Pr(\text{damage}|1.0 \leq \log P \leq 2.0) = 0.15$, $\Pr(\text{damage}|2.0 \leq \log P \leq 3.0) =$

0.9. Therefore, the probability of having damage within the next 12 months is Pr (damage) = $0.01 \cdot 0.827 + 0.15 \cdot 0.147 + 0.9 \cdot 0.026 = 0.054$, or 5.4%.

A.1.3 Distribution Function and Bimodality

The cumulative distribution function (CDF), $F(x)$, gives the probability of choosing at random an event with $X \leq x$, i.e. $F(x) = \Pr(X \leq x)$. The survival function is defined as $S(x) = 1 - F(x) = \Pr(X \geq x)$. The probability density of a continuous random variable (PDF) is a function that describes the relative likelihood of $X = x$, and $f(x) = F'(x)$ or $F(x) = \int_{-\infty}^{x} f(u)\,du$. If events are mutually exclusive and collectively exhaustive, then $\Pr(x) = \int_{-\infty}^{\infty} f(x)\,dx = 1$, and $\Pr(a \leq x \leq b) = F(b) - F(a) = \int_{a}^{b} f(x)\,dx$. If X is a discrete random variable that takes on specific values $x_1, x_2, \ldots$, then the probability density (or mass) function is $f(x_i) = \Pr(X = x_i)$, i.e. the probability that a discrete random variable is exactly equal to some value. The value of the cumulative distribution function for a discrete random variable is the sum of the probabilities of all x_i that are less than or equal to x, $\sum_{x_i \leq x} f(x_i)$ and $\sum_u f(x_i) = 1$, as u runs through the set of all possible values of X (Fig. A.1).

The mode of a probability distribution is the value at which its probability density function takes its maximum value. For a unimodal distribution, the median and the mean lie within $\sqrt{3/5}$, and the median and the mode lie within $\sqrt{3}$ standard deviations of each other. The mode is not necessarily unique, since the probability density function may take the same maximum value at several points. The extreme case being the uniform distributions, where all values occur equally frequently.

A bimodal distribution is a probability distribution with two modes. It most commonly arises as a superposition of two distinct unimodal distributions. A mixture of two normal distributions with equal standard deviations is bimodal only if their means differ by at least twice the common standard deviation, see Fig. A.2. Note that it is difficult to recognise bimodality by inspecting the cumulative distribution function and that one needs to examine the probability density function (Kijko & Stankiewicz, 1987; Lasocki et al., 2000; Lasocki & Papaditriou, 2006).

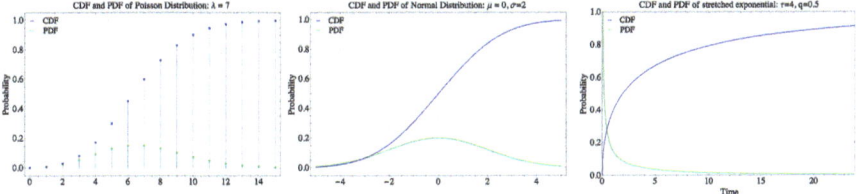

Fig. A.1 CDF and PDF of Poisson, Gaussian, and stretched exponential distributions

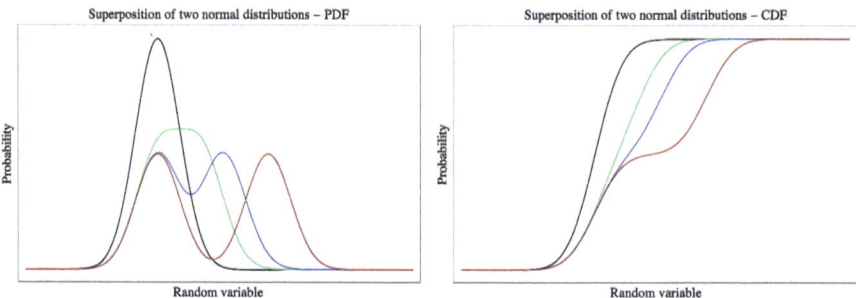

Fig. A.2 Mixture of two equally weighted PDFs (left) and CDFs (right) of normal distributions with common standard deviation, $s_{d1} = s_{d2}$, and with means $\mu_2 = 1 \cdot s_{d1}$ (black), $3 \cdot s_{d1}$ (green), $4 \cdot s_{d1}$ (blue), and $6 \cdot s_{d1}$ (red)

A.1.4 Conditional Probability and Hazard Function

The conditional probability that an event $\geq P$ will occur at some time t between t_1 and $t_1 + \Delta T$, where t_1 is the time since the last event of this size is,

$$\Pr(\geq P; t_1 \leq t \leq t_1 + \Delta T \mid t_1 < t) = \int_{t_1}^{t_1+\Delta T} f(t)\, dT \Big/ \int_{t_1}^{\infty} f(t)\, dt, \quad (A.1)$$

where $f(t) = dF(t)/dt$ is the probability density and $F(t)$ the cumulative distribution function of the recurrence times. It is the ratio of the probability that an event will occur within the time interval t_1 and $t_1 + \Delta T$, to the probability that the event will eventually occur, given that at the time t_1 it has not yet happened, which gives

$$\Pr(\geq P; t_1 \leq t \leq t_1 + \Delta T \mid t_1 < t) = \frac{[F(t_1 + \Delta T) - F(t_1)]}{[1 - F(t_1)]} = \frac{\Delta F}{S(t_1)}. \quad (A.2)$$

Figure A.3 illustrates the conditional probability expressed via the probability density function $f(t)$. The time interval of interest is from t_1, which is the present time, to $t_1 + \Delta T$. The ΔF is equal to the area marked in Fig. A.3 (right) in dark grey, and the survivor function, $S(t_1) = 1 - F(t_1)$, at time t_1 is equal to the area marked by light grey in Fig. A.3 (left). The conditional probability is the ratio of these two areas.

The hazard function gives the instantaneous rate of failure at time t_1 conditional upon no event having occurred up to time t_1. The hazard function is not a density or a probability, $h(t_1) > 1$ is acceptable, and however, one can think of it as the probability of having an event in a very small period of time between t_1 and $t_1 + \Delta T$, when $\Delta T \rightarrow 0$, given that there was no event before time t_1. The hazard function is

A Basic Concepts in Probability and Statistics

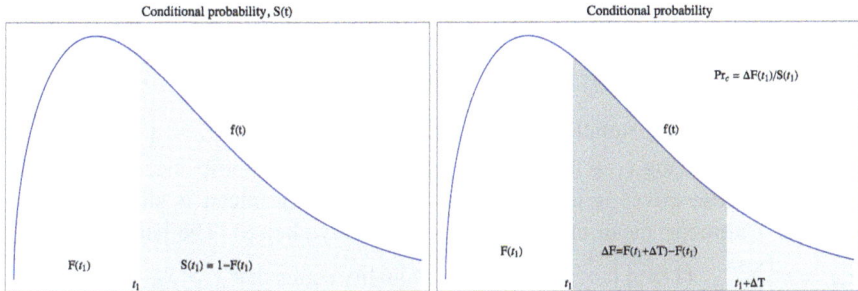

Fig. A.3 Illustration of conditional probability

also defined as $dF(t_1)/dt_1 = d[1 - S(t_1)]/dt = -d[S(t_1)]/dt = h(t_1)S(t_1)$, and therefore

$$h(t_1) = \frac{f(t_1)}{1 - F(t_1)} = \frac{f(t_1)}{S(t_1)}, \quad (A.3)$$

where $f(t_1)$ is the probability density function, $F(t_1) = \Pr(\leq t_1)$ is the cumulative distribution function, i.e. the probability that the event will happen before time t_1, and $S(t_1) = \Pr(\geq t_1) = 1 - F(t_1)$ is the survival function, i.e. the probability that the event will not happen until time t_1. The hazard function measures the probability that an event will occur at time t_1, under the assumption that it did not occur until this time. The hazard function is defined as

$$h(t_1) = \lim_{\Delta T \to 0} \frac{\Pr(\geq P; t_1 \leq t \leq t_1 + \Delta T \mid t_1 < t)}{dT}, \quad (A.4)$$

where the numerator is the conditional probability that the event of this size will occur in the time interval t_1 and $t_1 + \Delta T$ given that at the time t_1 it has not yet happened, and the denominator, dT, is the width of the interval.

A.2 Basic Concepts in Statistics

Statistics are a usage of mathematical procedures for summarising and interpreting observations. Random variables are uncertain numbers subject to variations due to chance. The expected value, $E[X]$, of a random variable $X(x_1, ..., x_j, ...)$ is a weighted average of its possible outcomes. For a discrete random variable, each outcome is weighted by its probability of occurrence, $E[X] = \sum_{x: f(x) > 0} x_j f(x_j)$, where $f(x)$ is the probability $\Pr(X = x_j)$, and for a continuous random variable, $E[X] = \int_{-\infty}^{\infty} x f(x) \, dx$.

For a sample of observed data, $x_1,...,x_n$, each of the individual x_j has associated probability $\Pr(x_j) = f(x_j) = 1/n$, and the sample mean is $\bar{x} = \sum_{j=1}^{n} x_j (1/n) = (1/n) \sum_{j=1}^{n} x_j$, which is called the arithmetic mean. The geometric mean of a sample for non-negative real numbers, $\bar{x}_G = \left(\prod_{j=1}^{n} x_j\right)^{1/n}$, and $\log \bar{x}_G = (\log x_1 + \log x_2, ..., \log x_n)/n$. The geometric mean is used for variables whose effect is multiplicative. The geometric mean is slightly smaller than the arithmetic mean unless the data are highly skewed. The harmonic mean, $\bar{x}_H = n \left[\sum_{j=1}^{n} (1/x_j)\right]^{-1}$. The general inequality states that $\bar{x} \geq \bar{x}_G \geq \bar{x}_H$.

A median of the distribution is any real number $\tilde{x}$ that satisfies the inequalities: $\Pr(\leq \tilde{x}) \geq 0.5$ and $\Pr(\geq \tilde{x}) \geq 0.5$. The median of the data set is the numerical value separating the higher half of a data sample from the lower half: It is the middle value of a set of data containing an odd number of values, or the average of the two middle values of a set of data with an even number of values, $x_m = x_{(n+1)/2}$ when n is odd and $x_m = 0.5(x_{n/2} + n_{n/2+1})$ when n is even. The median is less sensitive to outliers than the mean.

The mode of a set of data is the value which occurs most frequently. An empirical relation between the mean, median, and mode which appears to hold for unimodal distributions of moderate asymmetry is: $mean - mode \approx 3 (mean - median)$.

A.2.1 Variance, Coefficient of Variation, and Covariance

Variance is the average squared deviation of a random variable from its mean, $Var(x) = E\{(X - E[X])^2\}$. For a discrete random variable, $Var(X) = \sum_{x: f(x)>0} (x_j - E[X])^2 f(x_j)$, and for a continuous case, $Var(X) = \int_{-\infty}^{\infty} (x - E[X])^2 f(x) dx$. The variance of a sample is $\overline{Var} = s^2 = (1/n) \sum_{j=1}^{n} (x_j - \bar{x})^2$. The standard deviation measures the average distance of a random variable X from its mean, $\sqrt{Var(X)}$, and for a sample $s_d = (\overline{Var})^{1/2}$. The standard error of the sample mean, $s_d(\bar{x}) = s_d/\sqrt{n}$, is an estimate of how far the sample mean is likely to be from the population mean.

The coefficient of variation is a normalised measure of dispersion of a probability distribution or a frequency distribution, $C_v(X) = \sqrt{Var(X)}/E(X)$, and for a sample, $C_v(x) = s_d(x)/\bar{x}$. In a finite sample of n non-negative numbers with a real zero, the coefficient of variation can take value between 0 and $\sqrt{n-1}$, the maximum is when all values but one are equal to zero. If the underlying data is normally distributed, its standard error $\hat{sd}(C_v) = \hat{C}_v/\sqrt{2n}$.

Covariance measures the degree to which two variables differ from their mean, $Cov(X, Y) = E\{(X - E[X])(Y - E[Y])\}$. The covariance of a variable with itself is its variance, $Cov(X, X) = Var(X)$. The covariance of a sample is $\overline{Cov} = (1/n) \sum_{j=1}^{n} (x_j - \bar{x})(y_j - \bar{y})$. Note that covariance is a dimensional quantity; therefore, it depends on the units of measurement for X and Y. Correlation measures the degree of linear relationship between random variables,

$Corr(X, Y) = Cov(X, Y) / [\sqrt{Var(X)}\sqrt{Var(Y)}]$. The correlation of a variable with itself is always one, $Corr(X, X) = 1$.

Covariance and correlation measure the degree of linear relationship, and if their values are zero, then either the relationship between X and Y can be assumed to be of some non-linear type, or else the variables are independent.

A.3 Statistical Dependence

Having a number of parameters that may be relevant to the subject at hand, for example, stability analysis described in Chap. 3, we would like to select a subset that is most informative, mainly those that are statistically independent. However, some of the relevant parameters may not be totally independent, and then we need to select a set of parameters that are least dependent.

Two random variables are said to be statistically independent if the outcome of one random variable does not affect the conditional probability of the other. Dependence applies to any statistical relationship, whether causal or not, between two random variables or two sets of data.

It is well known that the correlation coefficient measures the degree of the linear relation between variables, and, in some cases, it may be zero, while there is a strong non-linear relation between these variables.

A.3.1 Correlation Coefficient

Correlation is a measure of a monotonic association between two variables. In correlated data, the change in the magnitude of one variable is associated with a change in the magnitude of another variable, either in the same or in the opposite direction. A linear relationship between two variables is a special case of monotonic relation. The degree to which the change in one variable is associated with a change in another continuous variable can mathematically be described in terms of the covariance. Covariance is similar to variance, but it measures how two variables vary together. The most popular correlation measure is the correlation coefficient of the two variables X and Y and is obtained by dividing their covariance by the square root of the product of their standard deviations, $\text{cor}(X, Y) = \text{cov}(X, Y) / \sqrt{\text{var}(X)\,\text{var}(Y)}$. The sample correlation coefficient, r_{xy}, for n measurements of x_i and y_i and the sample means $\bar{x}$ and $\bar{y}$ is

$$r_{xy} = \frac{\sum (x_i - \bar{x})(y_i - \bar{y})}{\sqrt{\sum (x_i - \bar{x})^2}\sqrt{\sum (y_i - \bar{y})^2}}, \tag{A.5}$$

where the summations go from $1, \ldots, n$. It is symmetric and defined only if both standard deviations are finite and nonzero. It ranges from $-1 \leq r_{xy} \leq 1$, and it is 1 when all data points lie on a line for which Y increases as X increases, indicating a perfect positive linear relation, and it is -1 in the opposite case. The case $r_{xy} = 0$ does not mean two variables are independent, i.e. it does not measure dependence. The correlation coefficient is limited to linear associations and is very sensitive to outliers. It is called the Pearson coefficient although its development was initiated by Galton and then formalised by Pearson.

A.3.2 Hoeffding Test

Unlike the Pearson correlation, r_{xy}, the Hoeffding (1948) D_H statistic tests whether two random variables, X and Y, are independent. If $F(X, Y) = \Pr(X < x, Y < y)$ is a joint cumulative distribution function and their marginal distribution functions are $F_X(x) = \Pr(X < x)$ and $F_Y(y) = \Pr(Y < y)$ and continuous, the null hypothesis that these two variables are independent is $F \equiv F_X F_Y$. Hoeffding proposed the following statistic:

$$D_H = \int [F(x, y) - F_X(x) F_Y(y)]^2 \, dF(x, y), \quad (A.6)$$

and $D_H = 0$ if X and Y are independent and $D_H > 0$ if the joint distribution is dependent and continuous. It is a non-parametric rank-based statistics that measures the dependence by the ranking of n paired samples (X_i, Y_i). It measures the quantity Q_i which is the number of bivariate observations (x_j, y_j) for which $x_j < x_i$ and $y_j < y_i$, where $Q_i = \sum_{j=1}^{n} \phi(x_j, x_i) \phi(y_j, y_i)$ with $\phi(a, b) = 1$ if $a < b$ and $\phi(a, b) = 0$ otherwise. If $D_{H1} = \sum_{i=1}^{n} Q_i (Q_i - 1)$ and $D_{H2} = \sum_{i=1}^{n} (R_i - 1)(R_i - 2)(S_i - 1)(S_i - 2)$ and $D_{H3} = \sum_{i=1}^{n} (R_i - 2)(S_i - 2) Q_i$, where R_i and S_i are ranks x_i and y_i, respectively, then

$$D_H = \frac{(n-2)(n-3) D_1 + D_2 - 2(n-2) D_3}{n(n-1)(n-2)(n-3)(n-4)}. \quad (A.7)$$

The parameter D_H lies on the interval $(-1/2, 1)$, but the positive or negative signs have no interpretation, the larger the value the higher the dependence. The D_H decreases as the number of duplicate values in the data increases. The D_H test is valid for the data drawn from a continuous distribution. Blum et al. (1960) proposed an improved method to overcome this problem.

A.3.3 Distance Correlation

Distance correlation, developed by Szekely et al. (2007), is a relatively new measure of statistical dependence between two random variables of arbitrary dimension. It

A Basic Concepts in Probability and Statistics

is obtained by dividing their distance covariance by the product of their distance standard deviations,

$$dCorr(X, Y) = \frac{dCov(X, Y)}{\sqrt{dVar(X) \, dVar(Y)}}, \quad (A.8)$$

which is zero only if X and Y are independent and 1 for a perfect linear dependence. In case of non-linear dependence, $0 < dCorr(X, Y) < 1$. The value of $dCorr(X, Y) = 0$ implies $r_{xy} = 0$ and $\|r_{xy}\| = 1$ only if $dCorr(X, Y) = 1$, and therefore $dCorr(X, Y) = 0$ only if X and Y are independent.

The distance correlation for a sample of n measurements of x_i and y_i can be estimated in the following way: (1) Define distance matrices, $A_{ij} = \|x_i - x_j\|$ and $B_{ij} = \|y_i - y_j\|$, for $i, j = 1, ..., n$. (2) Double centre the distance matrix A_{ij} so all rows and columns sum to zero, $\tilde{A}_{ij} = A_{ij} - \bar{A}_{j.} - \bar{A}_{.j} - \bar{A}_{..}$, where $\bar{A}_{j.}$ is the mean of row i, $\bar{A}_{.j}$ is the mean of column i, and $\bar{A}_{..}$ is the overall mean of A. (3) Double centre the distance matrix B_{ij} the same way. (4) Compute the estimate of the distance covariance, $\widehat{dCov}(X, Y) = \sqrt{(1/n^2) \sum_{i=1}^{n} \sum_{j=1}^{n} \tilde{A}_{ij} \tilde{B}_{ij}}$. Taking $\widehat{dVar}(X) = \widehat{dCov}(X, X)$ and $\widehat{dVar}(Y) = \widehat{dCov}(Y, Y)$ gives the estimate of Eq. (A.8).

Distance correlation coefficient measures any kind of dependence including non-linear or non-monotone dependencies (Edelmann et al., 2021).

A.3.4 Information Entropy

Information entropy is a measure of the unpredictability of information content and, for a random variable X with n outcomes $\{x_1, ..., x_n\}$, is defined as

$$H(X) = \sum_{i=1}^{n} \Pr(x_i) \log_2 [1/\Pr(x_i)] = -\sum_{i=1}^{n} \Pr(x_i) \log_2 [\Pr(x_i)], \quad (A.9)$$

where $\Pr(x_i)$ is the probability that a random variable X associated with x_i (Shannon, 1948a first paper). For $\log_2$, the units of entropy are called bits, and for $\log_{10}$ nats. In Eq. (A.9), $\log_2 [1/\Pr(x_i)]$ is the information content of event x_i with probability $\Pr(x_i)$, and if this probability is high, then knowledge that event x_i occurred gives very little information, since it had a high probability of occurrence to start with, see Fig. A.4 left. Information entropy does not depend on the actual values taken by the random variable X but only on the probabilities. If one of the outcomes has probability one, $\Pr(x_k) = 1$, then $\Pr(x_i) = 0$ for all $x_i \neq x_k$, and such a system conveys no information, thus $H = 0$. Note that $\lim_{x \to 0} x \log x = 0$, and therefore, adding terms of zero probability does not change the entropy. If,

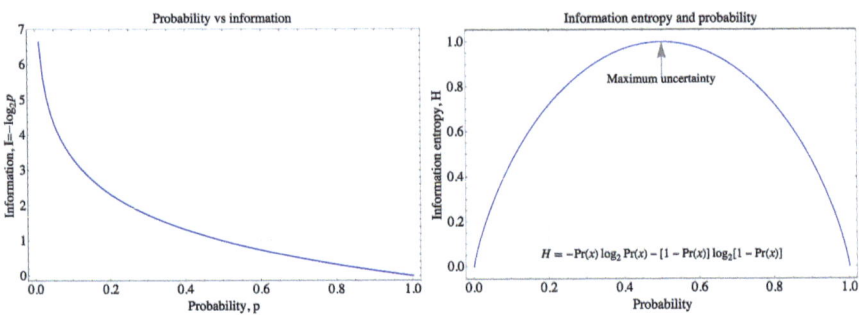

Fig. A.4 Probability vs. information *(left)*. Binary information entropy, i.e. the variable X that takes the value one with probability p and the value zero with probability $(1 - p)$, *(right)*

however, all possible outcomes are equally likely, $\Pr(x_i) = 1/n$ for $i = 1, ..., n$, then the uncertainty is maximal and $H = \log(n)$, see Fig. A.4 right.

Events with either very high or very low probabilities do not contribute significantly to the information entropy. H reaches its maximum value if all states are equally probable, and it decreases as the uniformity of the probability distribution is being eroded. In this sense, it is a measure of randomness.

The bounds on information entropy are $0 \leq H(X) \leq \log_2 n$, and it is zero when X is constant and $\log_2 n$ when X has a uniform probability distribution because the uncertainty is maximal when all possible events are equiprobable.

Information entropy can be used to test the variability of the underlying probability distribution. A sharply peaked distribution with low variance will have low entropy, and flat distributions will have high entropy since all potential outcomes have similar probabilities and the outcome of any particular one is very uncertain.

The name information entropy may be confusing since by definition information is always a measure of the decrease of uncertainty at a receiver. When Shannon developed his theory, he wanted to call it "information", then "uncertainty", eventually he asked the famous mathematician John von Neumann for advice. Von Neumann replied "You should call it entropy, for two reasons. In the first place your uncertainty function has been used in statistical mechanics under that name, so it already has a name. In the second place, and more important, nobody knows what entropy really is, so in a debate you will always have the advantage".

The joint entropy of a pair of discrete random variables (X, Y) with a joint distribution $\Pr(x, y)$ is defined as

$$H(X, Y) = -\sum_x \sum_y \Pr(x, y) \log \Pr(x, y).$$

The conditional entropy $H(Y \mid X)$ is defined as

$$H(Y \mid X) = \sum_y \Pr(x) H(Y \mid X = x) = -\sum_x \sum_y \Pr(x, y) \log \Pr(y \mid x),$$

and therefore, the entropy of a pair of random variables is the entropy of one plus the conditional entropy of the other,

$$H(X, Y) = H(X) + H(Y \mid X).$$

In summary, the entropy is a measure of the amount of information required on the average to describe the random variable, and the relative entropy is a measure of the distance between two distributions.

A.3.5 Mutual Information

Mutual information between two variables, $I(X, Y)$, introduced by Shannon (1948b), is the relative entropy between the joint distribution, $\Pr(x, y)$, and the product distribution, $\Pr(x)\Pr(y)$,

$$I(X; Y) = \sum_x \sum_y \Pr(x, y) \log_2 \left[\frac{\Pr(x, y)}{\Pr(x)\Pr(y)} \right], \quad (A.10)$$

and $I(X; Y) = H(X) - H(X \mid Y)$ and by symmetry $I(X; Y) = H(Y) - H(Y \mid X)$, and therefore X says as much about Y as Y says about X. Since $H(X, Y) = H(X) + H(Y \mid X)$, then $I(X; Y) = H(X) + H(Y) - H(X, Y)$ and $I(X; X) = H(X) - H(X \mid X) = H(X)$, i.e. the mutual information of a random variable with itself is the entropy of the random variable, also called self-information.

The relationship between these entropies is illustrated in Fig. A.5, where mutual information $I(X; Y)$ corresponds to the intersection of the information in X with the information in Y.

Mutual information measures the reduction in uncertainty of one variable due to knowledge of another. If knowledge of Y reduces uncertainty of X, then Y

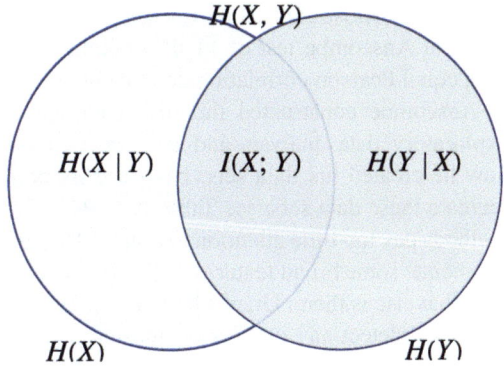

Fig. A.5 Illustration of mutual information expressed in a Venn diagram

carries information about X. The most important property of $I(X;Y)$ is that it is always non-negative, and is zero if and only if X and Y are independent, since for independent variables $\Pr(x, y) = \Pr(x)\Pr(y)$ and $\log 1 = 0$ gives $I(X;Y) = 0$.

The bounds on mutual information are $0 \leq I(X;Y) \leq \min[H(X), H(Y)]$, and it is zero when X and Y are independent and $\min[H(X), H(Y)]$ when XY is constant. Therefore, one can normalise mutual information by $I_N(X;Y) = I(X;Y)/\min[H(X), H(Y)]$. Alternatively, similar to the information coefficient of correlation introduced by Linfoot (1957),

$$ICC(X;Y) = \sqrt{1 - \exp[-2I(X;Y)]}, \qquad (A.11)$$

which for the bivariate normal distribution is equal to the classical correlation coefficient by Pearson.

The mutual information is a general measure of dependence of two variables, and, unlike correlation coefficient that measures the linear dependence, it places no assumptions or models on the variables under consideration and is sensitive to any possible relationships, including non-linear effects and effects in high-order statistics of the distributions. It measures the deviation of a given process from an uncorrelated one. Or, it is the average value of the logarithmic measure of distance from independence.

To estimate mutual information from data is nontrivial. The main problem is to estimate the joint distribution $\Pr(x, y)$ that can be done by binning the data over a rectangular grid (x, y). Then $\Pr(x, y)$ can be estimated as the fraction of data points falling into the respective bin. The efficient way is given by Kraskov et al. (2004).

Mutual information can also be used as a kind of non-linear autocorrelation function to determine when the values measured at time t and $t + \Delta t$ in a time series are independent enough of each other.

A.3.6 Example: Anscombe Test

This section illustrates the performance of different measures of dependence in the classical Anscombe test of 11 data points that give the same linear regression all with equal Pearson correlation coefficient.

Anscombe constructed the four data sets to demonstrate the importance of exploratory data analysis and the effect of outliers. Anscombe did not explain how he created his data sets, but since its publication some authors managed to recreate these data sets, see Table A.1. He stated that most textbooks on statistical methods pay too little attention to graphs. He said that graphs help us to perceive and appreciate some broad features of the data and to look behind these broad features to see what else is there. Graphs help to clarify the nature of the relationships between variables, detect any outliers or unusual patterns in the data, and facilitate further analysis.

A Basic Concepts in Probability and Statistics

Table A.1 Anscombe data sets 2, 3, and 4

x1	10	8	13	9	11	14	6	4	12	7	5
y1	8.04	6.95	7.58	8.81	8.33	9.96	7.24	4.26	10.84	4.82	5.68
x2	10	8	13	9	11	14	6	4	12	7	5
y2	9.14	8.14	8.74	8.77	9.26	8.1	6.13	3.1	9.13	7.26	4.74
x3	9.14	8.14	8.74	8.77	9.26	8.1	6.13	3.1	9.13	7.26	4.74
y3	7.46	6.77	12.74	7.11	7.81	8.84	6.08	5.39	8.15	6.42	5.73
x4	8	8	8	8	8	8	8	19	8	8	8
y4	6.58	5.76	7.71	8.84	8.47	7.04	5.25	12.5	5.56	7.91	6.89

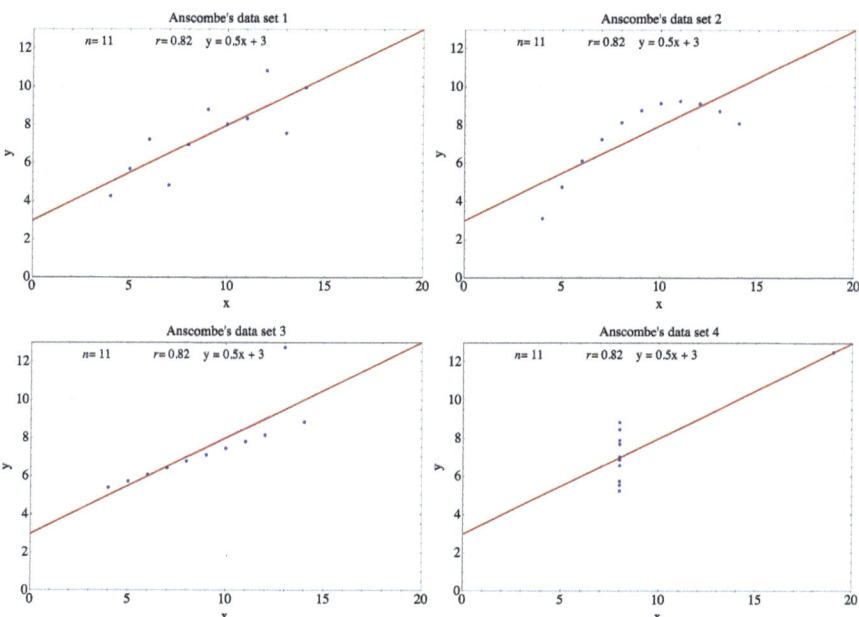

Fig. A.6 Distance correlation, $dCorr$, mutual information -based coefficient of correlation, ICC, and a simple linear correlation coefficient r for three Anscombe data sets

Figure A.6 shows the data sets 1, 2, 3, and 4 of the Anscombe's quartet that have almost identical sample mean and variance: $\bar{x} = 9\, s_x^2 = 11$ and $\bar{y} = 7.5\, s_y^2 = 4.12$, i.e. the same correlation, $r_{xy} = 0.816$, the same linear regression $y = 0.5x + 3$ and the same coefficient of determination $R^2 = 0.6667$, but different distributions (Anscombe, 1973).

Figure A.6 top left shows a simple linear relationship of two correlated variables with distributed noise. Figure A.6 top right shows the data that indicates a clean non-linear relation, and therefore, the linear Pearson correlation coefficient is not applicable. Such a data set should be fitted with a polynomial. Figure A.6 bottom left shows the data that indicates a perfect linear relation except for one outlier along the y-axis that is enough to lower the Pearson correlation coefficient r_{xy} from what would be 1 to 0.82. To fit such data set, one should use the first norm that would ignore the outlier. Figure A.6 bottom right shows an example where data lies along the straight line on the y-axis except for one outlier far in the x direction which is enough to produce a high linear correlation coefficient.

Anscombe test shows that the summary statistics, e.g. r, D_H, I, or $dCorr$, cannot capture all characteristics of a given data set. One should always visualise the data before fitting.

A.3.7 Example: Noisy Data

This section demonstrates the performance of different dependence measures in the presence of noise.

Figure A.7 shows a data set that represents a parabola, $y = 0.25x^2 + RN$, marked by a red line with added random noise $RN = 0.01$ (left) and $RN = 0.05$ (right). The distance correlation, $dCorr$, indicates correctly a non-linear relation, while the mutual information, I, and the Hoeffding statistic, D_H, drop with increasing noise. As expected, the Pearson coefficient, r, indicates no linear relation between variables and also drops slightly with increased noise.

Figure A.8 shows the power law, $y = x^{-\beta}$, for $\beta = 0.25, 0.5, 1.0, 1.25,$ and 1.5, marked by red lines, all with added random noise $RN = 0.1$. As the exponent β

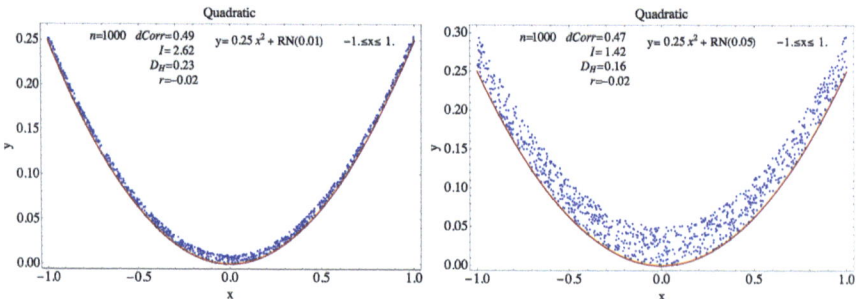

Fig. A.7 Distance correlation, $dCorr$, mutual information, I, Hoeffding D_H statistics, and the Pearson correlation coefficient, r, for the parabola, $y = 0.25x^2 + RN$, with random noise 0.01 and 0.05

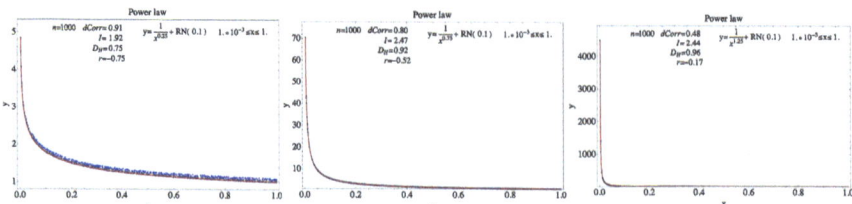

Fig. A.8 Distance correlation, $dCorr$, mutual information, I, Hoeffding D_H statistics, and the Pearson correlation coefficient, r, for the power law data sets with random noise, $y = x^{-\beta} + RN$, for increasing exponents β from 0.25 to 1.25

A Basic Concepts in Probability and Statistics   307

increases, the $dCorr$ decreases indicating a stronger non-linear relation, the Pearson coefficient, r, also decreases indicating less linear correlation, but the Hoeffding statistic, D_H, increases. Mutual information, I, however, tends to increase for β between 0.25 and 0.75 but then stays more or less the same.

A.3.8 Example: Source Parameters

Seismic events are routinely quantified by the following four independent parameters derived from recorded waveforms: the time of the event, t, location, $X = x, y, z$, seismic potency, P, and the radiated seismic energy, E. From seismic potency and energy, one can derive apparent stress, $\sigma_A = E/P$ (Aki, 1966), and apparent volume, $V_A = \mu P^2/E$, (Mendecki, 1993). Seismic potency is derived from a single measurement of amplitude at the low frequency asymptote of the displacement source spectrum, and seismic energy is derived from the integral of the velocity-squared source spectrum. Consequently, potency scales with the lowest frequency of seismic radiation, and since the bulk of seismic energy is released past the predominant frequency, it scales with high frequencies of seismic radiation. However, the fact that these two parameters are derived from independent measurements does not necessarily imply independence. We know that seismic energy increases with seismic potency and in general scales $\log E = d \log P + c$, and here d is the slope and c the intercept that measures the $\log E$ released by a seismic event with $\log P - 0$. In such a case, apparent stress scales with potency as $\log \sigma_A = (d-1) \log P + c$, and for $d = 1$, apparent stress is independent of potency, $\sigma_A = 10^c$. For $d > 1$, apparent stress would increase with an increase in seismic potency.

Seismic sources associated with a weak geological structure will yield slowly under lower differential stress producing larger seismic potency and radiating less seismic energy, resulting in a low apparent stress event. The opposite applies to a source associated with a strong geological feature or hard patch in the rock mass. However, in a relatively competent homogeneous rock, one would expect less scatter in $\log E$ vs. $\log P$ plot than in inhomogeneous rock, i.e. they should be more dependent. Here we will test the degree of dependence between P, E, σ_A, and V_A for a particular data set.

Example As an example, we used the same data set of 6073 events in the $\log P$ ranged from -4.45 to 2.61 recorded over 60 days or 1438 hours with a spatial span of 2171 metres, coefficient of variation in time of 1.32, and space 0.84.

Figure A.9 top row shows three scatter plots, E vs. P, σ_A vs. P, and V_A vs. P. The power law distribution of sizes makes these plots hardly useful, and most of the data points are squeezed in the left bottom corner or row. The bottom row shows the same data, but after logarithmic transformation: $\log E$ vs. $\log P$, $\log \sigma_A$ vs. $\log P$, and $\log V_A$ vs. $\log P$ that clearly demonstrates the power of logarithmic transformation that makes it easier to visualise and interpret.

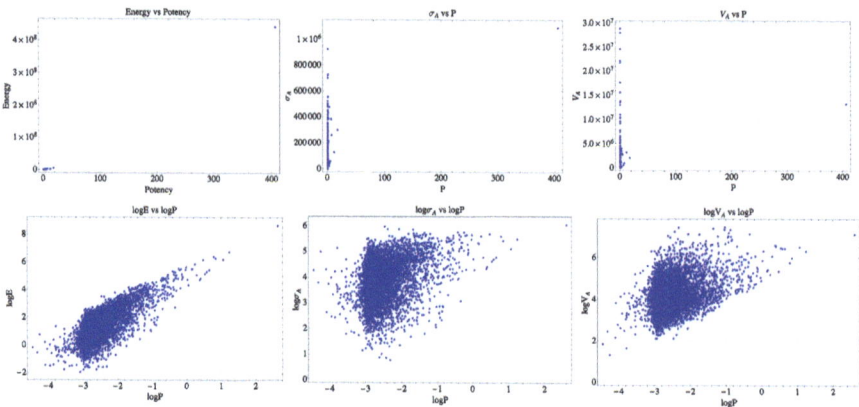

Fig. A.9 Simple scatter plots of source parameters tested for dependence

Table A.2 Dependence test for selected source parameters

	E vs. P	$\log E$ vs. $\log P$	σ_A vs. P	$\log \sigma_A$ vs. $\log P$	V_A vs. P	$\log V_A$ vs. $\log P$
$dCorr$	0.995	0.673	0.170	0.329	0.208	0.264
I	0.000	0.396	0.000	0.074	0.000	0.069
D_H	0.160	0.160	0.035	0.035	0.017	0.017
r	0.999	0.737	0.252	0.345	0.174	0.337

Table A.2 lists the four dependence measures: distance correlation, $dCorr$, mutual information, I, Hoeffding test, D_H, and the Pearson correlation, r, for the three data sets shown in Fig. A.9 before and after logarithmic transformation.

The distance correlation and Pearson correlation rate potency and energy as dependent, while mutual information as independent and Hoeffding test slightly dependent. The same two parameters after taking the log transform are measured slightly differently: Distance correlation indicates dependence but not necessarily linear at 0.673, mutual information shows slight dependence at 0.396, the Hoeffding test stays the same, and the linear Pearson coefficient drops from 0.999 to 0.737. Potency and apparent stress are rated as low dependence by all four measures, while after taking logs the dependence increased, but slightly. The same for potency and apparent volume. A logarithmic transformation is non-linear, and therefore, it will change the relationships between variables and modify the correlation between them.

A.4 Correlation Integral and Correlation Dimension

In mathematics, the dimension measures the space filling properties of a set. Regular objects, conforming to classical Euclidean geometry, can be characterised

A Basic Concepts in Probability and Statistics

by their Euclidean dimension. A point is a zero-dimensional object, a line is one dimensional, a smooth surface is two dimensional, and a cube is 3D. Therefore, the Euclidean dimension takes only integer values.

Since the geometry of most natural objects is highly irregular, their dimension may fall in between the integer values. One can say that the dimension of an object is the number of independent quantities needed to specify the positions of points on the object.

A fractal dimension is a statistical index that measures how detail in a pattern changes with the scale at which it is measured. It is also used as a measure of the space filling capacity, and, for three-dimensional geometries, the fractal dimension varies from 0 to 3. For example the fractal dimension of the coastline of Norway is 1.52, Britain is 1.25, Ireland is 1.22, Australia is 1.13, and South Africa is 1.02. The fractal dimension of the surface of broccoli is 2.7, cauliflower 2.8, and the human brain 2.79.

The correlation sum for the collection of n points is the fraction of all possible pairs of points which are closer than a given distance l,

$$C(l) = \frac{2}{n(n-1)} \sum_{i=1}^{n} \sum_{j=i+1}^{n} \{H[l - \|x_i - x_j\|]\}, \tag{A.12}$$

where H is the Heaviside step function. Theoretically, if $n \to \infty$ and l is very small, C would scale like a power law $C(l) \sim l^D$, where D is the correlation dimension (Grassberger & Procaccia, 1983).

The correlation dimension is invariant, i.e. unchanged under a specific transformation, but the correlation sum is not. If the system of points is a fractal set, the graph of $C(l)$ in logarithmic coordinates must be a linear function with slope D_l equal to the fractal correlation dimension of the system, $\partial \log C(l) / \partial \log l$. In the limit, $n \to \infty$, the correlation sum becomes the correlation integral and is equal to the probability that the distance separating two randomly chosen events is less than l.

The correlation dimension is a measure of clustering. A correlation dimension equal to the embedding space means that the data set covers it completely, i.e. no clustering, and a data set with low correlation dimension shows clustering. The correlation dimension is also related to randomness since random processes occupy all the available space.

The correlation dimension varies between 0 and the value of the embedding dimension, which for space is 3 and for time 1.

One can then define the space-time correlation integral,

$$C(l,t) = \frac{2}{n(n-1)} \sum_{i=1}^{n} \sum_{j=i+1}^{n} \{H[l - \|x_i - x_j\|] \cdot H[\tau - \|t_i - t_j\|]\},$$

where n is the number of observations and the sums run through all pairs whose spatial distance is less than l and time interval less than τ.

Likewise, the time correlation dimension and the space correlation dimension are

$$D_d = \frac{\partial \log C\,(l, t)}{\partial \log l} \qquad D_\tau = \frac{\partial \log C\,(l, t)}{\partial \log t}. \qquad (A.13)$$

There is a number of authors studying the fractal structure of seismicity, both natural and induced. Here we will mention only the first few papers on the subject we have read and learned from.

Kranz et al. (1995) showed the fractal structure of stope-scale seismicity in a silver mine in northern Idaho.

Eneva (1996) studied the generalised correlation integral that allows to test for multifractal behaviour of the data set. She counted numbers of events within balls of a certain radius centred at points in the data set. In the paper, she concluded that the appearance of multifractal features in the scaling properties of the spatial distribution of seismic activity in Creighton Mine in Ontario Canada is largely spurious in character. She also stated that it may be misleading to compare values of correlation dimensions estimated from data sets featuring differences in the number of events, size and geometry of the embedding space, scaling range used to evaluate the slopes, and location errors. Eneva (1998) also analysed the degree of non-randomness, the correlation dimension, and the time interval for occurrence of a constant number of events and compared them quantitatively with the times of larger events.

Tosi (1998) used the generalised correlation integral to analyse seismicity in Central Italy and found the background seismicity with higher spatial fractal dimension but dropping to a lower value before and during earthquake sequences. The first work on the space-time correlation integral applied to earthquakes known to the author was done by Tosi et al. (2004) and Tosi et al. (2008) that analysed changes in the space and time correlation dimensions of global seismicity over time.

Davidsen and Paczuski (2005) analysed the spatial distribution between successive earthquakes and concluded that the spatial distance and waiting time between subsequent earthquakes are uncorrelated with each other, which contradicts the theory of aftershock zone scaling with main shock magnitude.

Example As an example, we used the same data set we used in Sect. A.3.8, with 6073 events in the $\log P$ range from -4.45 to 2.61 recorded over 60 days or 1438 hours with spatial span of 2171 metres, coefficient of variation in time of 1.32 and space 0.84.

Figure A.10 left shows the PDF of the space correlation integral with two modes. The main one is between 80 and 130 metres, and the second smaller one between 350 and 380 metres.

Figure A.10 right shows the CDF of the space correlation integral with the first scaling regime up to 50 metres with the correlation dimension of 2.26. There may be a second one between 50 and 125 metres with a lower dimension of 1.7, but then scaling breaks down.

A Basic Concepts in Probability and Statistics   311

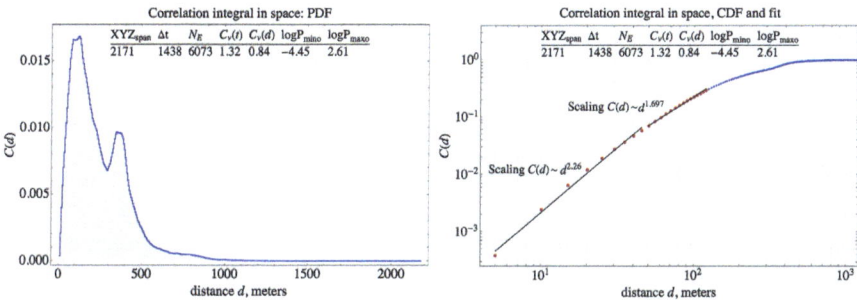

Fig. A.10 Histogram of correlation integral in time *(left)* and correlation integral with fit for correlation dimension D_t *(right)* of BCF dataSet1

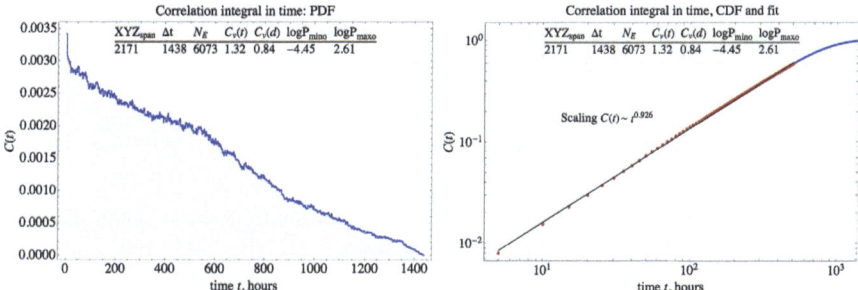

Fig. A.11 Histogram of correlation integral of distances *(left)* and correlation integral with fit for correlation dimension D_d *(right)* of BCF dataSet1

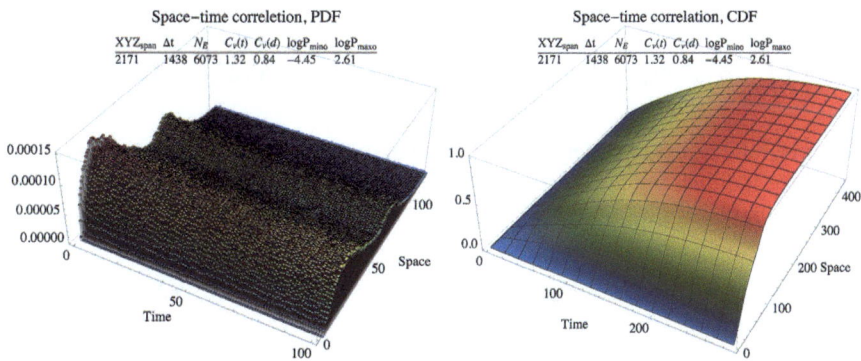

Fig. A.12 Histogram of space-time correlation integral (left) and space-time correlation integral (right) of BCF dataSet1

Figure A.11 left shows the PDF of the time correlation integral, and Fig. A.11 right shows the CDF of the time correlation integral with a well developed scaling regime up to 500 hours with a correlation dimension of 0.926.

Figure A.12 left shows the PDF of the space-time correlation integral with the two modes along the space axis and smooth along the time axis. Figure A.12

right shows the CDF of the space-time correlation integral which, as mentioned in Sect. A.1.3, smoothes over the second mode.

A.5 Random Walk and Diffusion

The degree of randomness in the spatial distribution of distances between consecutive seismic events may justify the interpretation of the seismicity in terms of a random walk. The following assumptions are made: (1) Isotropy, i.e. the next step can proceed from the previous one with equal probability in all directions. (2) Uniformity, i.e. the random walk proceeds from the last step with the same probability as it did in the previous step.

Let us consider the probability, $\Phi_n(\mathbf{X})$, that the random walker will reach the location $\mathbf{X}$ in the volume ΔV during the nth step,

$$\Phi_n(\mathbf{X}) = \iiint_{\Delta V} \phi(\mathbf{x}) \Phi_{n-1}(\mathbf{X} - \mathbf{x}) d^3 x, \tag{A.14}$$

where $\phi(\mathbf{x})$ is the probability of making a step $\mathbf{x}$. If the step $\mathbf{x}$ is small, the probability $\Phi_{n-1}(\mathbf{X} - \mathbf{x})$ can be expanded in 3D Taylor series,

$$\Phi_{n-1}(\mathbf{X} - \mathbf{x}) = \Phi_{n-1}(\mathbf{X}) - x_i \nabla_i \Phi_{n-1}(\mathbf{X}) + \frac{x_i x_j}{2} \nabla_i \nabla_j \Phi_{n-1}(\mathbf{X}) + \ldots$$

Integrating the first term gives

$$\iiint_{\Delta V} \phi(\mathbf{x}) \Phi_{n-1}(\mathbf{X}) d^3 x = \Phi_{n-1}(\mathbf{X}) \iiint_{\Delta V} \phi(\mathbf{x}) d^3 x = \Phi_{n-1}(\mathbf{X}).$$

The integral of the second term is proportional to the mean value of the step vector, and since the distribution $\phi(\mathbf{x})$ is isotropic, $\iiint_{\Delta V} x_i \phi(\mathbf{x}) d^3 x = \langle x_i \rangle = 0$. The third integral is proportional to the variance of $\mathbf{x}$,

$$\iiint_{\Delta V} x_i x_j \phi(\mathbf{x}) d^3 x = \frac{1}{3} \langle x_i x_j \rangle \delta_{ij} \quad \Rightarrow \quad \Phi_n(\mathbf{X}) = \Phi_{n-1}(\mathbf{X}) + \frac{1}{6} \langle x^2 \rangle$$

$$\nabla^2 \Phi_{n-1}(\mathbf{X}).$$

Assuming that the time intervals δt between consecutive steps are small and approximately the same, one can write

$$\frac{[\Phi_{n\delta t}(\mathbf{X}) - \Phi_{(n-1)\delta t}(\mathbf{X})]}{\delta t} = \frac{1}{6} \frac{\langle x^2 \rangle}{\delta t} \Phi_{(n-1)\delta t}.$$

A Basic Concepts in Probability and Statistics

Introducing the current time $t = n\delta t \approx (n-1)\delta t$ and identifying the left-hand side as the finite difference approximation to the partial derivative of $\Phi(X, t)$ with respect to time, one arrives at the following partial differential equation:

$$\frac{\partial \Phi}{\partial t} = D\nabla^2 \Phi \quad \text{where} \quad \nabla^2 \Phi \equiv \frac{\partial^2 \Phi}{\partial x^2} + \frac{\partial^2 \Phi}{\partial y^2} + \frac{\partial^2 \Phi}{\partial z^2}, \tag{A.15}$$

where D is the diffusivity, $\partial \Phi / \partial t$ is the change of Φ over time, and $\nabla^2 \Phi$ is called the Laplacian of Φ. The partial differential equation governing the time evolution of the spatial distribution is known in physics as the diffusion equation. It describes transport processes when a system of equilibrium is moving towards equilibrium at the rate governed by its distance from equilibrium. The solution in the 4D space of the initial value problem $\Phi(t=0, X) = \delta^3(X)$, where δ here is the Dirac delta function, is

$$\Phi(t, X) = \frac{1}{(4\pi Dt)^{3/2}} \exp\left(-\frac{X^2}{4Dt}\right). \tag{A.16}$$

If we compare with the standard Gaussian function, $g(x) = \exp\left(-x^2/\sigma^2\right)$, then the standard deviation $\sigma = \sqrt{4Dt}$, which points to the linear dependence of the variance with time, $\sigma^2 \sim t$, or $\langle d(t)^2 \rangle \propto t$, where $d(t)$ is the distance travelled over time t. This is the so-called normal diffusion, with linear dependence of the variance with time, $\langle d(t)^2 \rangle \propto t$, or equivalently with the number of steps, which is applicable to homogeneous and isotropic systems. However, in reality, this scaling is not always valid, and, for heterogeneous systems, the diffusion law becomes anomalous, $\langle d(t)^2 \rangle \propto t^\gamma$. The case $\gamma > 1$ signifies a super-diffusive and $\gamma < 1$ a sub-diffusive behaviour. In disordered systems, γ is frequently less than one. Plotting $\log \langle d(t)^2 \rangle$ vs. $\log t$ is an experimental way to determine the type of diffusion occurring in a given system.

A.5.1 Space-Time Distribution of Aftershocks

Most frequently aftershocks of larger events start close to the main shock and then migrate away diffusing stresses imposed by the main rupture. If we use the diffusion equation as a continuous analogue of the underlying discrete random walk and position the main shock at the centre of the coordinate system at $t=0$, then the probability of an aftershock to arrive at time t in an annulus defined by R_1 and R_2 is

$$\Pr(t; R, R_2) = \frac{4\pi}{(4\pi DT)^{\frac{3}{2}}} \int_{R_1}^{R_2} r^2 \exp\left(-\frac{r^2}{4DT}\right) dr. \tag{A.17}$$

Assuming that the interval $\Delta R = R_2 - R_1$ is small compared to $R = R_1$, the Eq. (A.17) gives

$$\Pr(t; R, \Delta R) \simeq \frac{1}{(4\pi Dt)^{3/2}} R^2 \exp\left(-\frac{R^2}{4Dt}\right) \Delta R. \tag{A.18}$$

The probability of an aftershock to occur in a small volume cut from a thin annulus around (R, θ, ϕ), where $\theta \in [\theta, \theta + \Delta\theta]$ and $\phi \in [\phi, \phi + \Delta\phi]$ in the interval of time between t_1 and $t + \delta t$ is

$$\Pr(t; R, \Delta R; \theta, \Delta\theta; \phi, \Delta\phi) \simeq \frac{1}{(4\pi Dt)^{3/2}} R \exp\left(-\frac{R^2}{4Dt}\right) R \sin(\theta) \Delta\theta \Delta\phi \Delta R. \tag{A.19}$$

If the small volume $\Delta V = R \sin(\theta) \Delta\theta \Delta\phi \Delta R$ about (R, θ, ϕ) is expressed in Cartesian coordinates, one gets

$$\Pr(t, R, \Delta V, \delta t) \simeq \frac{1}{(4\pi Dt)^{3/2}} \exp\left(-\frac{R^2}{4Dt}\right) R \Delta x \Delta y \Delta z,$$

where $R^2 = x^2 + y^2 + z^2$. Going back to a discrete random walk with mean time step $\delta t = t/n$ gives $4Dt = 2n\langle d^2 \rangle/3$, and therefore,

$$\Pr(n, R, \Delta V, \delta t) \simeq \frac{1}{\left(\frac{2}{3}\pi n \langle d^2 \rangle\right)^{\frac{3}{2}}} \exp\left(-\frac{3R^2}{2n\langle d^2 \rangle}\right) R \Delta x \Delta y \Delta z.$$

References

Aki, K. (1966). Generation and propagation of G waves from the Niigata earthquake of June 16, 1964. Part 2: Estimation of earthquake moment, released energy, and stress strain drop from the G-wave spectrum. *Bulletin Earthquake Research Institute Tokyo University, 44*, 73–88.

Anscombe, F. J. (1973). Graphs in statistical analysis. *American Statistician, 27*(1), 17–21.

Blum, J. R., Kiefer, J., & Rosenblatt, M. (1960). Distribution free tests of independence based on the sample distribution function. *Annals of Mathematical Statistics, 32*(2), 485–498.

Davidsen, J., & Paczuski, M. (2005). Analysis of the spatial distribution between successive earthquakes. *Physical Review Letters, 94*, 048501.

Edelmann, D., Mori, T. F., & Szekely, G. (2021). On relationships between the Pearson and the distance correlation coefficients. *Statistics and Probability Letters, 169*(108960). http://doi.org/10.1016/j.spl.2020.108960.

Eneva, M. (1996). Effect of limited data sets in evaluating the scaling properties of spatially distributed data: an example from mining-induced seismic activity. *Geophysical Journal International, 124*(3), 773–786.

Eneva, M. (1998). In search for a relationship between induced microseismicity and larger events in mines. *Tectonophysics*, 91–104.

Grassberger, P., & Procaccia, I. (1983). Characterization of strange attractors. *Physical Review Letters, 50*(5), 346–349.

Hoeffding, W. (1948). A non-parametric test of independence. *Annals of Mathematical Statistics, 19*(4), 546–557.

Kijko, A., & T. Stankiewicz (1987). Bimodal character of the distribution of extreme seismic events in Polish mines. *Acta Geophysica Polonica, 35*, 491–506.

Kranz, R. L., Coughlin, J. P., & Billington, S. (1995). Studies of stope-scale seismicity in a hard rock mine. *Tech. Rep. RI 9625*, USBM, Denver Colorado.

Kraskov, A., Sogbauer, H., & Grassberger, P. (2004). Estimating mutual information. *Physical Review E, 69*(066138).

Lasocki, S., & Papaditriou, E. E. (2006). Magnitude distribution complexity revealed in seismicity in Greece. *Journal of Geophysical Research, 111*(B11309).

Lasocki, S., Kijko, A., & Graham, G. (2000). Model-free seismic hazard estimation. In H. Gokcekus (Ed.), *Proceedings of the International Conference on Earthquake Hazard and Risk in the Mediterranean Region* (pp. 503–508).

Linfoot, E. H. (1957). An informational measure of correlation. *Information and Control, 1*, 85–89.

Mendecki, A. J. (1993). Real time quantitative seismology in mines: Keynote Address. In R. P. Young (Ed.), *Proceedings 3rd International Symposium on Rockbursts and Seismicity in Mines, Kingston, Ontario, Canada* (pp. 287–295). Balkema, Rotterdam.

Shannon, C. E. (1948a). A mathematical theory of communication. Part 1. *Bell System Technical Journal, 27*, 379–423.

Shannon, C. E. (1948b). A mathematical theory of communication. Part 2. *Bell System Technical Journal, 27*, 623–656.

Szekely, G. J., Rizzo, M. L., & Bakirov, N. K. (2007). Measuring and testing dependence by correlation of distances. *The Annals of Statistics, 35*(6), 2769–2794.

Tosi, P. (1998). Seismogenic structure behavior revealed by spatial clustering of seismicity in Umbria-Marche region (central Italy). *Annali di Geofisica, 41*(2), 215–224.

Tosi, P., Rubeis, V. D., Loreto, V., & Pietronero, V. (2004). Space-time combined correlation integral and earthquake interactions. *Annali di Geofisica, 47*(6), 1849–1854.

Tosi, P., Rubeis, V. D., Loreto, V., & Pietronero, L. (2008). Space–time correlation of earthquakes. *Geophysical Journal International, 173*, 932–941. http://doi.org/10.1111/j.1365-246X.2008.03770.x.

MIX
Papier aus verantwortungsvollen Quellen
Paper from responsible sources
FSC® C105338

If you have any concerns about our products,
you can contact us on
ProductSafety@springernature.com

In case Publisher is established outside the EU,
the EU authorized representative is:
**Springer Nature Customer Service Center GmbH
Europaplatz 3, 69115 Heidelberg, Germany**

Printed by Libri Plureos GmbH
in Hamburg, Germany